CONTACT DYNAMICS

Method of Differential Specific Forces

CONTACT DYNAMICS

Method of Differential Specific Forces

Nikolay Goloshchapov, PhD

APPLE ACADEMIC PRESS

Apple Academic Press Inc.
3333 Mistwell Crescent
Oakville, ON L6L 0A2
Canada

Apple Academic Press Inc.
1265 Goldenrod Circle NE
Palm Bay, Florida 32905
USA

First issued in paperback 2021

Exclusive worldwide distribution by CRC Press, a member of Taylor & Francis Group
No claim to original U.S. Government works

ISBN 13: 978-1-77463-172-0 (pbk)
ISBN 13: 978-1-77188-683-3 (hbk)

Library and Archives Canada Cataloguing in Publication

Goloshchapov, Nikolay, author
Contact dynamics : method of differential specific forces / Nikolay Goloshchapov, PhD.

Includes bibliographical references and index.
Issued in print and electronic formats.
ISBN 978-1-77188-683-3 (hardcover).--ISBN 978-1-351-16652-2 (PDF)

1. Contact mechanics. 2. Viscoelasticity. 3. Elastoplasticity. 4. Solids. 5. Surfaces. I. Title.

TA353.G65 2018 620.1'05 C2018-905277-5 C2018-905278-3

Library of Congress Cataloging-in-Publication Data

Names: Goloshchapov, Nikolay, author.

Title: Contact dynamics : method of differential specific forces / Nikolay Goloshchapov, PhD.

Description: Oakville, ON ; Waretown, NJ : Apple Academic Press, 2018. |
Includes bibliographical references and index.

Identifiers: LCCN 2018042866 (print) | LCCN 2018044037 (ebook) |
ISBN 9781351166522 (ebook) | ISBN 9781771886833 (hardcover : alk. paper)

Subjects: LCSH: Contact mechanics.

Classification: LCC TA353 (ebook) | LCC TA353 .G65 2018 (print) | DDC 620.1/05--dc23

LC record available at https://lccn.loc.gov/2018042866

ABOUT THE AUTHOR

Nikolay Goloshchapov, PhD, has more than 30 years of experience in the fields of physics, mechanical engineering, tribology, and materials science; particularly physics and mechanics of materials and polymers and elastomers. He is currently an independent private scientific consultant at NNG—Engineering & Physics Science Consultancy, London, United Kingdom. He is a specialist in applied physics and math-mechanics, tribology, and fatigue life (wear out and erosion of materials under collision of solids, particularly in optimizing the basic dynamic and mechanical qualities of viscoelastic materials and elastomers [rubbers] for their exploitation in gas jet or flow abrasive solids and substances in different temperatures and velocities of loading and impact). He has acted as a reviewer of articles for international journals on topics such as scientific instruments and nonlinear dynamics. He has developed a method to define the time of fatigue of the materials being in processes of a collision between solid bodies, and he has researched the processes of erosion and durability of polymers and elastomers and other materials being in gas jet and gas flow of abrasive particles under high and low temperatures.

CONTENTS

ABBREVIATIONS

DEM	discrete elements method
HM+D	Hertz Mindlin spring–dashpot model
LS+D	linear spring–dashpot model
MDR	method of dimensionality reduction
MDSF	method of the differential specific forces

PREFACE

This new book describes the application of the method of the differential specific forces (MDSF). By using this new method, the solutions to the problems of a dissipative viscoelastic and elastic-plastic contacts between curvilinear surfaces of two solid bodies have been found. The novelty is that the forces of viscosity and the forces of elasticity can be found by an integration of the differential specific forces acting inside an elementary volume of the contact zone. This volume shows that this method allows finding the viscoelastic forces for any theoretical or experimental dependencies between the distance of mutual approach of two curvilinear surfaces and the radii of the contact area. Also, the derivation of the integral equations of the viscoelastic forces has been delivered and the equations for the contact pressure have been obtained. The viscoelastic and elastic-plastic contacts at impact between two spherical bodies have been examined. The equations for work and energy in the phases of compression and restitution and at the rolling shear have been obtained. Approximate solutions for the differential equations of movement (displacement) by using the method of equivalent work have been derived.

This new method of differential specific viscoelastic forces allows us to find the equations for all viscoelastic forces. It is principally different from other methods that use Hertz's theory, the classical theory of elasticity and the tensor algebra. In this new method, how to find the elastic and viscous forces by an integration of the differential specific forces in the infinitesimal boundaries of the contact area is explained. This method will be useful in research of contact dynamics of any shape of contacting surfaces. It also can be used for determination of the dynamic mechanical properties of materials and in the design of wear-resistant elements and coverings for components of machines and equipment that are in harsh conditions where they are subjected to the action of flow or jet abrasive particles.

This volume will be useful for professional designers of machines and mechanisms as well as for the design and development of new advanced materials such as wear-resistant elastic coatings and elements for pneumatic and hydraulic systems, stop valves, fans, centrifugal pumps, injectors, valves, gate valves, gearings and in other installations.

CHAPTER 1

PROBLEMS IN CONTACT DYNAMICS BETWEEN SOLIDS

ABSTRACT

This chapter provides a brief review of problems in contact dynamics between solids. Basic rheological models, the discrete elements method (DEM) as well as the problems of definition of the viscoelastic forces in cases of viscoelastic, elastic-plastic and high-elastic contacts are considered in this chapter. The simplified models of a viscoelastic sliding and rolling friction between surfaces of two solids are proposed. Boundary value problems and specifics of the initial and boundary conditions of loading and unloading of contacting surfaces are considered here as well.

1.1 INTRODUCTION

The objective of this book is the application of the 'method of the differential specific forces' for solutions to the problems in contact dynamics between surfaces of two solid bodies. The new conception is proposed here, for example how to find the elastic and viscous forces by an integration of the differential specific forces in the boundaries of the contact area, which can be found by a consideration of the geometry of the contact.

As we know, the mechanics of an elastic contact problem between two smooth surfaces have been studied in the 19th century by Hertz (1882, 1896) and Boussinesq (1885), and then later, it was examined by many others researchers such as Bowden and Tabor (1939); Landau and Lifshits (1944, 1959); Timoshenko and Goodier (1951); Archard (1957); Galin (1961); Sneddon (1965); Greenwood and Williamson (1966); Johnson et al. (1971); Derjaguin et al. (1975); Bush et al. (1975); Tabor (1977a, 1977b); Johnson (1985); Webster and Sayles (1986); Stronge (2000); Persson et al. (2002); Wriggers (2006); Hyun and Robbins (2007). Also, a viscoelastic and an elastic-plastic contact between smooth and rough curvilinear surfaces of two solids have already been researched very widely, and their results are published in many different manuscripts (Mindlin,1949; Radok, 1957; Hunter,1960; Goldsmith, 1960; Galin, 1961; Lee and Radok, 1960; Lee, 1962; Graham, 1965; Ting, 1966; Greenwood and Williamson, 1966; Simon, 1967; Padovan and Paramadilok, 1984; Johnson, 1985; Brilliantov et al., 1996; Ramírez et al., 1999; Stronge, 2000; Barber and Ciavarella, 2000; Goloshchapov, 2001, 2003a, 2003b, 2014, 2015; Laursen, 2002; Dintwa, 2006; Carbone et al., 2009; Harrass et al., 2010; Persson, 2010; Cummins et al., 2012; Carbone and Putignano, 2013; Popov, 2010, Popov and Heβ, 2015). According to above-mentioned researches, we can allocate several main types of frictional contact as follows:

1. Purely elastic contact (in nature, practically does not exist): when only the elastic forces act between the contacting surfaces
2. Viscoelastic contact: when between the surfaces in contact, the dissipative forces of viscosity (also named as the forces of internal friction) begin to act as well
3. The elastic-plastic contact: when forces of viscosity are considerable and the contacting surfaces pass into a plastic state
4. Adhesive contact: when a significant adhesive force, which cannot be neglected, acts between surfaces in contact

But, as we know, in practice, the real contact between the contacting surfaces usually is implemented as a combination of these four basic types of contact.

On the other hand, we can allocate three main types of contact of the relative displacement between the contacting surfaces, such as a slip, a rolling and an impact.

In this book, I have tried to consider all these options, applying a method of differential specific forces, which is the further development of the method of specific forces and has already been developed (Goloshchapov, 2003b, 2015).

1.2 PROBLEMS OF VISCOELASTIC AND ELASTIC-PLASTIC CONTACTS

Also, many old papers and others published recently (Ferry, 1948, 1963; Mindlin, 1949; Goldsmith, 1960; Simon, 1967; Flügge, 1975; Moore, 1975,

1978; Johnson, 1985; Cundall and Strack,1979; Meyers,1994; Schafer et al., 1996; Lakes,1998; Ramírez et al., 1999; Menard,1999; Stronge, 2000; Roylance, 2001; Goloshchapov, 2001, 2003a, 2003b, 2014, 2015; Votsios, 2003; Makse et al., 2004; Hosford, 2005; Van Zeebroeck, 2005; Dintwa, 2006; Bordbar and Hyppänen, 2007; Schwager and Poschel, 2007, 2008; Cheng et al., 2008; Becker et al., 2008; Thornton, 2009; Antypov et al., 2011; Thornton et al., 2012; Menga et al. 2014) traditional theories of elasticity and viscoelasticity, and also theoretical rheological models and methods such as the 'linear spring–dashpot model'—(LS+D), the Hertz Mindlin spring–dashpot model—(HM+D), Maxwell model, Kelvin–Voigt model, the 'discrete elements model or method', and others have been used. But in many of these papers, for the definition of the elastic force, Hertz's theory of elastic contact between two surfaces (Landau and Lifshitz, 1944, 1959) is still used. Also, the coefficient of friction was taken as a constant for the purpose of finding the tangential forces. The review of these already known models and methods can be found in the monographs of the authors such as Stronge (2000); Van Zeebroeck (2005); Dintwa (2006) and Li (2006).

It is obvious that the dynamic contact between two smooth curvilinear surfaces can be considered like the process of their mutual collision at impact, which has two phases the compression or the loading and the restitution or unloading. And as well in the case of the contact between two rough surfaces, we can assume that the contact of two asperities is similar like between two smooth surfaces, see Figures 1.2 and 2.1. Generally, it does not matter whether it is a sliding or rolling contact, but anyway, we can say that in the process of contact the volumes of deformation of two contacting bodies always are involved in two phases of contact, that is, in the phase of compression and restitution.

Also, it is well known that, in the case of the viscoelastic contact, the viscoelastic forces can be found according to the Kelvin–Voigt rheological model, see Figure 1.1a, as the sums of the elastic forces and the viscous forces:

$$F_n = F_{bn} + F_{cn} \tag{1.1}$$

$$F_{\tau n} = F_{b\tau y} + F_{c\tau y} \tag{1.2}$$

$$F_{\tau z} = F_{b\tau z} + F_{c\tau z} \tag{1.3}$$

where F_{cn} is the normal elastic force, F_{bn} is the normal viscous force, $F_{b\tau y}$ is the tangential viscous force by axis Y, $F_{c\tau y}$ is the tangential elastic force by axis Y, $F_{b\tau z}$ is the tangential viscous force by axis Z and $F_{c\tau z}$ is the tangential elastic force by axis Z.

But on the other hand, in the case of the elastic-plastic contact, when the plastic deformation is increased in both the phases of compression and restitution we have to use the Maxwell model, see Figure 1.1b, where the elastic forces are equal to viscous (plastic) forces as follows:

$$F_n = F_{bn} + F_{cn} \tag{1.4}$$

$$F_{\tau n} = F_{b\tau y} + F_{c\tau y} \tag{1.5}$$

$$F_{\tau z} = F_{b\tau z} + F_{c\tau z} \tag{1.6}$$

The general tangential force can be found as

$$F_\tau = \sqrt{F_{\tau y}^2 + F_{\tau z}^2} \tag{1.7}$$

In the 'linear spring–dashpot models'—(LS+D), all viscoelastic forces can be represented simply like the functions:

$$F_{bn} = b_x \dot{x}_b \tag{1.8}$$

$$F_{cn} = c_x x_c \tag{1.9}$$

$$F_{b\tau y} = b_y \dot{y}_b \tag{1.10}$$

$$F_{c\tau y} = c_y y_c \tag{1.11}$$

$$F_{b\tau z} = b_z \dot{z}_b \tag{1.12}$$

$$F_{c\tau z} = c_z z_c \tag{1.13}$$

where $c_x, c_y, c_z, b_x, b_y, b_z$ are the parameters of viscoelasticity, which were taken as constants in the linear models of contact, and where $c_x, c_y, c_z, b_x, b_y, b_z$ were called as the stiffness and b_x, b_y, b_z were called as the damping parameter or the damping coefficient; x_c is an elastic deformation of the compression of an elastic element (spring); $\dot{x}_b$ is a velocity of deformations of the compression a viscous element—dash-pot; y_c, z_c are elastic deformations of the shear elastic elements; $\dot{y}_b, \dot{z}_b$ are the velocities of deformations of the shear a viscous elements.

Where for the Maxwell model, in the case of elastic-plastic contact $x_c + x_b = x$, $y_c + y_b = y$ and $z_c + z_b = z$, and for Kelvin–Voigt Model in the case of viscoelastic contact, $x_c = x_b = x$, $y_c = y_b = y$, and $z_c = z_b = z$. Or x,y,z are the relative displacements (deformations) between contacting surfaces; $\dot{x},\dot{y},\dot{z}$ are relative velocities of displacements (deformations) between contacting surfaces.

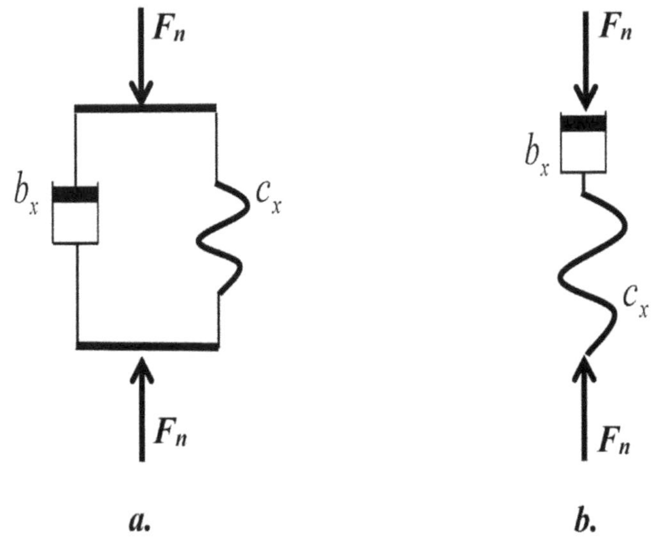

FIGURE 1.1 The basic rheological models of contact between two surfaces: (a) Kelvin–Voigt model, (b) Maxwell model.

The attempts have been made in the past to use some linear models for solutions of the contact's problems, for example, in the linear Winkler foundation model of an elastic foundation (see, for example Kerr, 1964), it is assumed that the contact stress σ at any point on the surface of the foundation is proportional to the deflection x,

$$\sigma_n = k\,x \tag{1.14}$$

where k is the coefficient called the modulus of the foundation, which is taken as a constant value and usually found in an experimental way.

Or by using the 'linear theory of viscoelasticity' viscoelastic stresses are described by exprηessions as

$$\sigma_{cn} = \varepsilon_{cx} E \tag{1.15}$$

and

$$\sigma_{bn} = \dot{\varepsilon}_{bx}\,\eta \tag{1.16}$$

where E is the Young's modulus, ε_{cx} is the relative deformation of the elastic element (spring), η is dynamic viscosity $\dot{\varepsilon}_{bx}$ is the velocity of the relative deformation of the viscous element (dashpot), and for the Kelvin–Voigt model $\varepsilon_x = \varepsilon_{cx} = \varepsilon_{bx}$ and for the linear Maxwell model $\varepsilon_x = \varepsilon_{cx} + \varepsilon_{bx}$; Equations 1.15 and 1.16 are valid only for linear deformations.

It is well known that these models give an oversimplified and inadequate description of the local contact interaction between curvilinear surfaces of two solid bodies, and lead to erroneous conclusions as they can be used only for linear deformations. But, as we will see further in this paper or as you can find in already published literature (Goloshchpov, 2003a, 2014, 2015), in the case of viscoelastic contact when $\varepsilon_x = \varepsilon_{cx} = \varepsilon_{bx}$, all contact deformations are reversible, it is possible to use the Kelvin–Voigt model in the case of the linear infinitesimal deformations, which arise in the infinitesimal volume.

Some authors use the 'Maxwell model' or so-called 'Maxwell body' for research of the viscoelastic contact, but it is not the correct way because according to this model, when $\varepsilon_x = \varepsilon_{cx} + \varepsilon_{bx}$, the viscous contact deformation between contacting surfaces is irreversible, since this viscous deformation always totally increases in both phases of the contact, in the phase of loading (compression or shear) and unloading (restitution). Therefore, this model can be used for an elastic-plastic contact deformation but only not in the case of a viscoelastic contact, when the plastic deformation has not placed. This is the right way to use 'Maxwell model' for research of vehicle damper systems, which include a dashpot and a spring, or in researches of sandwich composites, for example under a cyclic deformation of the extension—compression or at the shear under the action of an induced force.

The main problem is that all viscoelastic forces in Equations 1.8–1.13 are not the linear functions relative to displacements *x*, *y* and *z*. But indeed, it is known that (Brilliantov et al. 1996; Goloshchapov, 2003a, 2014; Jonson, 1985; Ramírez et al. 1999; Schafer et al. 1996; Schwager and Poschel, 2008; Stronge, 2000; Thornton, 2009, Popov and Heβ, 2015) the parameters $c_x, c_y, c_z, b_x, b_y, b_z$ are the variable functions depending on the displacement *x*, *y* and *z*, where, for example (Goloshchapov, 2003a, 2014, 2015) b_x is the effective parameter of viscosity (damping parameter), c_x is the effective parameter of elasticity (stiffness); b_y, b_z are the effective parameters of viscosity at a shift; c_y, c_z are the effective parameters of elasticity (stiffness) at a shift. Thus, the most basic problems in finding solutions for Equations 1.8–1.13 are that the dynamic contact between two curvilinear surfaces is a nonequilibral, nonlinear process of deformations and in this case, all mechanical dynamic parameters of viscoelasticity (c_x, c_y, b_x, b_y) are not the constant values. They all are variable magnitudes because all dynamic mechanical and physical properties of materials depend on dynamic conditions of loading (displacements, a velocity and frequency) and temperature. But, Hertz theory allows only finding the normal elastic force. The existing methods still cannot give the complete answer, how these nonlinear parameters of viscoelasticity can be found for the practical application by using the dynamic modulus of elasticity and viscosity, which usually can be found by using the known methods (Ferry, 1948, 1963; Moore, 1975; Van Krevelen, 1972; Nilsen, 1978, Nilsen and Landel, 1994). For example, as we know, according to Hertz theory (Landau and Lifshtz, 1944, 1959) for the contact between two spherical surfaces, $c_x = \frac{4}{3} ER^{1/2} x^{1/2}$, where R is the effective radius of contact curvature; E is the reduced Young's modulus of elasticity for static or quasi-static conditions of contact, but indeed in the case of dynamic contact, it is the effective elasticity modulus (Goloshchapov, 2003a, 2014, 2015). As we can see, the stiffness cx is the nonlinear function of displacement x. But still, the problem exists in the definition of the parameters c_y, c_z, b_x, b_y, b_z, which are the variable nonlinear functions, too.

It is necessary to mention here that the researches in the field of the collision of viscoelastic particles (granules) with identical mechanical properties have been made by Brilliantov et al. (1996). They have obtained the equation for the normal viscous force with variable viscosity parameter

$$F_{bn} = F_{dis}^{N} = \frac{Y}{(1-\nu^2)} \sqrt{R^{eff}}\, A \sqrt{\xi} \dot{\xi} \tag{1.17}$$

where $\xi = x$, $R^{eff} = R$, Y is Young's modulus or the elasticity modulus, ν is the Poisson's ratio, $A = \frac{1}{3} \frac{(3\eta_2 - \eta_1)^2}{(3\eta_2 + 2\eta_1)} \left[\frac{(1-\nu^2)(1-2\nu)}{Y\nu^2} \right]$ is the damping viscous parameter, where η_1 and η_2 are the viscous constants. But, Equation 1.17 can be used only for the contact of the bodies with the same physical-mechanical properties, and in this case, we have the problem of finding the viscous constants 'η_1' and 'η_2'. If the contacting surfaces have different physical-mechanical properties, this conception does not give the answer, because this is yet a more difficult problem.

Also, according to a well-known Hertzian contact solution, the connection between displacement x and normal elastic force can be written as

$$x^{3/2} = \frac{3}{4} \frac{F_n}{ER^{1/2}} \tag{1.18}$$

where E is known as the reduced Young's modulus of elasticity since it is less than the combined elasticity of the two contacting solids

$$\frac{1}{E} = \frac{1-\nu_1^2}{E_1} + \frac{1-\nu_2^2}{E_2} \tag{1.19}$$

where E_1,E_2 are Young's modulus in the initial moment of the time, when $t=0$ and v_1,v_2 are the Poisson's coefficients of the contacting surfaces.

Lee and Radok (1960) have used Equation 1.18 for contact between a rigid sphere and viscoelastic semi-space of elastomer (rubber). But since a rigid body is, in many times, harder than a soft semi-space of rubber, and since obviously $E_1 << E_2$ [where E_1 is modulus for elastomer (rubber) and E_2 is the modulus for material of rigid sphere], they took $\frac{1}{E}=\frac{1-\nu_1^2}{E_1}$, and since $E_1 = 2(1 + v)G_1$ where G_1 is the shear modulus in the initial moment of the time, when t=0, they received that

$$x^{3/2}=\frac{3}{8}\frac{F_n(1-\nu_1)}{G_1R^{1/2}} \tag{1.20}$$

After this, using the 'Boltzmann superposition model', the $\frac{F_n}{2G_1}$ in Equation 1.20 was replaced by the integral operator as follows

$$\left[\frac{F_n}{2G_1}\right]\rightarrow\int_0^t J(t-t')\frac{dF_n(t)}{dt'}dt' \tag{1.21}$$

where $J(t)$ is the function of creep (compliance) of a viscoelastic medium under shear, t' is the variable of integration, and finally, they got that

$$x(t)^{3/2}=\frac{3(1-\nu_1)}{4R^{1/2}}\int_0^t J(t-t')\frac{dF_n(t)}{dt'}dt' \tag{1.22}$$

Equation 1.22 has already been applied very widely for description of the viscoelastic behaviour of time-dependent materials such as elastomers (rubber) by many authors, for example Farine (1985); Cheng et al. (2005); Huang (2007); Carbone et al. (2009); Rodríguez (2012) and as well by many other authors. Different choices for the load function $F_n(t)$ and the creep compliance $J(t)$ can be made here, for example some researches use the Maxwell model, where $J(t)=\frac{1}{G_1}+\frac{t}{\eta_1}$; η_1 is the viscosity of the material of soft contacting surface; and some use the Kelvin–Voigt Model, where $J(t)=\frac{1}{G_1}[1-\exp(-t/\tau)]$; τ is the retardation time. These two models are the basic models, which are already used as elements for many complicated theoretical models in many researches. But nevertheless, the problems still exist, the first is that the viscoelastic force $F_n(t)$ and the creep compliance $J(t)$ are not independently linear functions, but they are dependent on each other, and also, the problem is how to find the $J(t)$, specifically in application to contact dynamics, because the creep compliance depends on the time of relaxation or the time of retardation, but on another hand, as we will see further, the time of relaxation or the time of retardation depends on the times of loading and unloading. Also indeed, in the initial moment of the time of dynamic viscoelastic contact at impact $t=0$, the normal and tangential contact forces as well as the contact stresses and deformations do not increase instantly, but they are equal to zero.

Moreover, using models of the single relaxation for an isotropic medium or the plural relaxations for anisotropic medium, we again have problems in their application as finding the relaxation or retardation times are not simple. In all these researches for finding viscoelastic forces and stresses, traditional theories and methods, for example the Radok and Lee's Equation 1.22, usually have been applied, but there are still many problems in the application of this theory for high speeds and frequencies of dynamic contact, because, as a rule, the creep is the process of slow deformation under static load, which takes the long period of time, but if the time of loading is reduced the time of relaxation is reduced, too. Theoretical expressions and equations obtained in them are usually difficult to use in a practical application because they are very complicated, and there still do not exist simple, convenient methods, which allow the finding of dynamic-mechanical parameters, which are contained in these expressions and equations. There is a problem in applying them in the case of a high-speed dynamic contact, using the modulus of elasticity and viscosity, and the Poisson's ratio (it is taken usually $\nu=0.5$ for rubber surfaces), which is obtained for the dead or quasi-static loads but gives wrong results. Also, in the finding of solutions by using the linear viscoelastic stress-strain relations, we have the problem in the definition of the relaxation modulus in realistic temperature/time/speed/frequency conditions of dynamic viscoelastic contact.

Also, we have to understand that the time of relaxation is the time of transition of a system of particles or elements of some volume of substance from the one equilibrium state into another. The physical relaxation is not the relaxation, which is taken in the mechanical oscillation. It is obvious that some molecular, submolecular or subatomic structures have their own times of relaxations, but the total relaxation time or the effective relaxation time is the time of relaxation, which matches the time when all relaxation processes inside of a substance are completely finished. Therefore, this total time of relaxation has to be taken into account in both the cases, when a volume of contact is isotropic or anisotropic medium. In contact dynamics, it is very important to know the time of relaxation process in the phase of viscoelastic compression between two surfaces, because at the point of maximum compression the velocity of mutual approach between surfaces equals zero, and all kinetic energy transforms into potential energy of deformations, and part of kinetic energy dissipates in result of an internal friction. This is similar to the new equilibrium state. Thus, the time of the phase of compression (or time the maximum mutual approach) and the effective time of relaxation in this phase have a common physical nature; and they are dependent on each other.

Also the method—'the reduction method of dimensionality (mapping theorem)' (Sneddon, 1965; Zhou, 2011; Heβ, 2012) or the 'method of dimensionality reduction (MDR)' has been used by Popov (2010), Popov and Heβ (2015). In the method (MDR) for the contact between a rigid spherical indenter with the radius R_1 and a half-space, it was obtained for the normal elastic force as

$$F_{cn} = \frac{4}{3} E^* a\delta \tag{1.23}$$

where $E^* = E$ is known as the reduced Young's modulus of elasticity; the radius of the contact area was the taken $a = \sqrt{2R_1\delta}$, where $\delta = x$ is the depth of indentation or it is the displacement x in the centre of the contact area. Thus, it was received as follows

$$F_{cn} = \frac{4}{3} E^* \sqrt{2R_1\delta}\delta \tag{1.24}$$

Then, the effective radius of curvature $R = 2R_1$, and of course, the same formula according to Hertz theory was obtained as

$$F_{cn} = \frac{4}{3} E^* \sqrt{2R\delta}\delta \tag{1.25}$$

However, as we know, according to Hertz theory (Landau and Lifshitz, 1965), the geometry of contact between two curvilinear surfaces (Goloshchapov, 2014) follows $\frac{1}{R} = \frac{1}{R_1} + \frac{1}{R_2}$. But, as we can see that in the case of contact between a rigid spherical indenter, having the radius R_1 and a semi-space and having the radius $R_2 = \infty$, we get $R = R_1$. But, why has this replacement $R = 2R_1$ been done by Popov (2010), if indeed $R = R_1$?

Maybe, it was taken, because the author of this replacement thinks that in the case of contact between a rigid body and surface of elastomer, Equation 1.24 more correctly reflects the specific of contact?

Also, Popov and Heβ (2015) considered the case when the smooth cylindrical indenter is pressed into a linearly viscous half-space (viscosity η) with a constant force F_n they obtained the equations for the constant normal forces in an elastic contact as

$$F_{cn} = 8Ga\delta \tag{1.26}$$

and for a viscous contact

$$F_{bn} = 8\eta a\dot{\delta} \tag{1.27}$$

where $\delta = x$, η is viscosity (it was not specified what kind of viscosity was taken here, but obviously it is the viscosity at the shear); a is considered as the instantaneous value of the contact radius. But on the other hand, as we can see, Equation 1.27 can be obtained by simple replacements such as $G \rightarrow \eta$ and $\delta \rightarrow \dot{\delta}$. It was proposed that Equation 1.27 is valid for any arbitrary-axially symmetric indenter, a is considered to be the instantaneous value for the contact radius. For example, in the case when the rigid axially-symmetric paraboloid is pressed into a viscous half-space with a constant force F_n, and since in this case was taken according to the Equation 1.27 and Hertz theory that $a = R^{1/2}\,\delta^{1/2}$, it was received by them for an incompressible body of elastomers with the viscosity η, that

$$F_{bn} = F_N = 8\eta R^{1/2}\delta^{1/2}\dot{\delta} \tag{1.28}$$

But since Equation 1.27 can be used for any arbitrary-axially symmetric indenter, it is logical that Equation 1.26, where $F_{cn} = 8Ga\delta$, it can be used for the same. If it is so, and since $G = E^*/4$ and Poisson's ratio was taken $\nu=0.5$ for an elastomer surface, we get the elastic force as follows

$$F_{cn} = 2E^* a\delta = 2E^* R^{1/2}\delta^{2/3} \tag{1.29}$$

But as we remember, the other result was given in the Equation 1.24.

The problem also is that generally in many of these researches the internal friction of materials and particularly into elastomers (in rubbers) is primitively considered like the shear between layers of material inside the block of elastomer. Therefore, the modulus and viscosity at shear are used by them for the description of viscous deformations problems. But as we know, indeed, in reality, the viscosity is a dissipative process of the changing of space-conformations between macromolecules or between atoms (ions) in crystal grids. In a dependency of the type of deformations (extension or tension/compression, bulk compression and shear), there are six kinds of modules and of viscosities in the descriptive dynamic mechanical properties of materials (Ferry, 1948, 1963; Moore, 1975; Van Krevelen, 1972; Nilsen, 1978; Nilsen and Landel, 1994). For example the main constitutive relations between them for the Kelvin–Voigt model are known as:

$$E' = 3K'(1-2\nu) = 2G'(1+\nu) \tag{1.30}$$

$$E'' = 3K''(1-2\nu) = 2G''(1+\nu) \tag{1.31}$$

and

$$E' = \omega\eta_E'' \tag{1.32}$$

$$K' = \omega\eta_K'' \tag{1.33}$$

$$G' = \omega\eta_G'' \tag{1.34}$$

$$E'' = \omega\eta_E' \tag{1.35}$$

$$K'' = \omega\eta_K' \tag{1.36}$$

$$G'' = \omega\eta_G' \tag{1.37}$$

where E'—the dynamic elasticity modulus (or the storage modulus), which is equal to Young's modulus of elasticity in the Kelvin–Voigt model; η_E' is the dynamic viscosity; G' is effective dynamic elasticity modulus at the shear (or it is the effective storage modulus at the shear); η_G' is the effective dynamic viscosity at the shear. Then, according to Equations 1.21 and 1.22

$$\eta_E' = 3\eta_K'(1-2\nu) = 2\eta_G'(1+\nu) \tag{1.38}$$

$$E'' = 3K''(1-2\nu) = 2G''(1+\nu) \tag{1.39}$$

Thus, as we can see in the common case, the expression for the normal viscoelastic can be written simply as

$$F_{bn} = \mathrm{k_b}\eta R^{1/2} x^{1/2} \dot{x} \tag{1.40}$$

Indeed interestingly, what is the value of this constant $\mathrm{k_b}$ in reality? Thus, as we can see, many problems still exist now in these research areas. Therefore, especially for solving these problems such as the definition of the normal viscous force and all tangential viscoelastic forces, and for the finding of the kinematic and the dynamic mechanical parameters between two contacting surfaces, such as the dynamic elasticity modulus, the viscosity modulus and viscoelastic pressure, the theoretical and experimental ways have been developed by Goloshchapov (2001, 2003a, 2014) and also represented in this book below. For example, the consideration of the contact between two spherical surfaces allows getting the equation for the viscous forces as

$$F_n = 4k_p \eta'_E R^{1/2} x^{1/2} \dot{x} \quad (1.41)$$

where k_p is the correlation coefficient.

And also in many already existing researches, the coefficient of friction usually is taken like a constant value for the purpose of finding the tangential forces, but according to Equations 1.1–1.3 and 1.8–1.13, for viscoelastic contact, it can be defined as follows:

$$f = \frac{F_\tau}{F_n} = \frac{\sqrt{(F_{b\tau y} + F_{c\tau y})^2 + (F_{b\tau z} + F_{c\tau z})^2}}{F_{cn} + F_{bn}} = \frac{\sqrt{(b_y \dot{y} + c_y y)^2 + (b_z \dot{z} + c_z z)^2}}{c_x x + b_x \dot{x}} \quad (1.42)$$

As we can see from this equation, the coefficient of friction is not always a constant value, but it changes during the time of contact because the dynamic contact between two bodies, in particular during collision at impact, is a nonequilibrial process, and all dynamic mechanical and physical properties of materials depend on dynamic conditions of loading and temperature. For example, the dynamic elasticity modules are very yieldable to the changing of velocity and temperature of the matter in the area of deformations (Ferry, 1948, 1963; Lee, 1962; Van Krevelen, 1972; Moor, 1978; Nilsen, 1978, 1994; Lakes, 1998; Meyers, 1994; Menard, 1999; Roylance, 2001; Goloshchapov, 2003a, 2014; Hosford 2005).

1.3 BOUNDARY VALUE PROBLEMS IN MECHANICS OF CONTACTS

Some researchers assume that quasi-static conditions of loading when $v<c$, where v is the speed of loading and c is the sound speed, but this is obviously wrong, because in the case of $v>c$ it is hypersonic deformations and all displacements of medium are out of field of the limit of strength in this case. The quasi-static state is the analogue of the static state when the acting forces are not dependant on speeds of loading or the duration of action of loading. Approximately, we can take it when $v<<c$. It is usually if the Mach number $M<0.01$. But, in dynamic analysis, we can lead a dynamic system to a quasi-static state, if we enter the forces of inertia. But, we must be aware that in this case, we cannot use mechanical parameters and characteristics of the properties of materials for static conditions, but we must take into account their dynamic mechanical properties, which are dependent on the time, speed and temperature. Particularly, this is very important in a dynamic contact between two solids, for example, at impact.

For finding the correct solutions to the problems of the contact mechanics, it is very important to set the boundary value problems and to specify the initial and boundary conditions of loading and unloading of contacting surfaces. First of all, it is necessary to consider the geometry of the contact area and to find its geometric characteristics such as the radius of contact area and a depth of indentation. Then, we have to find the basic constitutive relations between elastic and viscous parameters of materials of the contacting bodies. First of all, we have to find relations between the dynamic modulus of elasticity and dynamic viscosity. And then, using this information, the displacements and stresses that the surfaces of the contact are subjected to, can be found, too. All these found characteristics, such as the displacement and stresses, should satisfy the equilibrium equations, constitutive relations, compatibility conditions and initial and boundary conditions. Usually, as rule, the purpose of formulating and solving a boundary value problem is to:

1. To ensure that the viscoelastic stresses are within prescribed limits in the boundary of the contact area and their functions of distribution are known, where displacements are specified. This condition corresponds to the Neumann boundary condition and it simply can be written as:

$$\sigma_{nm}(x,t) \le \sigma_{n\lim}(x_{\lim}, t_{x\lim}) \tag{1.43}$$

$$\sigma_{ym}(y,t) \le \sigma_{y\lim}(y_{\lim}, t_{y\lim}) \tag{1.44}$$

$$\sigma_{zm}(z,t) \le \sigma_{z\lim}(z_{\lim}, t_{z\lim}) \tag{1.45}$$

where σ_{nm} is the maximal normal stress inside of the contact area or the volume of deformation, which corresponds to maximum value of the displacements $x(t)$, σ_{ym} is the maximal tangential stress inside of the contact area or the volume of deformation, which corresponds to maximum value of the displacements $y(t)$, σ_{zm} is the maximal tangential stress inside of the contact area or the volume of deformation, which corresponds to maximum value of the displacements $z(t)$; $x_{\lim}(t_{x\lim})$, $x_{\lim}(t_{x\lim})$, $x_{\lim}(t_{x\lim})$ are the permissible displacements or deformations; $\sigma_{n\lim}(x_{\lim}, t_{x\lim})$, $\sigma_{y\lim}(y_{\lim}, t_{y\lim})$, $\sigma_{z\lim}(z_{\lim}, t_{z\lim})$ are the permissible stresses; $t_{xlim,}$ $t_{ylim,}$ t_{zlim} are the limits of the time when physical-mechanical properties of contact medium are changed.

2. According to so-called Dirichlet boundary condition, to ensure that the displacements are within prescribed limits:

$$0 \leq x(t) \leq x_{\lim}(t_{x\lim}) \tag{1.46}$$

$$0 \leq x(t) \leq x_{\lim}(t_{x\lim}) \tag{1.47}$$

$$0 \leq x(t) \leq x_{\lim}(t_{x\lim}) \tag{1.48}$$

3. To ensure that the corresponding relationships between the times in the phases of loading (compression) and unloading (restitution) and their relaxation times have been found.
4. To ensure that the chosen dissipative model of contact corresponds to the viscoelastic or elastic-plastic behavior of the medium of the volume deformation in the area of contact, and also to be sure of the point of the boundary in which the viscoelastic model can be changed for the elastic-plastic model.

1.4 SIMPLIFIED MODEL OF DYNAMIC CONTACT AND FRICTION BETWEEN SOLIDS

In the case of a high size of contacting deformation, when the depth of the penetration of body into the surface of a semi-space reaches one order with the body size, the tangential force becomes commensurable with the normal force and the coefficient of friction also reaches large values. Thus, in the case of viscoelastic contact, we should take into account that the action of the tangential force, the area of contact between curvilinear surfaces can be deformed from an elliptical shape to an oval shape. Particularly, for example, it will be seen in the case of a viscoelastic deformation in the initial phase of compression and shear as it is depicted schematically in Figure 1.2. If we slowly apply the tangential force to a body with a very small velocity, step by step, keeping the normal force at constant value, we will see that the area of contact will be turning and in some moment, when the deformation of shear will reach the maximum at the point A, the body will start to slide or roll by the surface of a semi-space. It is obvious that the position of the contact area in moment of starting of sliding or rolling of a body by the surface of a semi-space will start to turn back, when the volume of deformation will start moving from the point A to the point C, (the volume of deformation will enter in the phase of restitution) and obviously the difference between size x_B

and x_C will decrease. And it is obvious that after when the velocity of sliding or rolling will reach the constant value, the difference between sizes x_B and x_C definitely will be equal to zero. And it is obvious that during the time of sliding or rolling motion, the volume of deformation *Vd* will be involved at the same time in the phase of the compression and the shear between the points *B* and *A* and into the phase of restitution between the points *A* and *C*, and work of elastic deformations at the compression and the shear will be fully returned to a body in the phase of restitution.

For example, let a rigid body having the weight $P = mg$ and the radius of curvature in the contact area *R* be put on the horizontal viscoelastic surface of a semi-space with the initial vertical velocity of compression $\dot{x}(t) = 0$. It is obvious that, in this case, the volume of deformation begins to oscillate together with a body along the axis *X*. Thus, we get nonequilibrial damping oscillations under an action of the normal force $F_n = c_x x + b_x \dot{x}$, which very soon will go down, and in some moment of the time at $t = t_0$ a body will stop oscillating and its vertical velocity will be equal to zero, $\dot{x}(t) = 0$, and also obviously the normal force will reach the constant equilibrium magnitude of $F_{n0} = c_x x_0 = mg$, where x_0 is the equilibrium static size of indentation of a body into a semi-space, see Figure 1.2.

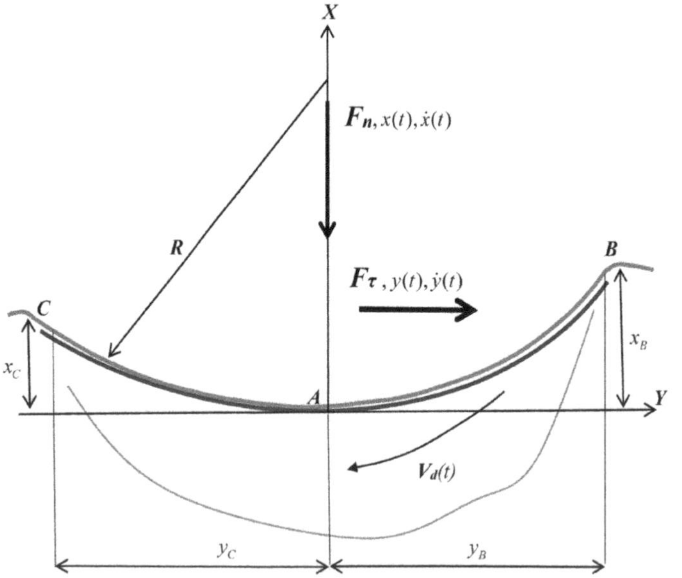

FIGURE 1.2 (See color insert.) Illustration of a sliding or rolling contact between the surface of the rigid body and a semi-space.

Then after short period of time, let us try to push a body along the tangential axis Y with the initial horizontal velocity of a body $\dot{y}(t) = V_y$. In this new moment of the time $t = 0$, we will begin to push a body, and the tangential force of friction $F_\tau(y, \dot{y}) = c_y y + b_y \dot{y}$ will begin to act as well. We will get the new nonequilibrium of the process of shear. A body will start to shear on the surface in a semi-space and the volume of deformation $V_d(t)$ will begin to move from the point B to the point A, and in the instant of the time $t = t_1$, the deformations will reach the maximum of a shear y_B and the maximum compression y_B. It will be the tangential shear phase and it will be the normal compression phase at the same time. It is obvious that at the point B, the velocity of a body $\dot{y}$ always equals to the velocity of shear $\dot{y}_B$ and the velocity of sliding always equals zero. But the velocities of a compression of the volume of deformation at the point B can be found from proportion $t_1 = \frac{y_B}{\dot{y}_B} = \frac{x_B}{\dot{x}_B}$ as

$$\dot{x}_B = \frac{x_B \dot{y}_B}{y_B} \tag{1.49}$$

Thus, in the boundary moment of the time $t = t_1$, at the point A, the deformations of the volume $V_d(t)$ will reach maximal sizes $y = y_B$ and $x_m = x_B$. Also, it is obvious that in duration of the time t_1 the velocity of sliding at the point A will reach maximum magnitude $\dot{y}_b = \dot{y}_{bA}$ and the velocities of shear deformation and compression will reach the minimum magnitudes $\dot{y} = \dot{y}_A = 0$ and $\dot{x} = \dot{x}_A = 0$. Thus, we can write for the maximums of the normal and the tangential forces in this moment of the time as

$$F_{\tau m} = c_y y_B \tag{1.50}$$

$$F_{nm} = c_x x_B \tag{1.51}$$

This maximal tangential force, F_{nm}, is usually named as the friction force of rest. The coefficient of friction will reach in this moment and the magnitude as follows

$$f_{sm} = \frac{F_{\tau m}}{F_{nm}} = \frac{c_y y_B}{c_x x_B} \tag{1.52}$$

Then, if we continue to push a body, the volume of deformation $V_d(t)$ will begin to be moved from point A to point C, and at the point C the deformation of restitution and velocity will reach values $y = -y_C$, $\dot{y} = -\dot{y}_C$, respectively, and the volume of deformation will start to move constantly relative to point $x = x_0$. It is obvious that the volume of deformation will be moving from point B to point A in the compression phase and at the same time it will be moving from point A to point C in the restitution phase. Also, it is obvious that the work of the normal elastic force in the compression phase will be returned to the body in the restitution phase. Therefore, we can find the average normal elastic force as the sum of the constant elastic force $\tilde{F}_{cn0} = c_x x_0$ and the relative normal elastic force of the moving of the volume of deformation $\tilde{F}_{cnV} = c_x(|x_B| - |x_C|)$ as

$$\tilde{F}_{cn} = \tilde{F}_{cn0} + \tilde{F}_{cnV} = c_x(x_0 + |x_B| - |x_C|) \tag{1.53}$$

And also, since the average velocity of deformation of the compression can be expressed as $V_x = \frac{1}{2}(|\dot{x}_B| + |\dot{x}_C|)$, we can find the average normal viscous force as

$$\tilde{F}_{bn} = b_x V_x = \frac{b_x}{2}(|\dot{x}_B| + |\dot{x}_C|) \tag{1.54}$$

Thus, the average viscous normal force can be expressed as

$$\tilde{F}_n = \tilde{F}_{cn} + \tilde{F}_{bn} = c_x(x_0 + |x_B| - |x_C|) + \frac{b_x}{2}(|\dot{x}_B| + |\dot{x}_C|) \tag{1.55}$$

But, on the other hand, when the body will continue moving with a constant velocity, the energy of the tangential elastic force in the shear (compression) phase will be returned to the body in the restitution phase. It is obvious that the volume of deformation will be moving from point B to point A in the shear phase and in the same time it will be moving from point A to point C in the restitution phase; thus, we can find the magnitude of the elastic friction force as follows

$$\tilde{F}_{cs} = c_y(|y_B| - |y_C|) \tag{1.56}$$

And since the energy of the tangential viscous force in the compression and restitution phases dissipates, we can find the average viscous frictional force as

$$\tilde{F}_{bs} = b_y\left(\frac{|\dot{y}_B| + |\dot{y}_C|}{2}\right) \tag{1.57}$$

and; thus, we can write the expression for the average frictional forces as follows

$$\tilde{F}_{\tau s} = c_y(|y_B| - |y_C|) + b_y\left(\frac{|\dot{y}_B| + |\dot{y}_C|}{2}\right) \tag{1.58}$$

Finally, the equation for the average coefficient of sliding friction can be written as

$$f_s = \frac{\tilde{F}_{\tau s}}{\tilde{F}_n} = \frac{2c_y(|y_B| - |y_C|) + b_y(|\dot{y}_B| + |\dot{y}_C|)}{2c_x(|x_B| + |x_C|) + b_x(|\dot{x}_B| + |\dot{x}_C|)} \tag{1.59}$$

In the case of viscoelastic deformations, in the stationary process of sliding with the constant velocity V_y, the deformations in phases of shear and compression and restitution are equal, $|y_B| = |y_C|$; it follows $\tilde{F}_{cs} = 0$. Also, since $V_y = |\dot{y}| = |\dot{y}_C| = |\dot{y}_B|$, $|\dot{x}_B| = |\dot{x}_C| = V_x$, $|x_B| = |x_C| = |x_0|$, where x_0 is the size of compression at the point A, and taking into account Equation 1.54, we get

$$\tilde{F}_{cn} = c_x x_0 \tag{1.60}$$

$$\tilde{F}_n = c_x x_0 + b_x V_x \tag{1.61}$$

$$\tilde{F}_{\tau s} = b_y V_y \tag{1.62}$$

Since in case of sliding with a stationary speed $V_y = \dot{y}_B$, $\dot{x}_B = V_x$, $x_B = x_0$, according to Equation 1.49, it can be also written as

$$V_x = \frac{x_0 V_y}{y_B} \tag{1.63}$$

and taking into account Equation 1.54, we get

$$\tilde{F}_{bn} = b_x \frac{x_0}{y_B} V_y \tag{1.64}$$

and that

$$\tilde{F}_n = c_x x_0 + b_x \frac{x_0 V_y}{y_B} \tag{1.65}$$

Thus, the average coefficient of sliding friction can be expressed as

$$f_s = \frac{\tilde{F}_{\tau s}}{\tilde{F}_n} = \frac{b_y V_y}{x_0 \left(c_x + \dfrac{b_x V_y}{y_B} \right)} \tag{1.66}$$

As in the case of rolling between the contacting surfaces, the instantaneous centre of rotation—the centre of velocities is placed at the point A, see Figures 1.2 and 1.3. It is obvious that the velocity of rotation of the point B around the instantaneous centre of velocities can be found as $V_B = \omega_A R_B$; since $\omega_A = \dot{y}/R$, it follows

$$V_B = \frac{\dot{y}R_B}{R} \tag{1.67}$$

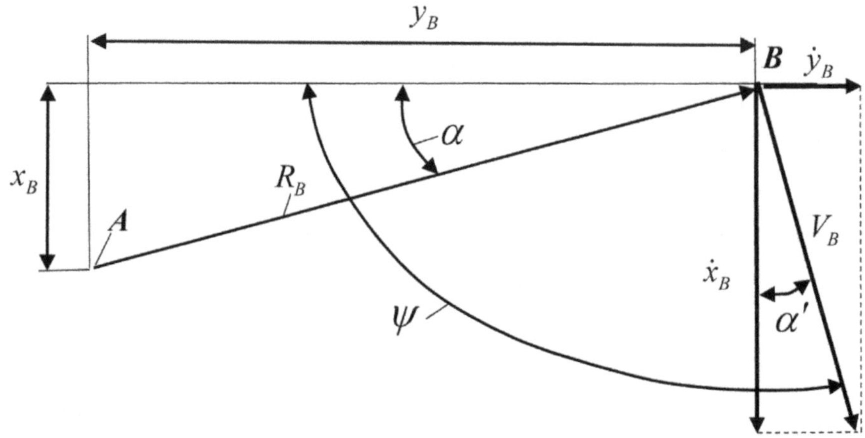

FIGURE 1.3 Kinematics of rotation the point B around the centre of velocities at the point A.

Also since $\dot{x}_B = V_B \cos\alpha'$, $\dot{y}_B = V_B \sin\alpha'$ and $R_B = x_B / \sin\alpha = y_B / \cos\alpha$; and since angle $\psi = \pi/2 + \alpha = \pi/2 + \alpha'$, see Figure 1.3, it follows that $\alpha = \alpha'$, and in case the rolling motion we get the following

$$\dot{y}_B = \frac{x_B}{R}\dot{y} \tag{1.68}$$

$$\dot{x}_B = \frac{y_B}{R}\dot{y} \tag{1.69}$$

Since the deformations in the phases of shear, compression and restitution are equal, that is $|y_B| = |y_C|$, $|x_B| = |x_C| = |x_0|$ and $V_{yr} = |\dot{y}_C| = |\dot{y}_B|$, $|\dot{x}_B| = |\dot{x}_C| = V_x$, it follows

$$V_{yr} = \frac{x_0}{R}V_y \tag{1.70}$$

$$V_x = \frac{y_B}{R}V_y \tag{1.71}$$

and; therefore, according to Equation 1.54, it follows

$$\tilde{F}_{bn} = b_x V_x = b_x \frac{y_B}{R} V_y \tag{1.72}$$

Thus, we can write finally that

$$\tilde{F}_{\tau r} = b_y V_{yr} = b_y \frac{x_0}{R} V_y \tag{1.73}$$

$$\tilde{F}_n = c_x x_0 + b_x V_y \frac{y_B}{R} \tag{1.74}$$

and in this case, the expression for the average magnitude of the coefficient of rolling friction can be written as follows

$$f_r = \frac{\tilde{F}_{\tau r}}{\tilde{F}_n} = \frac{b_y x_0 V_y}{c_x x_0 R + b_x y_B V_y} \tag{1.75}$$

It is obvious that usually in the case of viscoelastic sliding or rolling motion with a constant velocity $\tilde{F}_{cn} >> \tilde{F}_{bn}$; therefore, we can neglect the normal viscous force and respectively, it follows

$$\tilde{F}_n = \tilde{F}_{cn} = c_x x_0 \tag{1.76}$$

Thus, the average coefficient of sliding friction can be expressed as

$$f_s = \frac{\tilde{F}_{\tau s}}{\tilde{F}_n} = \frac{b_y V_y}{c_x x_0} \tag{1.77}$$

The coefficient of rolling friction can be expressed simply as

$$f_r = \frac{\tilde{F}_{\tau s}}{\tilde{F}_n} = \frac{b_y V_y}{c_x R} \tag{1.78}$$

As we can see here, the friction coefficient is the constant value for stationary conditions of sliding or rolling motion, when V_y is the constant value.

Also, for example, it is known according to Hertz theory, that in the case of contact between the spherical body and a semi-space $y_B = r = \sqrt{x_0 R}$. After substitution of this solution into Equation 1.63 or into 1.71, we will get the following

$$V_x = V_y \sqrt{\frac{x_0}{R}} \tag{1.79}$$

Thus, it is obvious that the friction can be considered as the dynamic cyclic process of deformations of the volume of deformation (Goloshchapov, 2003a). But, if a velocity of sliding or rolling is changed, the friction coefficient will be changed, too. Again, we have the problems here. How does a shape of contact area affect viscoelastic parameters c_x, b_x, c_y, b_y? What are the relations between the times of phases of compression/shear and restitution and the time of relaxation? Or better to say in other words, what is the effect of damping and relaxation on the process of sliding and rolling? Further, in this book, you will find answers and solutions to these problems. Also, in this book, the method of definition of the instant dynamic modules of visco-elasticity, using the time of compression as well as another novel theoretical and practical principals in dynamics of viscoelastic, high-viscoelastic and elastic-plastic contacts are proposed for your consideration.

KEYWORDS

- **rheological models**
- **boundary value problems**
- **elastic contact**
- **elastic-plastic contact**
- **viscoelastic contact**
- **Kelvin–Voigt model**
- **Maxwell model**

CHAPTER 2

METHOD OF THE DIFFERENTIAL SPECIFIC FORCES (MDSF)

ABSTRACT

The derivation of the basic principles of the 'method of the differential specific forces (MDSF)' in dynamics of contacts between smooth curvilinear surfaces of two solid bodies has been proposed in this chapter, and then the method of definition of the elastic and the viscous forces has been developed. The novelty is that, the forces of viscosity and the forces of elasticity have been found by integration of the differential specific forces acting inside an elementary volume of the contact zone. It is shown here that this method allows finding the viscoelastic forces for any theoretical or experimental dependencies between the distance of mutual approach of two contacting surfaces and the size of the contact area. Also, the derivation of integral equations of the viscoelastic forces has been given and the equations for contact pressure and general stresses have been obtained. The equations for work and energy for general viscoelastic forces in the phases of compression and restitution, and at the rolling shear have been obtained. Also, equations for kinematic and dynamic parameters have been derived in this chapter. The examples of solutions of contact problems between solids have been examined here.

2.1 INTRODUCTION

The majority of surfaces of solid bodies have very rigid structures, for example, metals and their alloys, ceramics, glass, crystal polymers and some amorphous polymers. In this chapter, the contact of highly elastic materials, such as elastomers is not considered.

The illustration of the mutual approach during viscoelastic contact between two smooth curvilinear surfaces of two solid bodies along axes X, Y

and Z relative to the initial point of contact A, is depicted in Figure 2.1. We can take the contact between two smooth surfaces as the kind of the contact, when $x >> R_{zi}$, where x is the deformation of compression between them and R_{zi} is the average height of roughness (asperity) of a more rough surface. Also, the contact surfaces of two rigid asperities can be considered as the contact between two smooth surfaces, too.

It is obvious that during the time of arbitrary mutual approach between the contacting surfaces, the shape of projection of the contact area on the tangential plane takes an elliptical or oval shape, see Figure 2.1.

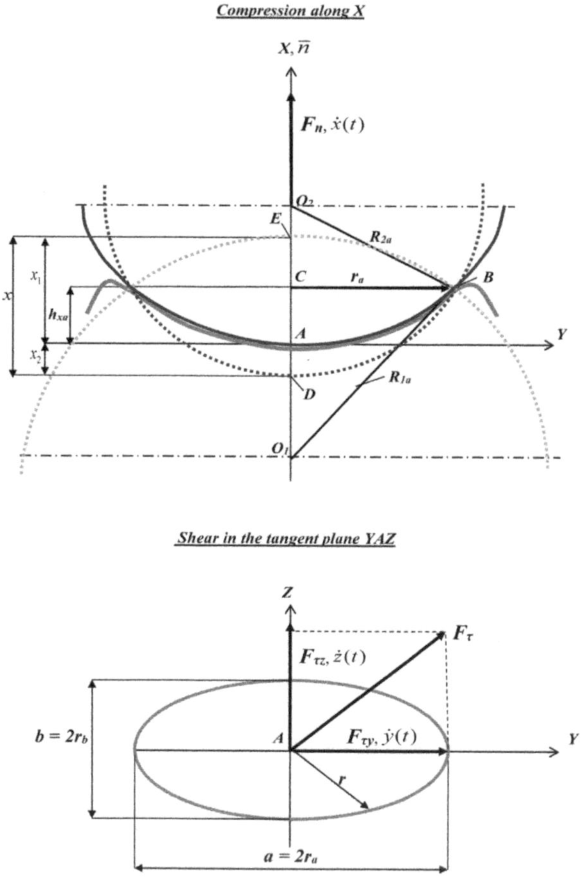

FIGURE 2.1 (See color insert.) Schematic illustration of the mutual approach between two curvilinear surfaces of two solid bodies along the axis X relative to the initial point of contact A.

Source: Reprinted from Goloshchapov, 2015b, http://dx.doi.org/10.5539/mer.v5n2p59 https://creativecommons.org/licenses/by/3.0/

You can say that in the process of sliding deformation under the action of tangential force, the area of contact does not take an elliptical shape, but is oval or maybe another asymmetric figure. Yes, indeed in the case of a fixed sliding contact between two surfaces, for example, when a hard indenter or a rigid asperity slides by the surface of a high elastic elastomer (if it is a very soft surface, when the modulus elasticity of rigid indenter is many times more than the modulus elasticity of this soft surface), the area of contact does not take an exact elliptical shape; it is oval. It was researched in many papers, for example by Putignano et al. (2016). The cases of sliding of the high viscoelastic and the elastic-plastic deformations will be considered further in this book in another chapter. But in the cases of rolling contact between the contacting surfaces or in the case of sliding contact between two rigid surfaces, we can take that the area of contact retains an elliptic shape. Moreover, in the method, which will be presented further in this chapter, the derivations of the viscoelastic forces are directly dependent on the minimum and maximum radius of curvature of the contact surface in planes *XAY* and *XAZ*, and therefore, it does not matter if the area of contact has an elliptical shape or an oval shape. In both cases, the derivation of expressions for sizes of the contact area and for all viscoelastic forces can be done in the same manner.

It is obvious that during the time of dynamic contact, at the initial point *A*, the curvilinear surfaces are moving relative to each other with the normal relative velocity $\dot{x}(t)$ and the tangential relative velocities of shear $\dot{y}(t)$ and $\dot{z}(t)$ between them under the action of normal force F_n and tangential shear force F_τ. Also, we can take that in the intimal moment of the time $t = 0$, the axis *X* coincides with the normal $\bar{n}$ between the contacting surfaces. Also let $F_{\tau y}$ and $F_{\tau z}$ be the tangential forces acting along the axes *Y* and *Z*, $x(t)$ is the relative deformations of compression between contacting surfaces along the axis *X*, $y(t)$, $z(t)$ are the relative deformations of shear between contacting surfaces along the axes *Y* and *Z* (the deformations of shear $y(t)$ and $z(t)$ are not shown here), r is the current radius of the contact area, h_{xa} is the depth of the contact surface in the plane *XAY* in the initial point of the contact *A*, or in other words, it is the depth of indentation of the surface of the harder body into the surface of the softer body, h_{xb} is the depth of the contact surface in the plane *XAZ* (this size is not shown in Figure 2.1), x_1 is the normal deformation of the compression of the surface of a softer body by *X* relative to the initial point of contact *A*, x_2 is the normal deformation of the compression of the surface of a harder body by *X* relative to the initial point of contact *A*, R_{1a} and R_{2a} are the radii of curvature of the contacting surfaces on the border of contact area in the plane *XAY*, a is the big axis of an elliptic or an oval contact area, b is the small axis of an elliptic or an oval contact area, O_1 and O_2 are the centres of curvature of the contacting surfaces in the initial point of contact A, R_{1b} and R_{2b} are the radii of curvature of the contacting surfaces on the border of the contact area in the plane *XAZ* (they are not shown in the Figure 2.1) Also, it is obvious that in the general case, the area of contact in the plane *YAZ* takes an oval or elliptical shape in the plane of *YAZ*, and at the same time, the contact surface takes the identical shape with the harder surface, and the surface of a soft body slides over the surface of the harder body. Also, let us take that the velocities $\dot{x}(t)$, $\dot{y}(t)$ and $\dot{z}(t)$ are less than the speed of waves of deformation (a sound speed) inside the volumes of deformations for both surfaces, and the depth of indentation is many times lesser than the radius of curvature of the contact area.

2.2 THE METHOD OF DIFFERENTIAL SPECIFIC FORCES (MDSF)

2.2.1 DIFFERENTIAL SPECIFIC FORCES

Let us assume that in the infinitesimal period of the time dt, when the mutual approach between surfaces of two bodies is the infinitesimal magnitude dx, see Figure 2.1, inside the elementary infinitesimal volume dV, which arises around the vicinity of the initial point of the contact A, see Figure 2.2, the differential specific elastic forces dF_{xic}, dF_{yic}, dF_{zic} and the differential specific viscous forces dF_{xib}, dF_{yib}, dF_{zib}, begin to act independently, it is very important that they are distributed in infinitesimal sizes da, db and dx parallel to axes X, Y and Z.

In other words, we can consider that these differential specific forces such as the infinitesimal distributed loads, placed by axes X, Y and Z and their dimension is [N/m]. (For remarks, please take in mind that here and further in this book, the index $i=1$ is used for forces in a softer body and $i=2$ is used for forces in a harder body, but equations for the effective forces will always be written without these indices.

Also, it is obvious that for infinitesimal period of the contact time dt, all deformations inside the volume of deformation are developing linearly and therefore, all differential specific forces are changing linearly, too, and thus, according to the linear theory of elasticity, the equations for these infinitesimal differential specific forces can be written as a series of linear functions as follows:

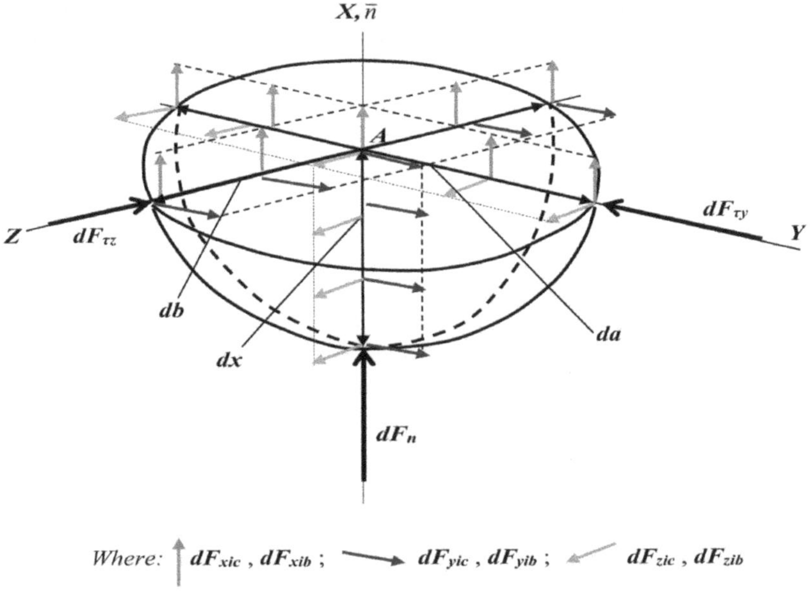

FIGURE 2.2 (See color insert.) Illustration of the action of the infinitesimal differential specific elastic and viscous forces inside the infinitesimal volume dV in the vicinity of point A.

Source: Adapted from Goloshchapov, 2015, with permission from SAGE Publications. http://journals.sagepub.com/doi/abs/10.1177/1056789514560912).

For the differential specific forces of a softer body, we can write as follows

$$dF_{x1c} = E_1' dx_1 \tag{2.1}$$

$$dF_{x1b} = \eta_{1E}' d\dot{x}_1 \tag{2.2}$$

$$dF_{y1c} = G_1' dy_1 \tag{2.3}$$

$$dF_{y1b} = \eta_{1G}' d\dot{y}_1 \tag{2.4}$$

$$dF_{z1c} = G_1' dz_1 \tag{2.5}$$

$$dF_{z1b} = \eta_{1G}' d\dot{z}_1 \tag{2.6}$$

where y_1 is the normal deformation of the shear of the surface of a softer body by axis Y relative to the initial point of contact A, z_1 is the normal deformation of the shear of the surface of a softer body by Z relative to the initial point of contact A, $\dot{x}_1$ is the velocity of the normal deformation of the compression of the surface of a softer body by X relative to the initial point of contact A, $\dot{y}_1$ is the velocity of the normal deformation of the shear of the surface of a softer body by axis Y, relative to the initial point of contact A, $\dot{z}_1$ is the velocity of the normal deformation of the shear of the surface of a softer body by Z relative to the initial point of contact A, F_{x1c} is the normal specific elastic force of a softer body, F_{x1b} is the normal specific viscous force of a softer body, F_{y1c} and F_{z1c} are the tangential elastic specific forces of the surface of a softer body, F_{y1b} and F_{z1b} are the tangential viscous specific forces of the surface of a softer body, E_1' is the dynamic elasticity module at the compression of the surface of a softer body, η_{1E}' is the dynamic viscosity at the compression of the surface of a softer body, G_1' is the dynamic elasticity module at the shear of the surface of a softer body and η_{1G}' is the dynamic viscosity at the shear of the surface of a softer body.

It should be remarked that the Young's elasticity moduli should be used for finding all the elastic forces. But, we also have to bear in mind that for the characterization of a viscoelastic contact between solids, we use the Kelvin–Voigt Model and according to this model, the Young's elasticity moduli E,E_1,E_2 and G,G_1,G_2 are equal to the dynamics elasticity moduli E', E_1', E_2' and G', G_1', G_2', which is also named as the dynamic storage moduli.

For the differential specific forces of a harder body, we can write as follows

$$dF_{x2c} = E_1' dx_2 \tag{2.7}$$

$$dF_{x2b} = \eta_{2E}' d\dot{x}_2 \tag{2.8}$$

$$dF_{y2c} = G_1' dy_2 \tag{2.9}$$

$$dF_{y2b} = \eta_{2G}' d\dot{y}_2 \tag{2.10}$$

$$dF_{z2c} = G_2' dz_2 \tag{2.11}$$

$$dF_{z2b} = \eta_{2G}' d\dot{z}_2 \tag{2.12}$$

where y_2 is the normal deformation of the shear of the surface of a harder body by axis Y relative to the initial point of contact A, z_2 is the normal deformation of the shear of the surface of a harder body by Z, relative to the initial point of contact A, $\dot{x}_2$ is the velocity of the normal deformation of the compression of the surface of a harder body by X relative to the initial point of contact A, $\dot{y}_2$ is the velocity of the normal deformation of the shear of the surface of a harder body by axis Y relative to the initial point of contact A, $\dot{z}_2$ is the velocity of the normal deformation of the shear of the surface of a harder body by Z relative to the initial point of contact A, F_{x2c} is the normal specific elastic forces of a harder body, F_{x2b} is the normal specific viscous force of a harder body, F_{x2c} and F_{x2c} are the tangential elastic specific forces of the surface of a harder body, F_{x2b} and F_{x2b} are the tangential viscous specific forces of the surface of a harder body, E_2' is the dynamic elasticity module at the compression of the surface of a harder body, η_{2E}' is the dynamic viscosity at the compression of the surface of a harder body, G_2' is the dynamic elasticity module at the shear of the surface of a harder body, η_{2G}' is the dynamic viscosity at the shear of the surface of a harder body.

For differential effective specific forces,

$$dF_{xc} = E'dx \tag{2.13}$$

$$dF_{xb} = \eta_E' d\dot{x} \tag{2.14}$$

$$dF_{yc} = G'dy \tag{2.15}$$

$$dF_{yb} = \eta_G' d\dot{y} \tag{2.16}$$

$$dF_{zc} = G'dz \tag{2.17}$$

$$dF_{zb} = \eta_G' d\dot{z} \tag{2.18}$$

where F_{xc} is the normal effective specific elastic force, F_{xb} is the normal effective specific viscous force, F_{yc} and F_{zc} are the tangential effective specific elastic forces, F_{yb} and F_{zb} are the tangential effective specific viscous forces, E' is the effective dynamic elasticity module at compression, η_E' is the effective dynamic viscosity at compression, G' is the effective dynamic elasticity module at shear, η_G' is the effective dynamic viscosity at shear.

According to the Kelvin–Voigt model, in the case of a viscoelastic contact with parallel action of the elastic and the viscous forces, we can write:

$$dF_{x1} = dF_{x1c} + dF_{x1b} \tag{2.19}$$

$$dF_{x2} = dF_{x2c} + dF_{x2b} \tag{2.20}$$

$$dF_{x} = dF_{xc} + dF_{xb} \tag{2.21}$$

$$dF_{y1} = dF_{y1c} + dF_{y1b} \tag{2.22}$$

$$dF_{y2} = dF_{y2c} + dF_{y2b} \tag{2.23}$$

$$dF_{y} = dF_{yc} + dF_{yb} \tag{2.24}$$

$$dF_{z1} = dF_{z1c} + dF_{z1b} \tag{2.25}$$

$$dF_{z2} = dF_{z2c} + dF_{z2b} \tag{2.26}$$

$$dF_z = dF_{zc} + dF_{zb} \tag{2.27}$$

where F_x is the normal effective specific viscoelastic force, F_y is the tangential effective specific viscoelastic force by axis Y, F_z is the tangential effective specific viscoelastic force by axis Z, F_{x1} and F_{x2} are the normal specific viscoelastic forces of the bodies by axis X, F_{y1} and F_{y2} are the tangential viscoelastic specific forces of the bodies by axis Y, F_{z1} and F_{z1} are the tangential viscoelastic specific forces by axis Z.

2.2.2 DEFINITION OF THE NORMAL DIFFERENTIAL SPECIFIC ELASTIC FORCES

Also, we can derive the equations for all differential specific forces in series of Equations 2.1–2.18) by another way. We can take that all deformations inside the infinitesimal volume of deformation change linearly for the infinitesimal period of the contact time *dt,* and therefore, all infinitesimal differential specific forces change linearly, too. Thus, equations for all infinitesimal differential specific forces can be written as the linear functions. Thus, the definition of these differential specific forces can be made by next method. Let us consider the contact of two infinitesimal by size Δh and absolutely identical cubes in the initial point A, which are in the process of compression under an action of the normal general differential specific force F_x, see Figure 2.3, and in the process of shear under an action of the tangential general specific force F_y and F_z, see Figure 2.4.

We can assume that in the infinitesimal period of the contact time, all deformations inside an infinitesimal volume are changing linearly. Based on this statement, the equations for all differential specific forces can be defined. First of all, the equations for the instant relative deformations in the compression phase can be written as follows (see Figure 2.3):

$$\varepsilon_{x1c} = \frac{\Delta h - (\Delta h - dx_1)}{\Delta h} = \frac{dx_1}{\Delta h} \tag{2.28}$$

$$\varepsilon_{x2c} = \frac{\Delta h - (\Delta h - dx_2)}{\Delta h} = \frac{dx_2}{\Delta h} \tag{2.29}$$

$$\varepsilon_{xc} = \frac{\Delta h - (\Delta h - dx)}{\Delta h} = \frac{dx}{\Delta h} \tag{2.30}$$

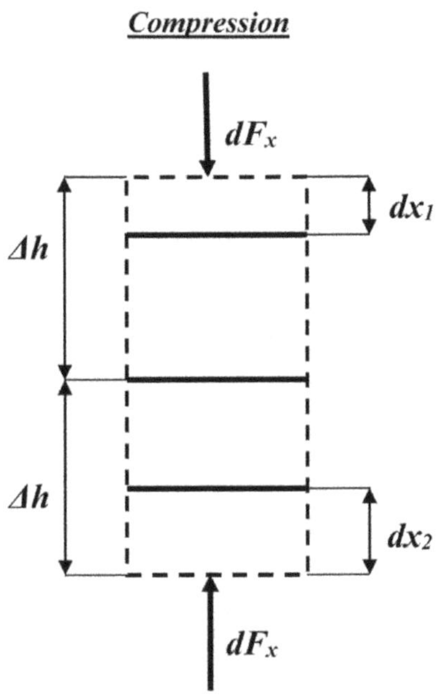

FIGURE 2.3 Illustration of the compression between two infinitesimal cubes of contacting surfaces in the initial point of contact *A*.

Source: Adapted from Goloshchapov, 2015a, with permission from SAGE Publications. http://journals.sagepub.com/doi/abs/10.1177/1056789514560912).

Then, according to the Hooke law the normal stresses of elasticity can be found as:

$$\sigma_{cn} = \frac{dx_1}{\Delta h} E_1' \tag{2.31}$$

$$\sigma_{cn} = \frac{dx_2}{\Delta h} E_2' \tag{2.32}$$

$$\sigma_{cn} = \frac{dx}{\Delta h} E' \tag{2.33}$$

where σ_{1cn} and σ_{2cn} are the normal stresses of elasticity of the contacting surfaces, σ_{cn} is the effective normal stress of elasticity of the contacting surfaces.

To represent the differential elastic specific forces such as $dF_{x1c} = \sigma_{1cn}\Delta h$, $dF_{x2c} = \sigma_{2cn}\Delta h$ and $dF_{xc} = \sigma_{cn}\Delta h$, the equations for normal differential elastic specific forces can be written as:

$$dF_{x1c} = E_2'dx_1 \tag{2.34}$$

$$dF_{x2c} = E_2'dx_2 \tag{2.35}$$

$$dF_{xc} = E'dx \tag{2.36}$$

2.2.3 DEFINITION OF THE NORMAL DIFFERENTIAL SPECIFIC VISCOUS FORCES

According to the Newton's formula for the stress of viscosity, we can write

$$\sigma_{1bn} = \eta_{1E}' \frac{d\dot{x}_1}{\Delta h} \tag{2.37}$$

$$\sigma_{2bn} = \eta_{2E}' \frac{d\dot{x}_2}{\Delta h} \tag{2.38}$$

$$\sigma_{bn} = \eta_{E}' \frac{d\dot{x}}{\Delta h} \tag{2.39}$$

where σ_{1bn} and σ_{2bn} are the normal stresses of viscosity of the contacting surfaces, σ_{bn} is the effective normal stress of viscosity of the contacting surfaces. If we take that $dF_{x1b} = \sigma_{1cn}\,\Delta h$, $dF_{x2b} = \sigma_{2bn}\,\Delta h$, $dF_{xb} = \sigma_{bn}\,\Delta h$, consequently, the equations for normal differential viscous specific forces can be written as:

$$dF_{x1b} = \eta_{1E}' d\dot{x}_1 \tag{2.40}$$

$$dF_{x2b} = \eta_{2E}' d\dot{x}_2 \tag{2.41}$$

$$dF_{xb} = \eta_{E}' d\dot{x} \tag{2.42}$$

2.2.4 DEFINITION OF THE TANGENTIAL DIFFERENTIAL SPECIFIC ELASTIC FORCES

The equations for the tangential viscoelastic forces of a sliding or rolling shear deformation can be found in the same way as for the compression. But, in the process of moving in a rolling shear along axis Y, the centres of mass of the contacting bodies have angular velocities $\vec{\omega}_{y1}$ and $\vec{\omega}_{y2}$ relative to the instant centre of velocities at the point of contact, and the relative rolling shear angular velocity $\vec{\Omega}_y$ can be found like the sum of these angular velocities, see Figure 2.4.

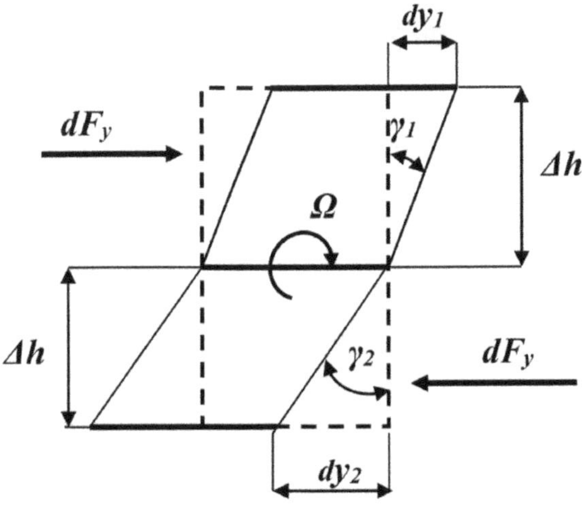

FIGURE 2.4 Illustration of the rolling shear between two infinitesimal cubes of contacting surfaces in the initial point of contact A.

And on the other hand, the tangential stresses of elasticity can be found by considering the stresses of a sliding or a rolling shear between two infinitesimal, identical by size cubes in the initial point of contact A as it follows below:

$$\sigma_{1cy} = G_1'\gamma_1 = G'\frac{dy_1}{\Delta h} \tag{2.43}$$

$$\sigma_{2cy} = G_2'\gamma_2 = G_2'\frac{dy_2}{\Delta h} \tag{2.44}$$

$$\sigma_{cy} = G_2'\gamma_2 = G_2'\frac{dy}{\Delta h} \tag{2.45}$$

where σ_{1cy} and σ_{2cy} are the tangential stresses of elasticity of the contacting surfaces, σ_{cy} is the effective tangential stress of elasticity of the contacting surfaces, γ_1 and γ_2 are the angles of shear. Analogically as for the differential elastic forces, if we take $dF_{y1c} = \sigma_{1cy}\,\Delta h$, $dF_{y2c} = \sigma_{2cy}\,\Delta h$, $dF_{yc} = \sigma_{cy}\,\Delta h$ the equation for tangential differential specific elastic forces can be written as:

$$dF_{y1c} = G_2'dy_1 \tag{2.46}$$

$$dF_{y2c} = G_2'dy_2 \tag{2.47}$$

$$dF_{yc} = G'dy \tag{2.48}$$

2.2.5 DEFINITION OF THE TANGENTIAL DIFFERENTIAL SPECIFIC VISCOUS FORCES

Analogically, we can find viscous specific forces using the known Newton's formula for viscous stress again:

$$\sigma_{1by} = \eta'_{1G} \frac{d\dot{y}_1}{dh} \tag{2.49}$$

$$\sigma_{2by} = \eta'_{2G} \frac{d\dot{y}_2}{dh} \tag{2.50}$$

$$\sigma_{by} = \eta'_{G} \frac{d\dot{y}}{dh} \tag{2.51}$$

where σ_{1by} and σ_{2bn} are the normal stresses of viscosity of the contacting surfaces, σ_{bn} is the effective normal stress of viscosity of contacting surfaces. And since we can take $dF_{y1b} = \sigma_{1by}\,\Delta h$, $dF_{y2b} = \sigma_{2by}\,\Delta h$, $dF_{yc} = \sigma_{by}\,\Delta h$ again, we get:

$$dF_{y1b} = \eta'_{1G}\, d\dot{y}_1 \tag{2.52}$$

$$dF_{y2b} = \eta'_{2G}\, d\dot{y}_2 \tag{2.53}$$

$$dF_{yb} = \eta'_{G}\, d\dot{y} \tag{2.54}$$

The model of a sliding or rolling shear along axis Z is the same as depicted in the Figure 2.4 for axis Y. Also, in the process of moving in a rolling shear along axis Z, the centres of mass of the contacting bodies have angular velocities $\vec{\omega}_{y1}$ and $\vec{\omega}_{y2}$ relative to the instant centre of velocities at the point of contact and the relative rolling shear angular velocity $\vec{\Omega}_z$ can be found by calculating the vector sum of these velocities. The equations for the tangential viscoelastic specific forces along axis Z can be found in the same way:

$$dF_{z1c} = G'_1 dz_1 \tag{2.55}$$

$$dF_{z2c} = G'_2 dz_2 \tag{2.56}$$

$$dF_{zc} = G' dz \tag{2.57}$$

$$dF_{z1b} = \eta'_{1G}\, d\dot{z}_1 \tag{2.58}$$

$$dF_{z2b} = \eta'_{2G}\, d\dot{z}_2 \tag{2.59}$$

$$dF_{zb} = \eta'_{G}\, d\dot{z} \tag{2.60}$$

2.2.6 THE EFFECTIVE DYNAMIC ELASTICITY MODULES AND DYNAMIC VISCOSITIES

According to 'Newton's Third Law,' the effective differential specific viscoelastic forces and the specific viscoelastic forces acting between two surfaces inside the elementary infinitesimal volume dV around the initial point of contact A, have to be equal:

$$dF_x = dF_{x1} = dF_{x2} \tag{2.61}$$

$$dF_y = dF_{y1} = dF_{y2} \tag{2.62}$$

$$dF_z = dF_{z1} = dF_{z2} \tag{2.63}$$

And we can write for specific forces as well,

$$F_x = F_{x1} = F_{x2} \tag{2.64}$$

$$F_y = F_{y1} = F_{y2} \tag{2.65}$$

$$F_z = F_{z1} = F_{z2} \tag{2.66}$$

In the case of viscoelastic contact, according to Kelvin–Voigt model, see Figure 2.5, when the specific elastic forces and the specific viscous forces act in parallel, the general specific forces can be found as the sum of the specific elastic forces and the specific viscous forces as follows:

$$F_x = F_{xb} + F_{xc} = F_{x1} = F_{x1b} + F_{x1c} = F_{x2} = F_{x2b} + F_{x2c} \tag{2.67}$$

$$F_y = F_{yb} + F_{yc} = F_{y1} = F_{y1b} + F_{y1c} = F_{y2} = F_{y2b} + F_{y2c} \tag{2.68}$$

$$F_z = F_{zb} + F_{zc} = F_{z1} = F_{z1b} + F_{z1c} = F_{z2} = F_{z2b} + F_{z2c} \tag{2.69}$$

We suppose that the elementary deformation between two bodies develops analogically like the deformation of the elementary discrete element, which is depicted in Figure 2.5a. It is a simple case of the linear model of deformations of elementary discrete elements, and instead of this model with four elements, we can use its analogy—the model with two effective elements depicted in Figure 2.5b. Also, the 'Elementary discrete elements model' for the normal forces can be used for the tangential forces in the same manner.

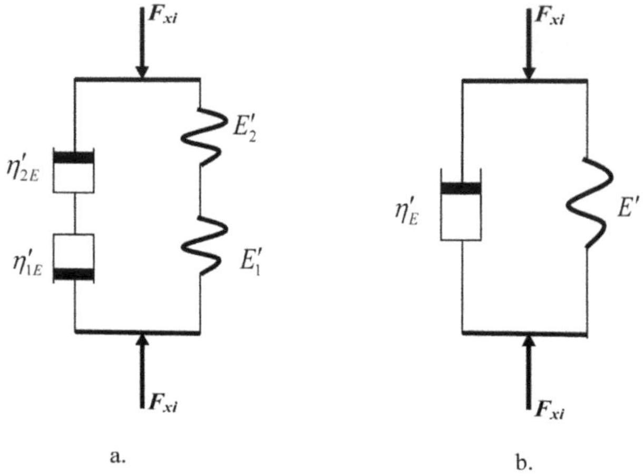

FIGURE 2.5 Illustration of the 'Elementary discrete elements Kelvin–Voigt model (EDEM)': (a) the elementary discrete element of the normal contact between two bodies; (b) the effective elementary discrete element of the normal contact.

Source: Reprinted from Goloshchapov, 2015a, with permission from SAGE Publications. http://journals.sagepub.com/doi/abs/10.1177/1056789514560912).

According to this model of contacts in Figure 2.5a and b, for the determination of the effective dynamic viscosity and the effective dynamic elasticity modules, we can summarize the elastic and viscous compliances as shown below:

$$\frac{1}{E'} = \frac{1}{E'_1} + \frac{1}{E'_2} \tag{2.70}$$

$$\frac{1}{\eta'_E} = \frac{1}{\eta'_{1E}} + \frac{1}{\eta'_{2E}} \tag{2.71}$$

$$\frac{1}{G'} = \frac{1}{G'_1} + \frac{1}{G'_2} \tag{2.72}$$

$$\frac{1}{\eta'_G} = \frac{1}{\eta'_{1G}} + \frac{1}{\eta'_{2G}} \tag{2.73}$$

Finally, the expressions for a calculation of the effective dynamic viscosities and the effective dynamic modules of elasticity can be written as:

$$E' = \frac{E'_1 E'_2}{E'_1 + E'_2} \tag{2.74}$$

$$\eta'_E = \frac{\eta'_{1E}\eta'_{2E}}{\eta'_{1E} + \eta'_{2E}} \tag{2.75}$$

$$G' = \frac{G'_1 G'_2}{G'_1 + G'_2} \tag{2.76}$$

$$\eta'_G = \frac{\eta'_{1G}\eta'_{2G}}{\eta'_{1G} + \eta'_{2G}} \tag{2.77}$$

We can derive Equations 2.74–2.77 in another way. It is obvious that the specific elastic forces are equal at the initial instance of the contact, when $t=0$, $x=0$ (at this point, they are equal to zero) and they are equal at the instance of the maximum compression between two bodies, when $x=x_m$ (at this point, they reach the maximum value). But at the same time, the specific viscous forces are equal at the initial instance of the contact, $t=0$, $x=0$ (at this point, they are equal to the maximum value) and they are equal at the instance of the maximum compression between two bodies, $x=x_m$ (at this point they are equal to zero). All of these forces are linear continuous functions and if they are equal for these two values of the argument x, they have to be equal for any other values also, or in other words, they are equal in any instance of the time of the contact. Thus, we can write:

$$F_{xc} = E'x = F_{x1c} = E'_1 x_1 = F_{x2c} = E'_2 x_2 \tag{2.78}$$

$$F_{xb} = \eta'_E \dot{x} = F_{x1b} = \eta'_{1E}\dot{x}_1 = F_{x2b} = \eta'_{2E}\dot{x}_2 \tag{2.79}$$

$$F_{yc} = G'y = F_{y1c} = G'_{1y1} = F_{y2c} = G'_{2y2} \tag{2.80}$$

$$F_{yb} = \eta'_G \dot{y} = F_{y1b} = \eta'_{1G}\dot{y}_1 = F_{y2b} = \eta'_{2G}\dot{y}_2 \tag{2.81}$$

$$F_{zc} = G'z = F_{z1c} = G'_1 z_1 F_{z2c} = G'_2 z_2 \tag{2.82}$$

$$F_{zb} = \eta'_G \dot{z} = F_{z1b} = \eta'_{1G}\dot{z}_1 = F_{y2b} = \eta'_{2G}\dot{z}_2 \tag{2.83}$$

According to Equations 2.78–2.83, since $x = x_1 + x_2$, $y = y_1 + y_2$, $z = z_1 + z_2$ we get the same equations again like in Equations 2.74–2.77.

Also, according to Equations 2.74, 2.76, 2.78, 2.80 and 2.82 we can write the next series of the equations:

For the normal deformations

$$x_1 = D_1 x \tag{2.84}$$

and

$$x_2 = D_2 x \tag{2.85}$$

where D_1 is the coefficient of normal deformation of the softer surface, D_2 is the coefficient of normal deformation of the harder surface, which can be calculated respectively as

$$D_1 = \frac{E_2'}{E_1' + E_2'} \tag{2.86}$$

$$D_2 = \frac{E_1'}{E_1' + E_2'} \tag{2.87}$$

For the tangential deformations, as

$$y_1 = D_{1\tau} y \tag{2.88}$$

$$z_1 = D_{1\tau} z \tag{2.89}$$

$$y_2 = D_{2\tau} y \tag{2.90}$$

$$z_2 = D_{2\tau} z \tag{2.91}$$

where $D_{1\tau}$ is the coefficient of the tangential elastic deformation of the softer surface, $D_{2\tau}$ is the coefficient of the tangential elastic deformation of the harder surface, which can be calculated respectively as

$$D_{1\tau} = \frac{G_2'}{G_1' + G_2'} \tag{2.92}$$

$$D_{2\tau} = \frac{G_1'}{G_1' + G_2'} \tag{2.93}$$

Further, since the effective differential specific viscoelastic force dF_{xc}, dF_{xb}, dF_{yc}, dF_{xb}, dF_{zc}, dF_{zb}, are distributed by sizes da, db and dx, as shown in Figure 2.2, we can write the twelve expressions for the effective differential viscoelastic forces acting along the axes X, Y and Z respectively as:

$$dF_{acn} = dF_{xc} da \tag{2.94}$$

$$dF_{abn} = dF_{xb} da \tag{2.95}$$

$$dF_{bcn} = dF_{cx} db \tag{2.96}$$

$$dF_{bbn} = dF_{xb} db \tag{2.97}$$

$$dF_{acz} = dF_{zc} da \tag{2.98}$$

$$dF_{abz} = dF_{zb} da \tag{2.99}$$

$$dF_{bcy} = dF_{yc}db \tag{2.100}$$

$$dF_{bby} = dF_{yb}db \tag{2.101}$$

$$dF_{xcz} = dF_{zc}dx \tag{2.102}$$

$$dF_{xbz} = dF_{zb}dx \tag{2.103}$$

$$dF_{xcy} = dF_{yc}dx \tag{2.104}$$

$$dF_{xby} = dF_{yb}dx \tag{2.105}$$

Where F_{acn}, F_{bcn}, F_{bcy}, F_{xcy}, F_{acz}, F_{xcz} are the elastic forces relative to the axes *da, db, dx*; F_{abn}, F_{bbn}, F_{bby}, F_{xby}, F_{abz}, F_{xbz} are the elastic forces relative to the axes *da, db, dx*, see Figure 2.2. Then, according to Equations 2.13–2.18, we get differential equations for the effective elastic and viscous forces:

$$dF_{acn} = E'dxda \tag{2.106}$$

$$dF_{abn} = \eta'_E d\dot{x}da \tag{2.107}$$

$$dF_{bcn} = E'dxdb \tag{2.108}$$

$$dF_{bbn} = \eta'_E d\dot{x}db \tag{2.109}$$

$$dF_{bcy} = G'dydb \tag{2.110}$$

$$dF_{bby} = \eta'_G d\dot{y}db \tag{2.111}$$

$$dF_{xcy} = G'dydx \tag{2.112}$$

$$dF_{xby} = \eta'_G d\dot{y}dx \tag{2.113}$$

$$dF_{xcz} = G'dydx \tag{2.114}$$

$$dF_{xbz} = \eta'_G d\dot{z}dx \tag{2.115}$$

$$dF_{acz} = G'dzda \tag{2.116}$$

$$dF_{abc} = \eta'_G d\dot{z}da \tag{2.117}$$

Now we can write the equations for the general differential viscoelastic forces, see Figure 2.2, respectively as

$$dF_n = dF_{cn} + dF_{bn} \tag{2.118}$$

$$dF_{\tau y} = dF_{cy} + dF_{by} \tag{2.119}$$

$$dF_{\tau z} = dF_{cz} + dF_{bz} \tag{2.120}$$

where,

$$dF_{cn} = dF_{acn} + dF_{bcn} \tag{2.121}$$

$$dF_{bn} = dF_{abn} + dF_{bbn} \tag{2.122}$$

$$dF_{cy} = dF_{bcy} + dF_{xcy} \tag{2.123}$$

$$dF_{by} = dF_{bby} + dF_{xby} \tag{2.124}$$

$$dF_{cz} = dF_{acz} + dF_{xcz} \tag{2.125}$$

$$dF_{bz} = dF_{abz} + dF_{xbz} \tag{2.126}$$

The series of Equations 2.78–2.83, 2.94–2.105, 2.106–2.117, 2.118–2.120 and 2.121–2.126 will be used further for finding the normal and tangential viscoelastic forces for both kinds of viscoelastic and elastic-plastic contacts.

2.2.7 GEOMETRY OF THE AREA OF CONTACT AND THE PRESSURE DISTRIBUTION

And now, the important moment, it can be seen that for finding the solutions for all Equations 2.106–2.117, we have to only know the equations or the formulas for $a=f(x)$, $b=f(x)$ and for $h_{xa}=f(x)$ and $h_{xb}=f(x)$. For example, we can use $r=(Rx)^{1/2}$ according to Hertz theory, but according to this theory, the area of contact is a flat surface and the depth of indentation (the depth of the contact surface) $h_x=0$. But in reality, the area of contact is usually not flat but is a curvilinear surface. In Hertz's theoretical models it has been taken that the contacting surfaces deform together without micro-sliding, but in reality, each surface deforms independently. Therefore, to find the radii of the contact area r_a and r_b in reality, first of all, let us consider the geometry of contact between two curvilinear surfaces in the normal section in the plane *XAY*, as it is depicted in the illustration in Figure 2.1. It is obvious that at the time of indentation of harder surface into a soft surface, the contact surface takes a curvilinear shape, where point *B* is a special point where the deformations always equal zero, and the border of the area of contact always passes through this point *B*. According to this statement, the distance O_2B between this point and the centre of curvature O_2 of the surface of the harder body will not be changed in the period of time of contact. This distance is always equal to the radius of curvature R_2. Also, the distance O_1B between this point and the centre of curvature O_1 of the surface of the lesser hard body will not be changed in the period of time of contact too. This distance is always equal to the radius of curvature R_1. Hence, it is obvious that $O_2B=O_2D=R_{2a}$ and $O_1B=O_1E=R_{1a}$, and we can also write that $O_1C + O_2C = (R_{1a} + R_{2a})$ and since $O_1C = (R^2_{1a} + r^2_a)^{1/2}$ and $O_2C = (R^2_{2a} + r^2_a)^{1/2}$ follows $[(R^2_{2a} + r^2_a)^{1/2} + (R^2_{1a} - r^2_a)^{1/2}]^2 = [(R_{1a} + R_{2a}) - x]^2$, then after simple calculation we respectively get, $r_a^2 - \frac{2r_a^2 x}{R_{1a}+R_{2a}} + \frac{r_a^2 x^2}{(R_{1a}+R_{2a})^2} = 2x\frac{R_{1a}R_{2a}}{R_{1a}+R_{2a}}$ $-x^2 + \frac{x^3}{R_{1a}+R_{2a}} - \frac{x^4}{4(R_{1a}+R_{2a})^2}$, and if we neglect the members of the smallest order, we get the equation for the radius of contact area in the plane *XAY as*

$$r^2_a = 2\, R_a x - x^2 \tag{2.127}$$

and then analogically in the same way we get the equation radius of the contact area in the plane *XAZ* respectively as

$$r^2_b = 2\, R_b x - x^2 \tag{2.128}$$

where $R_a = \frac{R_{1a}R_{2a}}{R_{1a}+R_{2a}}$ is the effective initial radius of contact curvature in the plane *XAY* and $R_b = \frac{R_{1b}R_{2ab}}{R_{1b}+R_{2b}}$ is the effective initial radius of contact curvature in the plane *XAZ*.

Equations 2.127 and 2.128 are not convenient for using and, therefore, let us rewrite them as

$$r^2_a = k^2_{pa} R_a x \tag{2.129}$$

$$r^2_b = k^2_{pb} R_b x \tag{2.130}$$

where k_{pb}, k_{pa} are the correlation coefficients, which can be calculated as

$$k_p = \sqrt{2 - \frac{x}{R}} \tag{2.131}$$

$$k_{pb} = \sqrt{2 - \frac{x}{R_b}} \tag{2.132}$$

Practically, for the solution of the contact problems, the correlation coefficient can be found by the method of iterations and a consecutive approximation.

Obviously from Equations 2.127 and 2.128, the section of the contact surface in the planes *XAY* and *XAZ* takes an elliptical shape, but for simplification, its shape can be approximated by parabolic functions as shown in Equations 2.129 and 2.130, respectively

$$x = \frac{1}{k_{pa}^2 R_a} r_a^2 \tag{2.133}$$

$$x = \frac{1}{k_{pb}^2 R_b} r_b^2 \tag{2.134}$$

Since the surface of the contact has an approximately parabolic shape, let us assume that the radial distribution of the pressure inside of this area changes analogically according to the parabolic function as given below

$$P = P_c \left(1 - \frac{r^2}{r_a^2}\right) \tag{2.135}$$

$$P = P_c \left(1 - \frac{r^2}{r_b^2}\right) \tag{2.136}$$

where r is the current radius of the contact area by axis Y, r_z is a current radius of the contact area by axis Z, P_c is the magnitude of the pressure in the centre of the contact area.

Since the area under functions in Equations 2.135 and 2.136 and the area under the linear function of the mean pressure P_m are equal, we can write that

$$P_c \int_0^{r_a} \left(1 - \frac{r_y^2}{r_a^2}\right) dr_y = P_a r_a \tag{2.137}$$

$$P_c \int_0^{r_b} \left(1 - \frac{r_z^2}{r_b^2}\right) dr_z = P_a r_b \tag{2.138}$$

Then, after the integration,

$$P_c \left(r_a - \frac{1}{3} r_a\right) = P_m r_a \tag{2.139}$$

$$P_c \left(r_b - \frac{1}{3} r_b\right) = P_m r_b \tag{2.140}$$

And, finally, the ratio between maximum and the mean pressure in the contact zone can be found as given below

$$P_c = \frac{3}{2} P_m \tag{2.141}$$

Now, let us define the depths of the contact surface in the planes *XAY* and *XAZ*. The expressions for the radii of contact area can also be found, see Figure 2.1, as follows

$$r_a^2 = R_{2a}^2 - (R_{2a}^2 - (x_2 + h_{xa}))^2 \tag{2.142}$$

$$r_b^2 = R_{2b}^2 - (R_{2b}^2 - (x_2 + h_{xb}))^2 \tag{2.143}$$

After simple algebraic transformations, neglecting members of the smallest order, we obtain the next equations for the radii of the contact area:

$$r_a^2 = 2R_{2a}(x_2 + h_{xa}) - (x_2 + h_{xa})^2 \tag{2.144}$$

$$r_b^2 = 2R_{2b}(x_2 + h_{xb}) - (x_2 + h_{xb})^2 \tag{2.145}$$

Then, after the comparison of Equations 2.144, 2.145 and 2.127, 2.128 we can write that

$$2R_{2a}(x_2 + h_{xa}) \approx 2R_a x \tag{2.146}$$

$$2R_{2b}(x_2 + h_{xb}) \approx 2R_b x \tag{2.147}$$

Finally, since $x_2 = D_2 x$, the formulas for h_{xa} and h_{xb} can be written as follows:

$$h_{xa} = \left(\frac{R_a - D_2 R_{2a}}{R_{2a}} \right) x = k_{ha} x \tag{2.148}$$

$$h_{xb} = \left(\frac{R_b - D_2 R_{2b}}{R_{2b}} \right) x = k_{hb} x \tag{2.149}$$

where k_{ha} and k_{hb} are the coefficients of depth of the contact surface, which can be found as the functions given below

$$k_{ha} = \left(\frac{R_a - D_2 R_{2a}}{R_{2a}} \right) \tag{2.150}$$

$$k_{hb} = \left(\frac{R_b - D_2 R_{2b}}{R_{2b}} \right) \tag{2.151}$$

Also, let us remark that in the case of a very small deformation, when $R_a >> x$ and $R_b >> x$, according to Equations 2.131 and 2.132, $k_{pa} \approx k_b \approx \sqrt{2}$ can be concluded, and we can write that

$$r_a = (2R_a x)^{1/2} \tag{2.152}$$

$$r_b = (2R_b x)^{1/2} \tag{2.153}$$

However, according to Hertz theory (Landau and Lifshitz, 1944, 1959), $r = (Rx)^{1/2}$, but it is possible only in the case when $k_{pa} = k_{pb} = 1$.

Since $a = 2r_a$ and $b = 2r_b$, according to Equations 2.129 and 2.130, we can write that

$$a = 2k_{pa} R_a^{1/2} x^{1/2} \tag{2.154}$$

$$b = 2k_{pb} R_b^{1/2} x^{1/2} \tag{2.155}$$

Then, incorporation of the derivatives da/dx and db/dx gives us the following equations

$$da = \frac{k_{pa} R_a^{1/2}}{x^{1/2}} dx \tag{2.156}$$

$$db = \frac{k_{pb} R_b^{1/2}}{x^{1/2}} dx \tag{2.157}$$

Now, since the expressions for da and db are already known, according to the series of Equations 2.106–2.117, we can write simple differential equations for all viscous and elastic forces as follows

$$dF_{acn} = k_{pa} R_a^{1/2} E' dx \frac{dx}{x^{1/2}} \tag{2.158}$$

$$dF_{abn} = k_{pa} R_a^{1/2} \eta_E' d\dot{x} \frac{dx}{x^{1/2}} \tag{2.159}$$

$$dF_{bcn} = k_{pb} R_b^{1/2} E' dx \frac{dx}{x^{1/2}} \tag{2.160}$$

$$dF_{bbn} = k_{pb} R_b^{1/2} \eta_E' d\dot{x} \frac{dx}{x^{1/2}} \tag{2.161}$$

$$dF_{bcy} = k_{pb} R_b^{1/2} G' dy \frac{dx}{x^{1/2}} \tag{2.162}$$

$$dF_{bby} = k_{pb} R_b^{1/2} \eta_G' d\dot{y} \frac{dx}{x^{1/2}} \tag{2.163}$$

$$dF_{xcy} = G' dy dx \tag{2.164}$$

$$dF_{xby} = \eta_G' d\dot{y} dx \tag{2.165}$$

$$dF_{acz} = k_{pa} R_a^{1/2} G' dz \frac{dx}{x^{1/2}} \tag{2.166}$$

$$dF_{abz} = k_{pa} R^{1/2} \eta'_G d\dot{z} \frac{dx}{x^{1/2}} \quad (2.167)$$

$$dF_{xcz} = G' dz dx \quad (2.168)$$

$$dF_{xbz} = \eta'_G d\dot{z} dx \quad (2.169)$$

2.2.8 DERIVATION OF EQUATIONS FOR THE VISCOELASTIC FORCES

Using the differential equations for dF_{acn} and dF_{bcn} from Equations 2.158 and 2.160, we get the effective normal elastic forces as given below

$$F_{acn} = k_{pa} E' R_a^{1/2} \int\int dx \frac{dx}{x^{1/2}} = \frac{2}{3} k_{pa} E' R_a^{1/2} x^{3/2} \quad (2.170)$$

$$F_{bcn} = k_{pb} E' R_b^{1/2} \int\int dx \frac{dx}{x^{1/2}} = \frac{2}{3} k_{pb} E' R_b^{1/2} x^{3/2} \quad (2.171)$$

Since $F_{cn} = F_{acn} + F_{bcn}$ follows that

$$F_{cn} = \frac{2}{3} E' x^{3/2} (k_{pa} R_a^{1/2} + k_{pb} R_b^{1/2}) = \frac{2}{3} E' \psi x^{3/2} \quad (2.172)$$

where $\psi = (k_{pa} R_a^{1/2} + k_{pb} R_b^{1/2})$, [m$^{1/2}$] is the parameter of curvature.

If $k_{pa} = k_{pb} = 1$ and if the contact area is a circle $R_a = R_b = R$, we have the same solution that has been obtained for the contact between spherical surfaces by using Hertz theory (Landau and Lifshitz, 1944, 1959) as follows

$$F_{cn} = \frac{4}{3} E' R^{1/2} x^{3/2} \quad (2.173)$$

Thus, it is obvious that the proposed method of finding the normal elastic forces is definitely valid and correct. It can be seen that if we know a functional dependency between r and x, we can always find the elastic force. But, if this method is correct for the definition of this force, hence, it should be valid for the definition of all viscoelastic forces in the series of Equations 2.158–2.169. The equation for the effective normal viscous forces can be found in the same way by integrating Equations 2.159 and 2.161:

$$F_{abn} = k_{pa} \eta'_E \int d\dot{x} \int \frac{R_a^{1/2}}{x^{1/2}} dx = 2 k_{pa} \eta'_E R_a^{1/2} \dot{x}(t) x^{1/2} \quad (2.174)$$

$$F_{bbn} = k_{pb}\eta'_E \int d\dot{x} \int \frac{R_b^{1/2}}{x^{1/2}} dx = 2k_{pb}\eta'_E R_b^{1/2} \dot{x}(t) x^{1/2} \tag{2.175}$$

And since $F_{bn} = F_{abn} + F_{bbn}$ follows that

$$F_{bn} = 2\eta'_E \dot{x} x^{1/2} (k_{pa} R_a^{1/2} + k_{pb} R_b^{1/2}) = 2\eta'_E \psi \dot{x} x^{1/2} \tag{2.176}$$

In the case when the contact area is a circle $k_{pa} = k_{pb} = k_p$ and $R_a = R_b = R$, we get,

$$F_{bn} = 4k_p \eta'_E R^{1/2} \dot{x} x^{1/2} \tag{2.177}$$

Since x and y are linearly independent and $h_{xa} = k_{ha}x, = h_{xb} = k_{hb}x$, after integration of the equations from series of Equations 2.162–2.169 for the effective tangential forces, their solutions can be written as follows:

$$F_{bcy} = k_{pb} R_b^{1/2} G' \int dy \int \frac{dx}{x^{1/2}} = 2k_{pb} G' R_b^{1/2} x^{1/2} y \tag{2.178}$$

$$F_{bby} = k_{pb} R_b^{1/2} \eta'_G \int d\dot{y} \int \frac{dx}{x^{1/2}} = 2k_{pb} \eta'_G R_b^{1/2} x^{1/2} \dot{y} \tag{2.179}$$

$$F_{xcy} = G' \int dy \int_0^{h_{xa}} dx = k_{ha} G' xy \tag{2.180}$$

$$F_{xby} = \eta'_G \int d\dot{y} \int_0^{h_{xa}} dx = k_{ha} \eta'_G x \dot{y} \tag{2.181}$$

$$F_{acz} = k_{pa} R_a^{1/2} G' \int dz \int \frac{dx}{x^{1/2}} = 2k_{pa} G' R_a^{1/2} x^{1/2} z \tag{2.182}$$

$$F_{abz} = k_{pa} R_a^{1/2} \eta'_G \int d\dot{z} \int \frac{dx}{x^{1/2}} = 2k_{pa} \eta'_G R_a^{1/2} x^{1/2} \dot{z} \tag{2.183}$$

$$F_{xcz} = G' \int dz \int_0^{h_{xb}} dx = k_{hb} G' xz \tag{2.184}$$

$$F_{xbz} = \eta'_G \int d\dot{z} \int_0^{h_{xb}} dx = k_{hb} \eta'_G x \dot{z} \tag{2.185}$$

The equations for the effective tangential viscoelastic forces can be finally written as the sum of the elastic and viscous tangential forces from the series of Equations 2.178–2.185:

$$F_{c\tau y} = F_{xcy} + F_{bcy} = G'P_{bx}y \quad (2.186)$$

$$F_{b\tau y} = F_{xby} + F_{bby} = \eta'_G P_{bx}\dot{y} \quad (2.187)$$

$$F_{c\tau z} = F_{xcz} + F_{acz} = G'P_{ax}z \quad (2.188)$$

$$F_{b\tau z} = F_{xbz} + F_{abz} = \eta'_G P_{ax}\dot{z} \quad (2.189)$$

where

$$P_{bx} = k_{ha}x + 2k_{pb}R_b^{1/2}x^{1/2} \quad (2.190)$$

$$P_{ax} = k_{hb}x + 2k_{pa}R_a^{1/2}x^{1/2} \quad (2.191)$$

Finally, using Equations 2.172, 2.176, 2.186–2.189, the next set of equations for the general viscoelastic forces can be respectively written as:

$$F_{cn} = F_{bn} + F_{cn} = 2\eta'_E \psi \dot{x}x^{1/2} + + \frac{2}{3}E'\psi x^{3/2} \quad (2.192)$$

$$F_{by} = F_{b\tau y} + F_{c\tau y} = \eta'_G P_{bx}\dot{y} + G'P_{bx}y \quad (2.193)$$

$$F_{\tau z} = F_{b\tau z} + F_{c\tau z} = \eta'_G P_{ax}\dot{z} + G'P_{ax}z \quad (2.194)$$

It is not possible to find the viscoelastic forces separately for each contacting body using Hertz theory and other existing theories, but it is possible by using the 'Method of the specific forces'. We can get the differential equations for the elastic and viscous forces separately for each contacting surface analogically as it was done for the series of Equations 2.106–2.117. Using Equations 2.1–2.6, the differential equations for the elastic and viscous forces for a softer surface can be expressed as:

$$dF_{acn1} = E'_1 dx_1 da \quad (2.195)$$

$$dF_{abn1} = \eta'_{1E} dx_1 da \quad (2.196)$$

$$dF_{acn1} = E'_1 dx_1 db \quad (2.197)$$

$$dF_{bbn1} = \eta'_{1E} dx_1 db \quad (2.198)$$

$$dF_{bcy1} = G'_1 dy_1 db \quad (2.199)$$

$$dF_{bby1} = \eta'_{1G} \dot{y}_1 db \quad (2.200)$$

$$dF_{xcy1} = G'_1 dy_1 dx \quad (2.201)$$

$$dF_{xby1} = \eta'_{1G} d\dot{y}_1 dx \quad (2.202)$$

$$dF_{xcz1} = G'_1 dy_1 dx \quad (2.203)$$

$$dF_{xbz1} = \eta'_{1G} d\dot{z}_1 dx \tag{2.204}$$

$$dF_{acz1} = G'_1 dz_1 da \tag{2.205}$$

$$dF_{abz1} = \eta'_{1G} d\dot{z}_1 da \tag{2.206}$$

And using Equations 2.7–2.12, we get differential equations for the elastic and viscous forces for the harder surface:

$$dF_{acn2} = E'_2 dx_2 da \tag{2.207}$$

$$dF_{abn2} = \eta'_{2E} d\dot{x}_2 da \tag{2.208}$$

$$dF_{bcn2} = E'_2 dx_2 db \tag{2.209}$$

$$dF_{bbn2} = \eta'_{2E} d\dot{x}_2 db \tag{2.210}$$

$$dF_{bcy2} = G'_2 dy_2 db \tag{2.211}$$

$$dF_{bby2} = \eta'_{2G} \dot{y}_2 db \tag{2.212}$$

$$dF_{xcy2} = G'_2 dy_2 dx \tag{2.213}$$

$$dF_{xby2} = \eta'_{2G} d\dot{y}_2 dx \tag{2.214}$$

$$dF_{xcz2} = G'_2 dz_2 dx \tag{2.215}$$

$$dF_{xbz2} = \eta'_{2G} d\dot{z}_2 dx \tag{2.216}$$

$$dF_{acz2} = G'_2 dz_2 da \tag{2.217}$$

$$dF_{abz2} = \eta'_{2G} d\dot{z}_2 da \tag{2.218}$$

Since $x = D_1^{-1} x_1 = D_2^{-1} x_2$, we can write that $\frac{dx}{x^{1/2}} = \frac{dx_1}{D_1^{1/2} x_1^{1/2}} = \frac{dx_2}{D_2^{1/2} x_2^{1/2}}$ and since $da = k_{pa} R_a^{1/2} \frac{dx}{x^{1/2}}$, $db = k_{pb} R_b^{1/2} \frac{dx}{x^{1/2}}$, we get

$$da = k_{pa} R_a^{1/2} \frac{dx_1}{D_1^{1/2} x_1^{1/2}} = k_{pa} R_a^{1/2} \frac{dx_2}{D_2^{1/2} x_2^{1/2}} \tag{2.219}$$

$$db = k_{pb} R_b^{1/2} \frac{dx_1}{D_1^{1/2} x_1^{1/2}} = k_{pb} R_b^{1/2} \frac{dx_2}{D_2^{1/2} x_2^{1/2}} \tag{2.220}$$

Substituting *da* and *db* from Equations 2.219 and 2.220 into Equations 2.195–2.206, we get the integral equations for a softer surface:

$$F_{acn1} = E'_1 \frac{k_{pa} R_a^{1/2}}{D_1^{1/2}} \int dx_1 \int \frac{dx_1}{x_1^{1/2}} \tag{2.221}$$

$$F_{abn1} = \eta'_{1E} \frac{k_{pa} R_a^{1/2}}{D_1^{1/2}} \int d\dot{x}_1 \int \frac{dx_1}{x_1^{1/2}} \tag{2.222}$$

$$F_{bcn1} = E'_1 \frac{k_{pb} R_b^{1/2}}{D_1^{1/2}} \int dx_1 \int \frac{dx_1}{x_1^{1/2}} \tag{2.223}$$

$$F_{bbn1} = \eta'_{1E} \frac{k_{pb} R_b^{1/2}}{D_1^{1/2}} \int d\dot{x}_1 \int \frac{dx_1}{x_1^{1/2}} \tag{2.224}$$

$$F_{bcy1} = k_{pb} R_b^{1/2} G'_1 D_1^{-1/2} \int dy_1 \int \frac{dx_1}{x_1^{1/2}} \tag{2.225}$$

$$F_{bby1} = k_{pb} R_b^{1/2} \eta'_{1G} D_1^{-1/2} \int d\dot{y}_1 \int \frac{dx_1}{x_1^{1/2}} \tag{2.226}$$

$$F_{xcy1} = G'_1 D_1^{-1} \int dy_1 \int_0^{h_{xa1}} dx_1 \tag{2.227}$$

$$F_{xby_1} = \eta'_{1G} D_1^{-1} \int d\dot{y}_1 \int_0^{h_{xa1}} dx_1 \tag{2.228}$$

$$F_{acz1} = k_{pa} R_a^{1/2} G'_1 D_1^{-1/2} \int dz_1 \int \frac{dx_1}{x_1^{1/2}} \tag{2.229}$$

$$F_{abz1} = k_{pa} R_a^{1/2} \eta'_{1G} D_1^{-1/2} \int d\dot{z}_1 \int \frac{dx_1}{x_1^{1/2}} \tag{2.230}$$

$$F_{xcz1} = G'_1 D_1^{-1} \int dz_1 \int_0^{h_{xb1}} dx_1 \tag{2.231}$$

$$F_{xbz1} = \eta'_{1G} D_1^{-1} \int d\dot{z}_1 \int_0^{h_{xb1}} dx_1 \tag{2.232}$$

Then, integration of the series of Equations 2.221–2.232 gives the following equations for elastic and viscous forces:

$$F_{acn1} = E'_1 \frac{2k_{pa} R_a^{1/2}}{3D_1^{1/2}} x_1^{3/2} \tag{2.233}$$

$$F_{abn1} = \eta'_{1E} \frac{2k_{pa} R_a^{1/2}}{D_1^{1/2}} \dot{x}_1 x_1^{1/2} \tag{2.234}$$

$$F_{bcn1} = E_1' \frac{2k_{pb}R_b^{1/2}}{2D_1^{1/2}} x_1^{3/2} \tag{2.235}$$

$$F_{bbn1} = \eta_{1E}' \frac{2k_{pb}R_b^{1/2}}{D_1^{1/2}} \dot{x}_1 x_1^{1/2} \tag{2.236}$$

$$F_{bcy1} = 2k_{pb}R_b^{1/2}G_1'D_1^{-1/2}y_1x_1^{1/2} \tag{2.237}$$

$$F_{bby1} = 2k_{pb}R_b^{1/2}\eta_{1G}'D_1^{-1/2}\dot{y}_1x_1^{1/2} \tag{2.238}$$

$$F_{xcy1} = G_1'D_1^{-1}k_{ha}y_1x_1 \tag{2.239}$$

$$F_{xby_1} = \eta_{1G}'D_1^{-1}k_{ha}\dot{y}_1x_1 \tag{2.240}$$

$$F_{acz1} = 2k_{pa}R_a^{1/2}G_1'D_1^{-1/2}z_1x_1^{1/2} \tag{2.241}$$

$$F_{abz1} = 2k_{pa}R_a^{1/2}\eta_{1G}'D_1^{-1/2}\dot{z}_1x_1^{1/2} \tag{2.242}$$

$$F_{xcz1} = G_1'D_1^{-1}k_{hb}z_1x_1 \tag{2.243}$$

$$F_{xbz1} = \eta_{1G}'D_1^{-1}k_{hb}\dot{z}_1x_1 \tag{2.244}$$

Since the normal elastic force $F_{cn1} = F_{acn1} + F_{bcn1}$ follows

$$F_{cn1} = \frac{2}{3}E_1'D_1^{-1/2}(k_{pa}R_a^{1/2} + k_{pb}R_b^{1/2})x_1^{3/2} = \frac{2}{3}E_1'D_1^{-1/2}\psi x_1^{3/2} \tag{2.245}$$

and since the normal viscous force $F_{bn1} = F_{abn1} + F_{bbn1}$, we get

$$F_{bn1} = 2\eta_{1E}'D_1^{-1/2}\dot{x}_1x_1^{1/2}(k_{pa}R_a^{1/2} + k_{pb}R_b^{1/2}) = 2\eta_{1E}'D_1^{-1/2}\psi\dot{x}_1x_1^{1/2} \tag{2.246}$$

The equations for the general tangential elastic forces can be finally written as:

$$F_{c\tau y1} = F_{xcy1} + F_{bcy1} = G_1'P_{bx1}y_1 \tag{2.247}$$

$$F_{c\tau z1} = F_{xcz1} + F_{acz1} = G_1'P_{ax1}z_1 \tag{2.248}$$

where

$$P_{bx1} = k_{ha}D_1^{-1}x_1 + 2k_{pb}R_b^{1/2}D_1^{-1/2}x_1^{1/2} \tag{2.249}$$

$$P_{ax1} = k_{hb}D_1^{-1}x_1 + 2k_{pa}R_a^{1/2}D_1^{-1/2}x_1^{1/2} \tag{2.250}$$

The equations for the general tangential viscous forces can be expressed respectively as

$$F_{b\tau y1} = F_{xby1} + F_{bby1} = \eta_{1G}'P_{bx1}\dot{y}_1 \tag{2.251}$$

$$F_{b\tau z1} = F_{xbz1} + F_{abz1} = \eta'_{1G} P_{ax1} \dot{z}_1 \tag{2.252}$$

Thus, the equations for the general viscoelastic forces for a softer surface can be obtained as the sums given below:

$$F_{n1} = F_{bn1} + F_{cn1} = 2\eta'_{1E} D_1^{-1/2} \psi \dot{x}_1 x_1^{1/2} + \frac{2}{3} E'_1 D_1^{-1/2} \psi x_1^{3/2} \tag{2.253}$$

$$F_{\tau y1} = F_{b\tau y1} + F_{c\tau y1} = \eta'_{1G} P_{bx1} \dot{y}_1 + G'_1 P_{bx1} y_1 \tag{2.254}$$

$$F_{\tau z1} = F_{b\tau z1} + F_{c\tau z1} = \eta'_{1G} P_{ax1} \dot{z}_1 + G'_1 P_{ax1} z_1 \tag{2.255}$$

Equations for the elastic and viscous forces for a harder surface are derived in the same order:

$$F_{acn2} = E'_2 \frac{2k_{pa} R_a^{1/2}}{3D_2^{1/2}} x_2^{3/2} \tag{2.256}$$

$$F_{abn2} = \eta'_{2E} \frac{2k_{pa} R_a^{1/2}}{D_2^{1/2}} \dot{x}_2 x_2^{1/2} \tag{2.257}$$

$$F_{bcn2} = E'_2 \frac{2k_{pb} R_b^{1/2}}{3D_2^{1/2}} x_2^{3/2} \tag{2.258}$$

$$F_{bbn2} = \eta'_{2E} \frac{2k_{pb} R_b^{1/2}}{D_2^{1/2}} \dot{x}_2 x_2^{1/2} \tag{2.259}$$

$$F_{bcy2} = 2k_{pb} R_b^{1/2} G'_2 D_2^{-1/2} y_2 x_2^{1/2} \tag{2.260}$$

$$F_{bby2} = 2k_{pb} R_b^{1/2} \eta'_{2G} D_2^{-1/2} \dot{y}_2 x_2^{1/2} \tag{2.261}$$

$$F_{xcy2} = G'_2 D_2^{-1} k_{ha} y_2 x_2 \tag{2.262}$$

$$F_{xby_2} = \eta'_{2G} D_2^{-1} k_{ha} \dot{y}_2 x_2 \tag{2.263}$$

$$F_{acz2} = 2k_{pa} R_a^{1/2} G'_2 D_2^{-1/2} z_2 x_2^{1/2} \tag{2.264}$$

$$F_{abz2} = 2k_{pa} R_a^{1/2} \eta'_{2G} D_2^{-1/2} \dot{z}_2 x_2^{1/2} \tag{2.265}$$

$$F_{xcz2} = G'_2 D_2^{-1} k_{hb} z_2 x_2 \tag{2.266}$$

$$F_{xbz2} = \eta'_{2G} D_2^{-1} k_{hb} \dot{z}_2 x_2 \tag{2.267}$$

Since $F_{cn2} = F_{acn2} + F_{bcn2}$ follows

$$F_{cn2} = \frac{2}{3} E'_2 D_2^{-1/2} (k_{pa} R_a^{1/2} + k_{pb} R_b^{1/2}) x_2^{3/2} = \frac{2}{3} E'_2 D_2^{-1/2} \psi x_2^{3/2} \tag{2.268}$$

and since $F_{bn2} = F_{abn2} + F_{bbn2}$

$$F_{bn2} = 2\eta'_{2E} D_2^{-1/2} \dot{x}_2 x_2^{1/2} (k_{pa} R_a^{1/2} + k_{pb} R_b^{1/2}) = 2\eta'_{2E} D_2^{-1/2} \psi \dot{x}_2 x_2^{1/2} \quad (2.269).$$

The equations for the general tangential elastic forces can be written as the sum:

$$F_{c\tau y2} = F_{xcy2} + F_{bcy2} = G' P_{bx2} y_2 \quad (2.270)$$

$$F_{c\tau z2} = F_{xcz2} + F_{acz2} = G' P_{ax2} z_2 \quad (2.271)$$

where

$$P_{bx2} = k_{ha} D_2^{-1} x_2 + 2k_{pb} R_b^{1/2} D_2^{-1/2} x_2^{1/2} \quad (2.272)$$

$$P_{ax2} = k_{hb} D_2^{-1} x_2 + 2k_{pa} R_a^{1/2} D_2^{-1/2} x_2^{1/2} \quad (2.273)$$

and the equations for general tangential viscous forces can be expressed as

$$F_{b\tau y2} = F_{xby2} + F_{bby2} = \eta'_{2G} P_{bx2} \dot{y}_2 \quad (2.274)$$

$$F_{b\tau z2} = F_{xbz2} + F_{abz2} = \eta'_G P_{ax2} \dot{z}_2 \quad (2.275)$$

The equations for general viscoelastic forces for a harder surface can be written as sums given below:

$$F_{n2} = F_{bn2} + F_{cn2} = 2\eta'_{2E} D_2^{-1/2} \psi \dot{x}_2 x_2^{1/2} + \frac{2}{3} E'_2 D_2^{-1/2} \psi x_2^{3/2} \quad (2.276)$$

$$F_{\tau y2} = F_{b\tau y2} + F_{c\tau y2} = \eta'_{2G} P_{bx2} \dot{y}_2 + G'_2 P_{bx2} y_2 \quad (2.277)$$

$$F_{\tau z2} = F_{b\tau z2} + F_{c\tau z2} = \eta'_{2G} P_{ax2} \dot{z}_2 + G'_2 P_{ax2} z_2 \quad (2.278)$$

Thus, we can find the viscoelastic forces separately for each body; hence, the viscoelastic stresses as well. But, it is not possible to do this by using Hertz theory.

As we can see, the method of the differential specific viscoelastic forces allows us to find the equations for all viscoelastic forces. The proposed method is principally different from others which use Hertz theory, the classical theory of elasticity and tensor algebra. In this method, new conception is proposed on how to find the elastic and viscous forces by the integration of the specific forces in the infinitesimal boundaries of the contact area. The radius of contact area can be taken according to Hertz theory or can be found by considering the geometry of the contact. This method can be used in researches of contact dynamics of any shape of contacting surfaces. Also, the method of solution of the differential equations of a movement has been proposed here and it has been solved. This method can also be used for the determination of the dynamic mechanical properties of materials, and it can be used in the design of wear-resistant elements and coverings for components of machines and equipment, which are utilized in harsh conditions where they are subjected to the action of flow or jet abrasive particles. Also, the theoretical and experimental results that are presented here can be useful in the design of elements and details of machines as well as mechanism present during the conditions of the dynamic contact. The method of the specific forces presented in this chapter can be used for the determination of the viscoelastic forces, contact stresses, durability and fatigue life for a wide spectrum of the tasks relevant to collisions between solid bodies under different loading conditions. Opportunities exist to use the obtained results practically in the design and development of new advanced materials, wear-resistant elastic coatings and elements for pneumatic and hydraulic systems, stop valves, fans, centrifugal pumps, injectors, valves, gate valves and in other installations. Also, the use of this theory gives an opportunity for the development of analytical and experimental methods, allowing the optimization of the basic dynamics and mechanical viscoelastic qualities of the already existing materials, and in the development of new advanced materials and elements of machines. This theory can be used not only for a viscoelastic contact but also for any other kind of contacts such as a high elastic contact as well as for an elasto-plastic contact.

2.3 WORK OF VISCOELASTIC FORCES IN A VISCOELASTIC SLIDING CONTACT BETWEEN CURVILINEAR SURFACES OF TWO SOLID BODIES

2.3.1 WORK OF THE NORMAL VISCOELASTIC FORCES

Since the equations for general viscoelastic forces have already been derived, we can now find the work of these forces as in the phase of loading as well as in the phase of unloading.

As we already know, the time period of the dynamic contact includes two principally different phases such as the phase of the deformation of a compression or a shear and the phase of the restitution. The work A_{xcm} of the normal elastic force F_{cn} in the compression phase can be found by integration as

$$A_{xcm} = \int_0^{x_m} F_{cn} dx = \frac{2}{3} E'\psi \int_0^{x_m} x^{3/2} dx = \frac{4}{15} E'\psi x_m^{5/2} \tag{2.279}$$

where x_m is the maximum magnitude of the compression (or the maximum of mutual approach) between contacting surfaces. And analogically, the work A_{xct} of the normal elastic force in the restitution phase, when $x = x_t = 0$, can be found as

$$A_{xct} = -\int_{x_m}^{0} F_{cn} dx = -\frac{2}{3} E'\psi \int_{x_m}^{0} x^{3/2} dx = \frac{4}{15} E'\psi x_m^{5/2} \tag{2.280}$$

It is obvious that $A_{xcm} = A_{xct}$ because the potential energy which has accumulated in the phase of compression fully returns in the phase of restitution.

Now, as $F_{bn} = 2\eta'_E \psi \dot{x} x^{1/2}$, we can write the next equation for the work of the dissipative viscous force in the phase of compression as follows:

$$A_{xbm} = \int_0^{x_m} F_{bn} dx = 2\eta'_E \psi \frac{\int_0^{x_m} x^{1/2} \int dx dx}{\int_0^{\tau_1} dt} = \frac{4\psi \eta'_E x_m^{5/2}}{5\tau_1} \tag{2.281}$$

where τ_1 is the time period of compression.
Analogically, the work of the dissipative viscous force in the restitution phase, when $x = x_t = 0$, can be written as follows

$$A_{xbt} = -\int_{x_m}^{0} F_{bn} dx = -2\eta'_E \psi \frac{\int_{x_m}^{0} x^{1/2} \int dx dx}{\int_{\tau_1}^{\tau_x} dt} = \frac{4\psi \eta'_E x_m^{5/2}}{5\tau_2} \tag{2.282}$$

where $\tau_x = \tau_1 + \tau_2$ is the time period of contact, τ_2 is the time period of restitution. As we can see that $A_{xbm} = A_{xbt,}$ but $A_{xbm} \neq Axbt$.

Now the equations for the work of the normal viscoelastic forces in the compression and the restitution can be written as follows:

$$A_{xm} = A_{xcm} + A_{xbm} = \frac{4}{15} \psi x_m^{5/2} \left(E' + \frac{3\eta'_E}{\tau_1} \right) \tag{2.283}$$

$$A_{xt} = A_{xct} - A_{xbt} = \frac{4}{15} \psi x_m^{5/2} \left(E' - \frac{3\eta'_E}{\tau_2} \right) \tag{2.284}$$

Also, the linear theory of elasticity (Ferry, 1948; Flügge, 1975; Moore, 1975; Van Krevelen, 1972; Nilsen, 1978; McCrum et al. 1997; Lakes, 1998; Menard, 1999; Hosford, 2005; Roylance, 2001) follows that

$$tg\beta_1 = \frac{E_1''}{E_1'} = \frac{G_1''}{G_1'} \tag{2.285}$$

$$tg\beta_2 = \frac{E_2''}{E_2'} = \frac{G_2''}{G_2'} \tag{2.286}$$

where E_2'' is the viscosity modulus of a harder surface, G_1'' is the viscosity modulus at shear of a softer surface, G_2'' is the viscosity modulus at shear of a harder surface, β_1 is the angle of mechanical losses of a softer surface, β_2 is the angle of mechanical losses of a harder surface, and also

$$\frac{E_1''}{\omega_x} = \eta_{1E}' \tag{2.287}$$

$$\frac{E_2''}{\omega_x} = \eta_{2E}' \tag{2.288}$$

$$\frac{G_1''}{\omega_y} = \eta_{1G}' \tag{2.289}$$

$$\frac{G_2''}{\omega_y} = \eta_{2G}' \tag{2.290}$$

where ω_x is the normal angular frequency of oscillations (frequency of loading) of the volume of deformation V_d by the axis X, ω_y is the tangential angular frequency of oscillations (frequency of loading) of a sliding contact by the axes Y and Z (Or further, ω_{yr} denotes the frequency of oscillation in the case of a rolling contact).

It is obvious that we can write analogically for the effective tangent of mechanical losses

$$tg\beta_E = \frac{E''}{E'} \tag{2.291}$$

$$tg\beta_G = \frac{G''}{G'} \tag{2.292}$$

where E'' is the effective viscosity modulus, E'' is the effective viscosity modulus at shear, β_E is the effective angle of mechanical losses, β_G is the effective angle of mechanical losses at a shear, and we can write that as

$$\frac{E''}{\omega_x} = \eta_E' \tag{2.293}$$

$$\frac{G''}{\omega_y} = \eta'_G \tag{2.294}$$

Substituting Equation 2.75 into 2.293 and Equation 2.77 into 2.294 we get

$$\frac{E''}{\omega_x} = \frac{\eta'_{1E}\dot{\eta}_{2E}}{\eta'_{1E} + \eta'_{2E}} \tag{2.295}$$

$$\frac{G''}{\omega_y} = \frac{\eta'_{1G}\dot{\eta}_{2G}}{\eta'_{1G} + \eta'_{2G}} \tag{2.296}$$

Then using Equations 2.287, 2.288 and 2.295 and also Equations 2.289, 2.290 and 2.296, we can obtain that

$$E'' = \frac{E''_1 E''_2}{E''_1 + E''_2} \tag{2.297}$$

$$G'' = \frac{G''_1 G''_2}{G''_1 + G''_2} \tag{2.298}$$

Further, substituting E' and G' from Equations 2.74, 2.76 and E'' and G'' from Equations 2.297, 2.298 into Equations 2.291, 2.292, the expressions for the tangents of angles of mechanical losses can be respectively written as

$$tg\beta_E = \frac{E''_1 E''_2 (E'_1 + E'_2)}{E'_1 E'_2 (E''_1 + E''_2)} \tag{2.299}$$

$$tg\beta_G = \frac{G''_1 G''_2 (G'_1 + G'_2)}{G'_1 G'_2 (G''_1 + G''_2)} \tag{2.300}$$

Taking into account Equations 2.285 and 2.286, we simply get

$$tg\beta_E = tg\beta_1 tg\beta_2 \frac{(E'_1 + E'_2)}{(tg\beta_1 E'_1 + tg\beta_2 E'_2)} \tag{2.301}$$

$$tg\beta_G = tg\beta_1 tg\beta_2 \frac{(G'_1 + G'_2)}{(tg\beta_1 G'_1 + tg\beta_2 G''_2)} \tag{2.302}$$

Thus, we have to be aware that usually $tg\beta_E \neq tg\beta_G$. They will be equal only in one case when $tg\beta_1 \neq tg\beta_2$.

Now taking into account Equations 2.291 and 2.293, Equations 2.283 and 2.284 can be rewritten as

$$A_{xm} = A_{xcm} + A_{xbm} = \frac{4}{15}\psi x_m^{5/2} E'\left(1 + \frac{3tg\beta_E}{\omega_x \tau_1}\right) \quad (2.303)$$

$$A_{xt} = A_{xct} - A_{xbt} = \frac{4}{15}\psi x_m^{5/2} E'\left(1 - \frac{3tg\beta_E}{\omega_x \tau_2}\right) \quad (2.304)$$

Since the works in phases of compression and restitution are already known, we can define the energetic coefficient of restitution e_x, which is equal to the square of the kinematic coefficient of restitution k_x (further, it will be simply named as the coefficient of restitution), as the ratio between A_{xt} and A_{xm}, specifically is

$$e_x = k_x^2 = \frac{A_{xt}}{A_{xm}} = \left(\frac{\omega_x \tau_2 - 3tg\beta_E}{\omega_x \tau_1 + 3tg\beta_E}\right)\frac{\tau_1}{\tau_2} \quad (2.305)$$

Also, we can find the coefficient of the mechanical losses of compression as given below

$$\chi_x = 1 - e_x = \frac{3\tau_x tg\beta_E}{(\omega_x \tau_1 + 3tg\beta_E)\tau_2} \quad (2.306)$$

2.3.2 WORK OF THE TANGENTIAL VISCOELASTIC FORCES AT A SLIDING CONTACT

Since $\frac{G''}{\omega_y} = \eta'_G$, see Equations 2.294, the work at a shear of the tangential viscoelastic forces $F_{b\tau y}$, $F_{c\tau y}$, and $F_{b\tau z}$, $F_{c\tau z}$, see Equations 2.186–2.189, in the phase of the compression can be found by integration as follows:

$$A_{ycm} = \int_0^{y_m} F_{c\tau y}\, dy = G' \int_0^{P_{bxm}} dP_{bx} \int_0^{y_m} y\, dy = \frac{G'}{2} P_{bxm} y_m^2 \quad (2.307)$$

$$A_{ybm} = \int_0^{y_m} F_{b\tau y}\, dy = \frac{G''}{\omega_y} \int_0^{P_{bxm}} dP_{bx} \int_0^{y_m} \dot{y}\, dy = \frac{G''}{\omega_y} P_{bxm} \frac{\int_0^{y_m} \int dy dy}{\int_0^{\tau_1} dt} = \frac{G''}{2\omega_y} P_{bxm} \frac{y_m^2}{\tau_1} \quad (2.308)$$

$$A_{zcm} = \int_0^{z_m} F_{c\tau z}\, dz = G' \int_0^{P_{axm}} dP_{ax} \int_0^{z_m} z\, dz = \frac{G'}{2} P_{axm} z_m^2 \quad (2.309)$$

$$A_{zbm} = \int_0^{z_m} F_{b\tau z} dz = \frac{G''}{\omega_y} \int_0^{P_{axm}} dP_{ax} \int_0^{z_m} \dot{z} dz = \frac{G''}{\omega_y} P_{axm} \frac{\int_0^{z_m} \int dz dz}{\int_0^{\tau_1} dt} = \frac{G''}{2\omega_y} P_{axm} \frac{z_m^2}{\tau_1} \quad (2.310)$$

where y_m and z_m are the relative shears between the contacting surfaces along axis Y and Z at the instance of time $t = \tau_1$,

$$P_{bxm} = P_{bx}(x_m) = k_{ha} x_m + 2k_{pb} R_b^{1/2} x_m^{1/2} \quad (2.311)$$

$$P_{axm} = P_{ax}(x_m) = k_{hb} x_m + 2k_{pa} R_a^{1/2} x_m^{1/2} \quad (2.312)$$

The work of all tangential viscoelastic forces $F_{b\tau y}$, $F_{c\tau y}$ and $F_{b\tau z}$, $F_{c\tau z}$ in the phase of the restitution can be found by integration as:

$$A_{yct} = -\int_{y_m}^{y_t} F_{c\tau y} dy = -G' \int_{P_{bxm}}^{0} dP_{bx} \int_{y_m}^{y_t} y dy = \frac{G'}{2} P_{bxm} (y_t^2 - y_m^2) \quad (2.313)$$

$$A_{ybt} = -\int_{y_m}^{y_t} F_{b\tau y} dy = -\frac{G''}{\omega_y} \int_{P_{bxm}}^{0} dP_{bx} \int_{y_m}^{y_t} \dot{y} dy = \frac{G''}{\omega_Y} P_{bxm} \frac{\int_{y_m}^{y_t} \int dy dy}{\int_{\tau_1}^{\tau_x} dt} = \frac{G''}{2\omega_y} P_{bxm} \frac{y_t^2 - y_m^2}{\tau_2} \quad (2.314)$$

$$A_{zct} = -\int_{z_m}^{z_t} F_{c\tau z} dz = -G' \int_{P_{axm}}^{0} dP_{ax} \int_{z_m}^{z_t} z dz = \frac{G'}{2} P_{axm} (z_t^2 - z_m^2) \quad (2.315)$$

$$A_{zbt} = -\int_{z_m}^{z_t} F_{b\tau z} dz = -\frac{G''}{\omega_y} \int_{P_{axm}}^{0} dP_{ax} \int_{z_m}^{z_t} \dot{z} dz = \frac{G''}{\omega_Y} P_{axm} \frac{\int_{z_m}^{z_t} \int dz dz}{\int_{\tau_1}^{\tau_x} dt} = \frac{G''}{2\omega_y} P_{axm} \frac{z_t^2 - z_m^2}{\tau_2} \quad (2.316)$$

where y_t and z_t are the relative deformations of shear between contacting surfaces along axes Y and Z at the instance of time $t = \tau_x$. The direction of the displacement in the restitution phase by the axis Y is not changing, therefore, the integrals for the works were taken with the plus sign.

The equations for work of the tangential viscoelastic forces can be found as

$$A_{yc} = A_{ycm} + A_{yct} = \frac{G'}{2} P_{bxm} y_t^2 \quad (2.317)$$

$$A_{yb} = A_{ybm} + A_{ybt} = \frac{G''}{2\omega_y} P_{bxm} \left(\frac{y_m^2 \tau_2 + \tau_1 (y_t^2 - y_m^2)}{\tau_1 \tau_2} \right) \quad (2.318)$$

$$A_{zc} = A_{zcm} + A_{zct} = \frac{G'}{2} P_{axm} z_t^2 \quad (2.319)$$

$$A_{zb} = A_{zbm} + A_{zbt} = \frac{G''}{2\omega_y} P_{axm} \left(\frac{z_m^2 \tau_2 + \tau_1 (z_t^2 - z_m^2)}{\tau_1 \tau_2} \right) \tag{2.320}$$

The dissipative energy W_b at the shear equals the dissipative work A_b and is equal to the sum of the work of the viscous tangential forces, and can be found as follows

$$W_b = A_b = A_{yb} + A_{zb} = \frac{G''}{2\tau_1\tau_2\omega_y} \left(P_{bxm} [y_m^2 \tau_2 + \tau_1 (y_t^2 - y_m^2)] + P_{axm} [z_m^2 \tau_2 + \tau_1 (z_t^2 - z_m^2)] \right) \tag{2.321}$$

The energy of elasticity W_c at the shear equals to the work of elasticity A_c and the sum of the work of the elastic tangential forces, and can be found as follows

$$A_c = W_c = A_{yc} + A_{zc} = \frac{G'}{2} (P_{bxm} y_t^2 + P_{axm} z_t^2) \tag{2.322}$$

Taking into account that $tg\beta_G = \frac{G''}{G'}$, the full energy, $W_t = W_c + W_b$, at the shear can be determined as

$$W_t = \frac{G'}{2} \left[(P_{bxm} y_t^2 + P_{axm} z_t^2) + \frac{tg\beta_G}{\tau_1 \tau_2 \omega_y} \left[P_{bxm} [y_m^2 \tau_2 + \tau_1 (y_t^2 - y_m^2)] + P_{axm} [z_m^2 \tau_2 + \tau_1 (z_t^2 - z_m^2)] \right] \right] \tag{2.323}$$

And now, the equations for the coefficient of mechanical losses can be written as

$$\chi_y = \frac{W_b}{W_t} = \frac{tg\beta_G \{ P_{bxm} [y_m^2 \tau_2 + \tau_1 (y_t^2 - y_m^2)] + P_{axm} [z_m^2 \tau_2 + \tau_1 (z_t^2 - z_m^2)] \}}{\tau_1 \tau_2 \omega_y (P_{bxm} y_t^2 + P_{axm} z_t^2) + tg\beta_G \{ P_{bxm} [y_m^2 \tau_2 + \tau_1 (y_t^2 - y_m^2)] + P_{axm} [z_m^2 \tau_2 + \tau_1 (z_t^2 - z_m^2)] \}} \tag{2.324}$$

2.4 GENERAL CONTACT STRESSES

2.4.1 INTRODUCTION

It is obvious that the contact stresses of a compression and shear arise under the action of the viscoelastic forces in the contact area during the time of a mutual approach between the contacting surfaces. The contact stresses of elasticity arise as a result of the compression and shear of structures of the volume of deformation V_d in the contact area, but at the same time, the contact stresses of viscosity arise in the process of the internal friction of these structures inside of the volume of deformation in this contact space.

According to the hypothesis of maximum tangential stresses, the equation for the general contact stresses of viscoelasticity in the centre of the contact area can be expressed as

$$\sigma_A = \sqrt{\sigma_n^2 + 4\sigma_{\tau c}^2} \tag{2.325}$$

where σ_n is the normal contact stresses of viscoelasticity and $\sigma_{\tau c}$ is the tangential contact stresses of viscoelasticity.

The most dangerous values of contact stresses occur at the points when the values of the elastic and viscosity forces reach their maximum. The problem is in finding these points and then in the calculation of the contact stress in them.

2.4.2 NORMAL GENERAL CONTACT STRESSES

The normal contact stresses of viscoelasticity in the centre of the contact area should be equal to the maximum of pressure $\sigma_n = P_c = \frac{3}{2}P_a$, see Equation 2.141, and since $P_a = \frac{F_n}{S_x}$, we get

$$\sigma_n = \frac{3F_n}{2S_x} \tag{2.326}$$

where $S_x = \frac{\pi}{4}ab$ is the square of the contact area, and according to Equations 2.154 and 2.155, we can conclude that $S_x = \pi k_{pa} k_{pb} R_a^{1/2} R_b^{1/2} x$, and the expression for general normal viscoelastic, known from Equation 2.192, follows

$$\sigma_n = \frac{\psi(3\eta_E' \dot{x} + E'x)}{\pi k_{pa} k_{pb} R_a^{1/2} R_b^{1/2} x^{1/2}} \tag{2.327}$$

2.4.3 TANGENTIAL GENERAL CONTACT STRESSES

It is obvious that the tangential contact stresses of viscoelasticity arise in the tangential section S_x (plane *YAZ*), in the tangential section S_y (plane *XAY*) and in the tangential section S_z (plane *XAZ*), see Figure 2.2. Therefore, the equation for the average tangential viscoelastic stresses in the section S_x, which act along axis *Y* can be expressed as

$$\sigma_{\tau by} = \frac{F_{by}}{S_x} = \frac{2(G'y + \eta'_G \dot{y})}{\pi k_{pa} R_a^{1/2} x^{1/2}} \tag{2.328}$$

where

$$F_{by} = F_{bcy} + F_{bby} = 2k_{pb} R_b^{1/2} x^{1/2} (G'y + \eta'_G \dot{y}) \tag{2.329}$$

see also Equations 2.178 and 2.179.

On the other hand, the equation for the average tangential viscoelastic stresses in the section S_y, acting along axis Y can be written as

$$\sigma_{\tau xy} = \frac{F_{xy}}{S_y} = \frac{3(G'y + \eta'_G \dot{y})}{4k_{ha}^{1/2} R_a^{1/2} x^{1/2}} \tag{2.330}$$

where

$$F_{xy} = F_{xcy} + F_{xby} = k_{ha} x (G'y + \eta'_G \dot{y}) \tag{2.331}$$

see Equations 2.178 and 2.179, and where S_y can be found by integration $dS_y = 2r_a dx$, see Figure 2.1 and 2.2,

$$S_y = 2\int_0^{h_{xa}} r_a dx = 2R_a^{1/2} \int_0^{h_{xa}} x^{1/2} dx = \frac{4}{3} R_a^{1/2} h_{xa}^{3/2} \tag{2.332}$$

and since $h_{xa} = k_{ha} x$, it follows

$$S_y = \frac{4}{3} k_{ha}^{3/2} R_a^{1/2} x^{3/2} \tag{2.333}$$

Now, the average tangential viscoelastic stresses, acting along the axis Y can be expressed as the sum

$$\sigma_{\tau y} = \sigma_{\tau by} + \sigma_{\tau xy} = \varsigma_y \frac{G'y + \eta'_G \dot{y}}{R_a^{1/2} x^{1/2}} \tag{2.334}$$

where

$$\varsigma_y = \left(\frac{2}{\pi k_{pa}} + \frac{3}{4k_{ha}^{1/2}} \right) \tag{2.335}$$

The equation for the average tangential viscoelastic stresses in the section S_x, which act along axis Z can be obtained as

$$\sigma_{\tau az} = \frac{F_{az}}{S_x} = \frac{2(G'z + \eta'_G \dot{z})}{\pi k_{pb} R_b^{1/2} x^{1/2}} \tag{2.336}$$

where

$$F_{az} = F_{acz} + F_{abz} = 2k_{pa} R_a^{1/2} x^{1/2} (G'z + \eta_G' \dot{z}) \tag{2.337}$$

see Equations 2.182 and 2.183.

On the other hand, the equation for the average tangential viscoelastic stresses in the section S_z, acting along axis Z can be expressed as

$$\sigma_{\tau xz} = \frac{F_{xz}}{S_z} = \frac{3(G'z + \eta_G' \dot{z})}{4k_{hb}^{1/2} R_b^{1/2} x^{1/2}} \tag{2.338}$$

where

$$F_{xz} = F_{xcz} + F_{xbz} = k_{hb} x (G'z + \eta_G' \dot{z}) \tag{2.339}$$

see Equations 2.184 and 2.185, and where S_z can be found by integration $dS_z = 2r_b dx$, see Figure 2.1 and 2.2,

$$S_z = 2\int_0^{h_{xb}} r_b dx = 2R_b^{1/2} \int_0^{h_{xb}} x^{1/2} dx = \frac{4}{3} R_a^{1/2} h_{xb}^{3/2} \tag{2.340}$$

and since $h_{xb} = k_{hb} x$, it follows

$$S_z = \frac{4}{3} k_{hb}^{3/2} R_b^{1/2} x^{3/2} \tag{2.341}$$

Now, the average tangential stresses acting along axis Z can be written as sum

$$\sigma_{\tau z} = \sigma_{\tau az} + \sigma_{\tau xz} = \varsigma_z \frac{G'z + \eta_G' \dot{z}}{R_b^{1/2} x^{1/2}} \tag{2.342}$$

where

$$\varsigma_z = \left(\frac{2}{\pi k_{pb}} + \frac{3}{4k_{hb}^{1/2}} \right) \tag{2.343}$$

The equation for the general average tangential viscoelastic stresses can be written as

$$\sigma_\tau = \sqrt{\sigma_{\tau y}^2 + \sigma_{\tau z}^2} \tag{2.344}$$

Substituting Equation 2.334 and 2.342 into Equation 2.344, we get

$$\sigma_\tau = \frac{1}{x^{1/2}} \sqrt{\varsigma_y^2 \frac{(G'y + \eta_G' \dot{y})^2}{R_a} + \varsigma_z^2 \frac{(G'z + \eta_G' \dot{z})^2}{R_b}} \tag{2.345}$$

Since the surface of the contact has an approximately parabolic shape and the magnitude of shear on the border of contact area equals to zero, we can take that the radial distribution of the tangential stress inside of this area changes analogically according to the parabolic function as

$$\sigma_\tau = \sigma_{\tau c}\left(1-\frac{r_y^2}{r_a^2}\right) \tag{2.346}$$

$$\sigma_\tau = \sigma_{\tau c}\left(1-\frac{r_z^2}{r_b^2}\right) \tag{2.347}$$

where r_y is the current radius of the contact area by axis Y, r_z is the current radius of the contact area by axis Z, P_c is the maximum magnitude of the stress in the centre of the contact area.

Further, since the square under these functions in Equations 2.346 and 2.347 and the square under the linear function of the mean stress in the contact area are equal, we can write that

$$\sigma_{\tau c}\int_0^{r_a}\left(1-\frac{r_y^2}{r_a^2}\right)dr_y = \sigma_\tau r_a \tag{2.348}$$

$$\sigma_{\tau c}\int_0^{r_b}\left(1-\frac{r_z^2}{r_b^2}\right)dr_z = \sigma_t r_b \tag{2.349}$$

Then, after the integration follows

$$\sigma_{\tau c}\left(r_a-\frac{1}{3}r_a\right)=\sigma_\tau r_a \tag{2.350}$$

$$\sigma_{\tau c}\left(r_b-\frac{1}{3}r_b\right)=\sigma_\tau r_b \tag{2.351}$$

and finally, the ratio between maximum and the mean tangential stress in the contact zone can be found as

$$\sigma_{\tau c} = \frac{3}{2}\sigma_\tau \tag{2.352}$$

Thus, we can write

$$\sigma_{\tau c} = \frac{3}{2x^{1/2}}\sqrt{\varsigma_y^2\frac{(G'y+\eta_G'\dot{y})^2}{R_a}+\varsigma_z^2\frac{(G'z+\eta_G'\dot{z})^2}{R_b}} \tag{2.353}$$

Finally, according to Equation 2.325, we get the equation for the general contact stresses of viscoelasticity.

$$\sigma_A = \frac{1}{x^{1/2}}\sqrt{\frac{\psi^2(E'x+\eta'_E\dot{x})^2}{\pi^2 k_{pa}^2 k_{pb}^2 R_a R_b} + 9\left(\varsigma_y^2 \frac{(G'y+\eta'_G\dot{y})^2}{R_a} + \varsigma_z^2 \frac{(G'z+\eta'_G\dot{z})^2}{R_b}\right)} \quad (2.354)$$

2.5 EXAMPLES OF SOLUTIONS OF CONTACT PROBLEMS BETWEEN SOLIDS

2.5.1 CONTACT BETWEEN TWO ELLIPSOIDS

As we already know, the viscoelastic forces are directly dependent on the shape of the contact area. Therefore, we have to consider the geometry of the contact area in each case of contact. For example, the case of an arbitrary direction of a compression and a shear between two ellipsoids is depicted in the Figure 2.6.

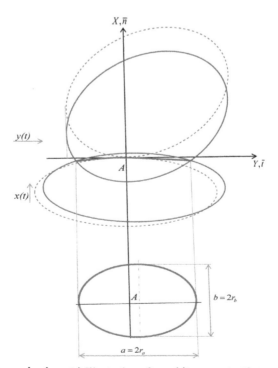

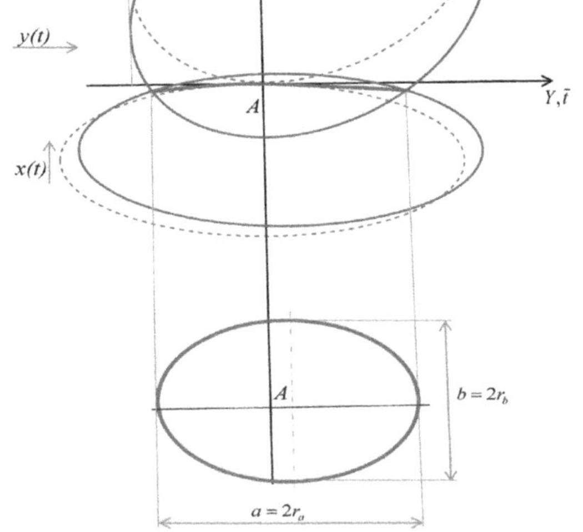

FIGURE 2.6 (See color insert.) Illustration of an arbitrary contact between two ellipsoids.

As we can see here, the area of contact in the tangential plane $\overline{t}$ takes the approximate shape of an oval, even if a displacement between the contacting surfaces has a big size. Also, as we know, if the viscoelastic deformation of the rigid body will exceed a few per cent, plastic deformation will begin to develop. We can use Equation 2.40 for finding the viscoelastic forces in this case. But, it is obvious that in any kind of contact (viscoelastic or elastic-plastic), the basic problem in the application of the MDSF is finding sizes of radii of curvature and sizes a and b of the contact area. For example, an elliptical contact has a place in the next case such as in the contact between spherical surface and cylindrical surface, in contact between the spherical and toroidal surface, for example, in rolling contact bearings, in contact between two cylinders with crossed axes. In all these cases, we can find the sizes of the contact area by using Equation 2.25 and the formulas $R_a = \dfrac{R_{1a}R_{2a}}{R_{1a}+R_{2a}}$, $R_b = \dfrac{R_{1b}R_{2ab}}{R_{1b}+R_{2b}}$ are the effective radii of contact curvature between convex surfaces, $R_a = \dfrac{R_{1a}R_{2a}}{R_{1a}-R_{2a}}$, $R_b = \dfrac{R_{1b}R_{2ab}}{R_{1b}-R_{2b}}$ are the effective radii of curvature in a case when one of the surfaces is concave (see also Figure 2.1).

2.5.2 VISCOELASTIC CONTACT BETWEEN TWO SPHERICAL BODIES

A contact between two spheres of the radii $R_1 = R_{1a}$ and $R_2 = R_{2a}$, and the initial velocities V_x and V_y, ($V_z=0$) come into viscoelastic contact at the initial instance of the time $t=0$, at the initial point of contact A, see Figure 2.1. Also, in the case of the contact between two spheres, when the contact area is a circle $R_a=R_b=R$, $r_a=r_b=r$, $a=b=2r$, $F_{\tau z}=0$, $F_\tau=F_{\tau y}$, $F_{b\tau}=F_{b\tau y}$, $F_{c\tau}=F_{c\tau y}$ follows that $k_p = k_{pa} = k_{pb}$, $k_x = k_{xa} = h_{xb}$, $k_h = k_{ha} = h_{hb}$, $\psi = 2k_pR^{1/2}$. Thus, according to Equations 2.172, 2.176 and 2.186, 87, we get the following equations for normal and tangential elastic and viscous forces:

$$F_{cn} = \frac{4}{3}E'R^{1/2}x^{3/2} \tag{2.355}$$

$$F_{bn} = 4k_p\eta_E'R^{1/2}\dot{x}x^{1/2} \tag{2.356}$$

$$F_{c\tau} = G'P_xy \tag{2.357}$$

$$F_{b\tau} = \eta_G'P_x\dot{y} \tag{2.358}$$

where $R = \dfrac{R_1 R_2}{R_1 + R_2}$ is the effective radius of curvature between convex surfaces, $R = \dfrac{R_1 R_2}{R_1 - R_2}$ is the effective radius of curvature in a case when one of the surfaces is concave, and where

$$P_x = k_h x + 2k_p R^{1/2} x^{1/2} \tag{2.359}$$

If Equations 1.1–1.3 are taken into account, we get the system of equations for the general viscoelastic forces:

$$F_n = F_{cn} + F_{bn} = \frac{4}{3} E' R^{1/2} x^{3/2} + 4k_p \eta'_E R^{1/2} \dot{x} x^{1/2} \tag{2.360}$$

$$F_\tau = F_{b\tau} + F_{c\tau} = \eta'_G P_x \dot{y} + G' P_x y \tag{2.361}$$

where the variable viscoelasticity parameters

$$b_x = 4\eta'_E R^{1/2} x^{1/2} \tag{2.362}$$

$$c_x = \frac{4}{3} E' R^{1/2} x^{1/2} \tag{2.363}$$

$$b_y = \eta'_G P_x \tag{2.364}$$

$$c_y = G' P_x \tag{2.365}$$

Also, for this case, the expression for the depth of the contact surface can be written as follows

$$h_x = k_h x \tag{2.366}$$

where

$$k_h = \left(\frac{R - D_2 R_2}{R_2} \right) \tag{2.367}$$

In the case of contact of a spherical body and a semi-space, since $R = R_2$.

$$k_h = 1 - D_2 = D_1 \tag{2.368}$$

2.5.3 VISCOELASTIC CONTACT BETWEEN TWO CYLINDERS WITH PARALLEL AXES

The contact area between two cylinders with parallel axes is a parallelogram, see Figure 2.7.

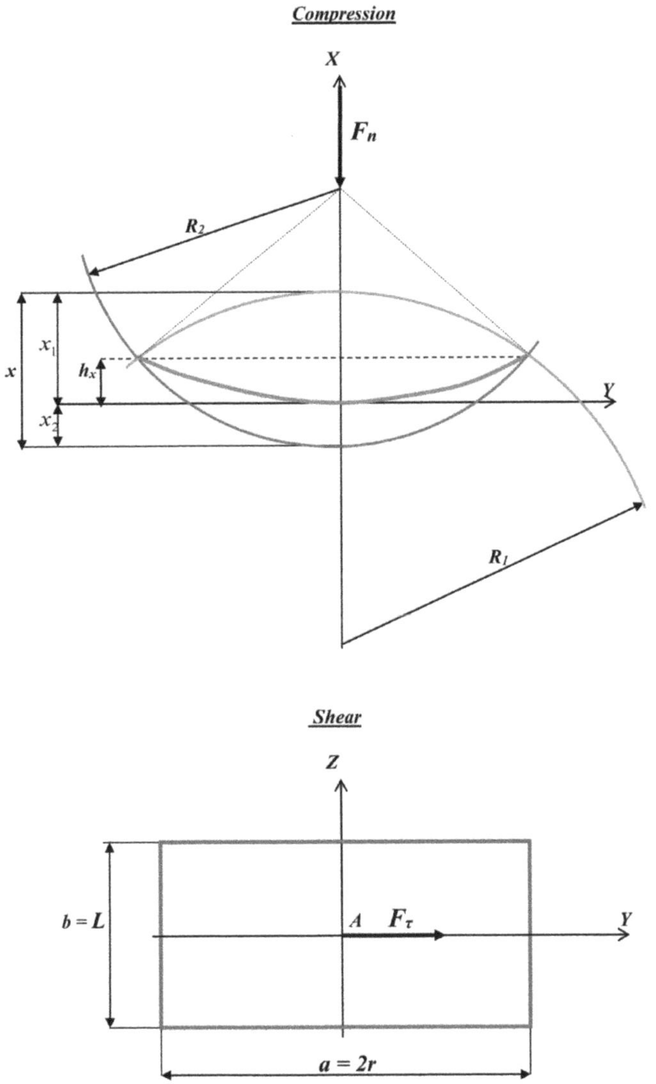

FIGURE 2.7 (See color insert.) Illustration of the contact between two cylinders with parallel axes.

In this case, let us denote that $L=b$ is the length of the contact area, a is the width of the contact area, and since $F_{\tau z}=0$, see Figure 2.1 and 2.7, x and y are linearly independent and $da=\frac{k_p R^{1/2}}{x^{1/2}}dx$, and $h_x=k_h x$, Equations 2.170, 2.171, 2.174, 2.175 as well as, in this case Equations 2.178–2.185 can be written in integral form as follows:

$$F_{acn}=k_{pa}E'R_a^{1/2}\int\int dx\frac{dx}{x^{1/2}}=\frac{2}{3}k_p E'R^{1/2}x^{3/2} \quad (2.369)$$

$$F_{bcn}=E'\int dx\int dL=E'xL \quad (2.370)$$

$$F_{abn}=k_{pa}\eta'_E\int^{\dot{x}} d\dot{x}\int^{x}\frac{R_a^{1/2}}{x^{1/2}}dx=2k_p\eta'_E R^{1/2}\dot{x}x^{1/2} \quad (2.371)$$

$$F_{bbn}=\eta'_E\int d\dot{x}\int dL=2\eta'_E\dot{x}L \quad (2.372)$$

$$F_{bcy}=G'\int dy\int dL=G'Ly \quad (2.373)$$

$$F_{bby}=\eta'_G\int d\dot{y}\int dL=\eta'_G L\dot{y} \quad (2.374)$$

$$F_{xcy}=G'\int dy\int_0^{h_{xa}}dx=k_h G'xy \quad (2.375)$$

$$F_{xby}=\eta'_G\int d\dot{y}\int_0^{h_{xa}}dx=k_h\eta'_G x\dot{y} \quad (2.376)$$

The viscous and elastic forces can be found after the summation of all partial elastic and viscous forces as follows:

$$F_{cn}=F_{acn}+F_{bcn}=\frac{2}{3}k_p E'R^{1/2}x^{3/2}+E'xL \quad (2.377)$$

$$F_{bn}=F_{abn}+F_{bbn}=2k_p\eta'_E R^{1/2}\dot{x}x^{1/2}+2\eta'_E\dot{x}L \quad (2.378)$$

$$F_{c\tau y}=F_{bcy}+F_{xcy}=G'Ly+k_h G'xy \quad (2.379)$$

$$F_{b\tau y}=F_{bby}+F_{xby}=\eta'_G L\dot{y}+k_h\eta'_G x\dot{y} \quad (2.380)$$

Thus, finally, according to Equations 1.1–1.3, we can get the next system of equations for general viscoelastic forces:

$$F_n = F_{bn} + F_{cn} = 2\eta'_E (k_p R^{1/2} x^{1/2} + L)\dot{x} + E'x\left(\frac{2}{3}k_p R^{1/2} x^{1/2} + L\right) \quad (2.381)$$

$$F_\tau = F_{b\tau y} + F_{c\tau y} = \eta'_G P_{xL}\dot{y} + G' P_{xL} y \quad (2.382)$$

where

$$P_{xL} = k_h x + L \quad (2.383)$$

Since usually L>>$k_h x$ and $L >> \frac{2}{3}k_p R^{1/2} x^{1/2}$, Equations 2.381 and 2.382 can be rewritten as

$$F_n = 2\eta'_E L\dot{x} + E'Lx \quad (2.384)$$

$$F_\tau = F_{b\tau y} + F_{c\tau y} = \eta'_G L\dot{y} + G'Ly \quad (2.385)$$

As we can see in this case that $F_{cn} = E'Lx$, this result completely matches the known formula for the contact between two cylinders with parallel axes.

2.5.4 VISCOELASTIC CONTACT BETWEEN AN AXIS-SYMMETRICAL CURVILINEAR SURFACE OF INDENTER OR ASPERITY AND A FLAT SEMI-SPACE

The contact area between an axis-symmetrical curvilinear surface of indenter or asperity and a flat surface of a semi-space is a circle, see Figure 2.8.

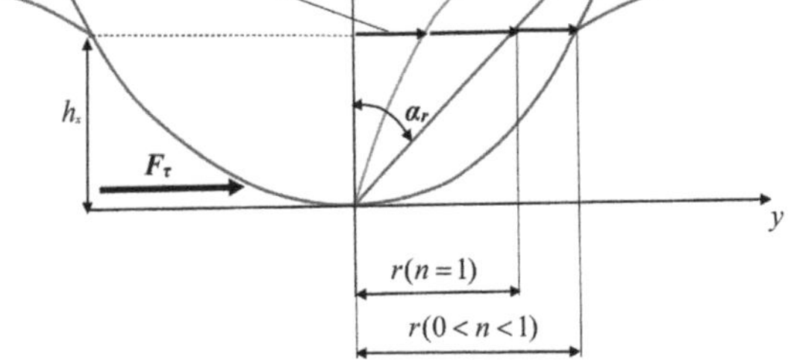

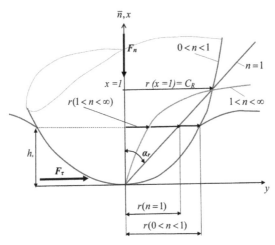

FIGURE 2.8 (See color insert.) Illustration of the contact between an axis-symmetrical curvilinear surface of indenter or asperity and the flat surface of a semi-space.

In a general case, the radius of generatrix of the curvilinear surface can be expressed as function

$$r = C_r x^{1/n} \tag{2.386}$$

where $0 < n < \infty$ and as we can see here, the dimension of C_r is $[m^{\frac{n-1}{n}}]$. If $1 < n < \infty$, we have the paraboloidal or hyperboloidal shapes of the indenter/asperity. On the other hand, if $1 < n < \infty$, Equation 2.386 describes the inverse paraboloidal or hyperboloidal shape of the indenter/asperity. In the case when $n = 1$, we get the conical shape of the indenter, where $C_r = tg\alpha_r$ and where α_r denotes the angle between the generatrix of the cone and the axis X, which coincidences with the contact normal $\bar{n}$, see Figure 2.8.

Also, since the contact area is a circle $r_a = r_b = r$, $a = b$, $h_x = h_{xa} = h_{xb}$ and since $a = 2r$, we can write that

$$a = 2r = 2C_r h_x^{1/n} \tag{2.387}$$

Then, as in the case of contact between a sphere and a semi-space if $h_x = k_h x$ is taken analogically, where $h_x = k_h x$, then according to Equation 2.368, it follows $h_x = D_1 x$ and

$$a = 2r = 2C_r D_1^{1/n} x^{1/n} \tag{2.388}$$

The taking of the derivatives da/dx gives us

$$da = \frac{2C_r D_1^{1/n}}{n} x^{\frac{1-n}{n}} dx \tag{2.389}$$

Also, since for the cyclic contact area $F_{\tau z} = 0$, $F_\tau = F_{\tau y}$, $F_{b\tau} = F_{b\tau y}$, $F_{c\tau} = F_{c\tau y}$, $da = db$, see Figure 2.1, according to Equations 2.106–2.113, also see Figure 2.2, we can write that

$$dF_{acn} = dF_{bcn} = \frac{2C_r D_1^{1/n} E'}{n} dxx^{\frac{1-n}{n}} dx \tag{2.390}$$

$$dF_{abn} = dF_{bbn} = \frac{2C_r D_1^{1/n} \eta'_E}{n} d\dot{x}x^{\frac{1-n}{n}} dx \tag{2.391}$$

$$dF_{bcy} = G' \frac{2C_r D_1^{1/n}}{n} dyx^{\frac{1-n}{n}} dx \tag{2.392}$$

$$dF_{bby} = \eta'_G \frac{2C_r D_1^{1/n}}{n} d\dot{y}x^{\frac{1-n}{n}} dx \tag{2.393}$$

$$dF_{xcy} = G' dy dx \tag{2.394}$$

$$dF_{xby} = \eta'_G d\dot{y} dx \tag{2.395}$$

After integration of these equations, we get

$$F_{acn} = F_{bcn} = \frac{2C_r D_1^{1/n} E'}{n} \iint dx x^{\frac{1-n}{n}} dx = \frac{2C_r D_1^{1/n}}{n+1} E' x^{\frac{n+1}{n}} \tag{2.396}$$

$$F_{abn} = F_{bbn} = \frac{2C_r D_1^{1/n} \eta'_E}{n} \int d\dot{x} \int x^{\frac{1-n}{n}} dx = 2C_r D_1^{1/n} \eta'_E \dot{x} x^{1/n} \tag{2.397}$$

$$F_{bcy} = G' \frac{2C_r D_1^{1/n}}{n} \int dy \int x^{\frac{1-n}{n}} dx = 2C_r D_1^{1/n} G' y x^{1/n} \tag{2.398}$$

$$F_{bby} = \eta'_G \frac{2C_r D_1^{1?n}}{n} \int d\dot{y} \int x^{\frac{1-n}{n}} dx = 2C_r D_1^{1/n} \eta'_G \dot{y} x^{1/n} \tag{2.399}$$

$$F_{xcy} = G' \int dy \int^{h_x} dx = G' y D_1 x \tag{2.400}$$

$$F_{xby} = \eta'_G \int d\dot{y} \int^{h_x} dx = \eta'_G \dot{y} D_1 x \tag{2.401}$$

The viscous and elastic forces can be found after the summation of all partial elastic and viscous forces as follows:

$$F_{cn} = F_{acn} + F_{bcn} = \frac{4C_r D_1^{1/n}}{n+1} E' x^{\frac{n+1}{n}} \tag{2.402}$$

$$F_{bn} = F_{abn} + F_{bbn} = 4C_r D_1^{1/n} \eta'_E \dot{x} x^{1/n} \tag{2.403}$$

$$F_{c\tau y} = F_{bcy} + F_{xcy} = G' P_{xn} y \tag{2.404}$$

$$F_{b\tau y} = F_{bby} + F_{xby} = \eta'_G P_{xn} \dot{y} \tag{2.405}$$

where

$$P_{xn} = 2C_r D_1^{1/n} x^{1/n} + D_1 x \tag{2.406}$$

Thus finally, according to Equations 1.1–1.3 the next system of equations for the general viscoelastic forces can be expressed as follows:

$$F_n = F_{cn} + F_{bn} = \frac{4C_r D_1^{1/n}}{n+1} E' x^{\frac{n+1}{n}} + 4C_r D_1^{1/n} \eta'_E \dot{x} x^{1/n} \tag{2.407}$$

$$F_\tau = F_{b\tau y} + F_{c\tau y} = \eta'_G P_{xn} \dot{y} + G' P_{xn} y \tag{2.408}$$

Thus, for example, in the case of a conical contact, when $n = 1$ and $C_r = tg\alpha$ we get

$$F_n = F_{cn} + F_{bn} = 2tg\alpha_r D_1 E'x^2 + 4tg\alpha_r D_1 \eta_E' \dot{x}x \tag{2.409}$$

$$F_\tau = F_{b\tau y} + F_{c\tau y} = \eta_G' P_{xc} \dot{y} + G'P_{xc} y \tag{2.410}$$

where

$$P_{xc} = D_1 x(2tg\alpha_r + 1) \tag{2.410*}$$

But on the other hand, in the case of a parabolic contact, when $n = 2$, we can write that

$$F_n = F_{cn} + F_{bn} = \frac{4C_r D_1^{1/2}}{3} E'x^{3/2} + 4C_r D_1^{1/2} \eta_E' \dot{x}x^{1/2} \tag{2.411}$$

$$F_\tau = F_{b\tau y} + F_{c\tau y} = \eta_G' P_{xp} \dot{y} + G'P_{xp} y \tag{2.412}$$

where

$$P_{xp} = 2C_r D_1^{1/2} x^{1/2} + D_1 x \tag{2.413}$$

2.5.5 VISCOELASTIC CONTACT BETWEEN AN ELLIPTICAL CONE AND A FLAT SURFACE OF SEMI-SPACE

The contact area between an elliptical cone and a flat surface of a semi-space has an elliptical form, see Figure 2.9.

In this case, since $h_x = D_{1x}$, we can write, see Figure 2.9, that

$$a = 2r_a = 2tg\alpha_a h_x = 2tg\alpha_a D_1 x \tag{2.414}$$

$$b = 2r_b = 2tg\alpha\, h_x = 2tg\alpha_b D_1 x \tag{2.415}$$

where α_a denotes the angle of the cone in the plane *XAY*, see Figure 2.9, and α_b denotes the angle of the cone in the plane *XAZ* (is not shown).

Taking derivatives *da/dx* and *db/dx* gives us

$$da = 2tg\alpha_a D_1 dx \tag{2.416}$$

$$db = 2tg\alpha_b D_1 dx \tag{2.417}$$

According to Equations 2.106–2.113, we get

$$dF_{acn} = 2tg\alpha_a D_1 E' dx dx \tag{2.418}$$

$$dF_{bcn} = 2tg\alpha_b D_1 E' dx dx \tag{2.419}$$

$$dF_{abn} = 2\eta'_E tg\alpha_a D_1 d\dot{x} dx \tag{2.420}$$

$$dF_{bbn} = 2\eta'_E tg\alpha_b D_1 d\dot{x} dx \tag{2.421}$$

$$dF_{bcy} = 2tg\alpha_b D_1 G' dy dx \tag{2.422}$$

$$dF_{bby} = 2tg\alpha_b D_1 \eta'_G d\dot{y} dx \tag{2.422a}$$

$$dF_{xcy} = G' dy dx \tag{2.423}$$

$$dF_{xby} = \eta'_G d\dot{y} dx \tag{2.424}$$

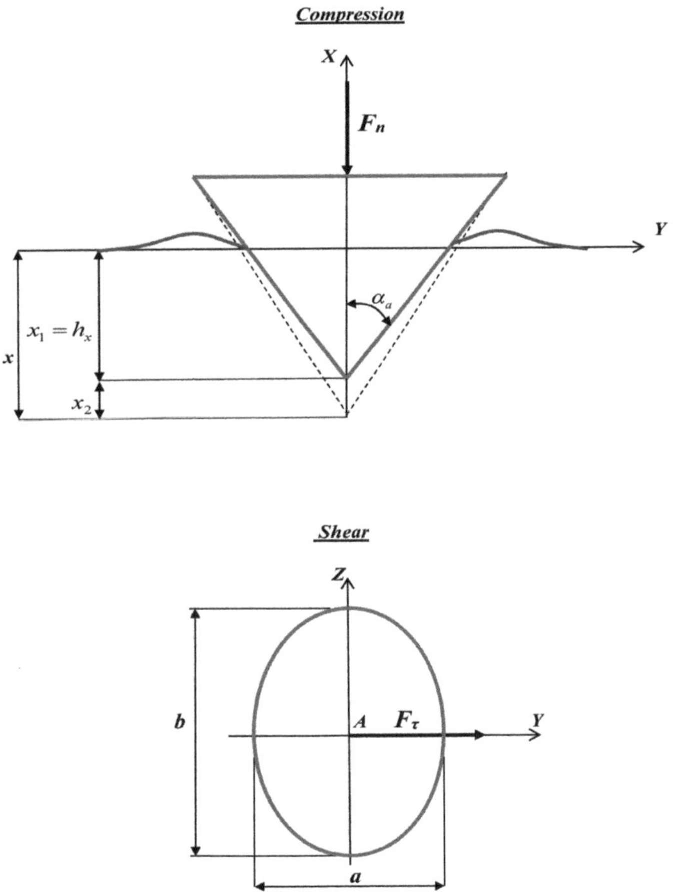

FIGURE 2.9 (See color insert.) Illustration of the contact between an elliptical cone and the flat surface of a semi-space.

After integration of these equations, we get:

$$F_{acn} = 2tg\alpha_a D_1 E' \iint dxdx = tg\alpha_a D_1 E' x^2 \tag{2.425}$$

$$F_{bcn} = 2tg\alpha_b D_1 E' \iint dxdx = tg\alpha_b D_1 E' x^2 \tag{2.426}$$

$$F_{abn} = 2tg\alpha_a D_1 \eta'_E \int d\dot{x} \int dx = 2tg\alpha_a D_1 \eta'_E \dot{x}x \tag{2.427}$$

$$F_{bbn} = 2tg\alpha_b D_1 \eta'_E \int d\dot{x} \int dx = 2tg\alpha_b D_1 \eta'_E \dot{x}x \tag{2.428}$$

$$F_{bcy} = 2tga_b D_1 G' \int d\dot{x} \int dx = 2tga_b D_1 G' yx \tag{2.429}$$

$$F_{bby} = 2tg\alpha_b D_1 \eta'_G \int d\dot{y} \int dx = 2tg\alpha_b D_1 \eta'_G \dot{y}x \tag{2.430}$$

$$F_{xcy} = G' \int dy \int^{h_x} dx = G' y D_1 x \tag{2.431}$$

$$F_{xby} = \eta'_G \int d\dot{y} \int^{h_x} dx = \eta'_G \dot{y} D_1 x \tag{2.432}$$

The viscous and elastic forces can be found after the summation of all partial elastic and viscous forces as follows:

$$F_{cn} = F_{acn} + F_{bcn} = D_1 E' x^2 (tg\alpha_a + tg\alpha_b) \tag{2.433}$$

$$F_{bn} = F_{abn} + F_{bbn} = 2D_1 \eta'_E \dot{x}x (tg\alpha_a + tg\alpha_b) \tag{2.434}$$

$$F_{c\tau y} = F_{bcy} + F_{xcy} = G' P_{xb} y \tag{2.435}$$

$$F_{b\tau y} = F_{bby} + F_{xby} = \eta'_G P_{xb} \dot{y} \tag{2.436}$$

where

$$P_{xb} = D_1 x (2tg\alpha_b + 1) \tag{2.436*}$$

Thus, finally, according to Equations 1.1–1.3 the next system of equations for general viscoelastic forces can be expressed as follows:

$$F_n = F_{cn} + F_{bn} = D_1 x (tg\alpha_a + tg\alpha_b)(E'x + 2\eta'_E \dot{x}) \tag{2.437}$$

$$F_\tau = F_{b\tau y} + F_{c\tau y} = P_{xb} (\eta'_G \dot{y} + G' y) \tag{2.438}$$

In the case if the contact area is a circle, $tg\alpha_b = tg\alpha_a$ and; therefore, it follows that

$$F_n = F_{cn} + F_{bn} = 2D_1 x tg\alpha_a (E'x + 2\eta'_E \dot{x}) \tag{2.439}$$

2.5.6 VISCOELASTIC CONTACT BETWEEN A PYRAMID AND A FLAT SURFACE OF A SEMI-SPACE

The contact area between a pyramid and the flat surface of a semi-space is a parallelogram, see Figure 2.10. L_1 and L_2 are the sizes of this contact area and α_a is the angle of pyramid equal to the angle of the equivalent cone α_{L1} in the plane *XAY*. (the shape of the contact area of the equivalent cone in Figure 2.10 is designated by blue line). α_{L2} is the angle of the pyramid in the plane *XAZ* and α_b is the angle of pyramid in the plane *XAZ* (not shown). The area of contact of the equivalent cone should be equal to the area of contact of the pyramid as the normal stresses should be the same in these areas. Thus, we can write that normal area of contact $S = L_1 L_2 = \frac{\pi ab}{4}$ and since $L_1 = a$, we get

$$b = \frac{4L_2}{\pi} \tag{2.440}$$

As well we can write $H = \frac{L_2}{2tg\alpha_{L2}} = \frac{b}{2tg\alpha_b}$ for an arbitrary height *H* of a cone and a pyramid, and taking into account Equation 2.440 we have the following

$$tg\alpha_b = \frac{4}{\pi} tg\alpha_{L2} \tag{2.441}$$

Substituting Equation 2.441 into Equations 2.437 and 2.438 gives us

$$F_n = F_{cn} + F_{bn} = D_1 x(tg\alpha_a + \frac{4}{\pi} tg\alpha_{L2})(E'x + 2\eta'_E \dot{x}) \tag{2.442}$$

$$F_\tau = F_{b\tau y} + F_{c\tau y} = P_{xn}(\eta'_G \dot{y} + G'y) \tag{2.443}$$

since Pxb = $2tg\alpha_b$ and $P_{xb} = 2tg\alpha_b + 1$, it follows

$$P_{xL} = D_1 x(\frac{8}{\pi} tg\alpha_{L2} + 1) \tag{2.444}$$

If the pyramid has a square base and since in this case $L_1 = L_2 = L$, it follows that the contact area of the equivalent cone should be a circle with diameter $a = b = 2r$, where *r* is the radius of this area, and therefore,

$$S = L^2 = \pi r^2 = \frac{\pi a^2}{4} \tag{2.445}$$

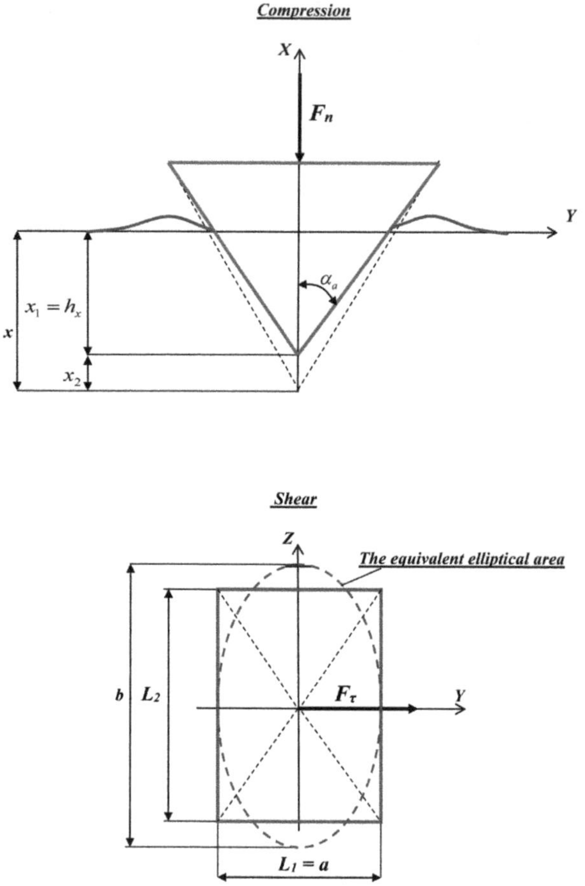

FIGURE 2.10 (See color insert.) Illustrations of the contact between a pyramid and the flat surface of a semi-space.

and respectively

$$L = \sqrt{\pi} r = \frac{\sqrt{\pi} a}{2} \tag{2.446}$$

As well we can write that $H = \dfrac{L}{2tg\alpha_L} = \dfrac{a}{2tg\alpha_a}$ for an arbitrary height H of a cone and a pyramid, then taking into account Equation 2.446, it follows that

$$tg\alpha_a = tg\alpha_b = \frac{2}{\sqrt{\pi}} tg\alpha_L \tag{2.447}$$

Substituting Equation 2.447 into 2.439 and 2.438 gives us

$$F_n = F_{cn} + F_{bn} = \frac{4}{\sqrt{\pi}} D_1 x tg\alpha_L (E'x + 2\eta'_E \dot{x}) \quad (2.448)$$

$$F_\tau = F_{b\tau y} + F_{c\tau y} = P_{xC}(\eta'_G \dot{y} + G'y) \quad (2.449)$$

where since $P_{xb} = 2tg\alpha_b + 1$ and $tg\alpha_a = tg\alpha_b = \frac{2}{\sqrt{\pi}} tg\alpha_L$, it follows

$$P_{xC} = D_1 x(\frac{4}{\sqrt{\pi}} tg\alpha_L + 1) \quad (2.450)$$

2.5.7 VISCOELASTIC CONTACT BETWEEN A WEDGE-SHAPED BLADE AND THE FLAT SURFACE OF A SEMI-SPACE

The contact area between a wedge-shaped blade and the flat surface of a semi-space is a parallelogram, as depicted in Figure 2.10. L_1 and L_2 are the sizes of this contact area, α_a is the angle of the wedge in the plane XAY and $L_2 = b = L$ is a constant length of the wedge. In this case, since $h_x = D_1 x$ x, see Figure 2.10, we can write that

$$a = 2r_a = 2tg\alpha_a h_x = 2tg\alpha_a D_1 x \quad (2.451)$$

Taking of the derivatives gives us

$$da = 2tg\alpha_a D_1 dx \quad (2.452)$$

$$db = dL \quad (2.453)$$

According to Equations 2.94–2.101 and 2.106–2.113, respectively

$$dF_{acn} = 2tg\alpha_a D_1 E' dx dx \quad (2.454)$$

$$dF_{bcn} = D_1 E' dL dx \quad (2.455)$$

$$dF_{abn} = 2\eta'_E tg\alpha_a D_1 d\dot{x} dx \quad (2.456)$$

$$dF_{bbn} = \eta'_E D_1 L d\dot{x} \quad (2.457)$$

$$dF_{bcy} = D_1 G' L dy \quad (2.458)$$

$$dF_{bby} = D_1 \eta'_G dL d\dot{y} \quad (2.459)$$

$$dF_{xcy} = G' dy dx \quad (2.460)$$

$$dF_{xby} = \eta'_G d\dot{y} dx \quad (2.461)$$

After integration of these equations, we get:

$$F_{acn} = 2tg\alpha_a D_1 E' \iint dxdx = tg\alpha_a D_1 E' x^2 \tag{2.462}$$

$$F_{bcn} = D_1 E' L \int dx = D_1 E' Lx \tag{2.463}$$

$$F_{abn} = 2tg\alpha_a D_1 \eta'_E \int d\dot{x} \int dx = 2tg\alpha_a D_1 \eta'_E \dot{x}x \tag{2.464}$$

$$F_{bbn} = D_1 \eta'_E L \int d\dot{x} = D_1 \eta'_E L\dot{x} \tag{2.465}$$

$$F_{bcy} = D_1 G' L \int dy = D_1 G' Ly \tag{2.466}$$

$$F_{bby} = D_1 G' L \int d\dot{y} = D_1 \eta'_G L\dot{y} \tag{2.467}$$

$$F_{xcy} = G' \int dy \int^{h_x} dx = G' D_1 yx \tag{2.468}$$

$$F_{xby} = \eta'_G \int d\dot{y} \int^{h_x} dx = \eta'_G D_1 \dot{y}x \tag{2.469}$$

The viscous and elastic forces can be found after the summation of all partial elastic and viscous forces as follows:

$$F_{cn} = F_{acn} + F_{bcn} = D_1 E' x(xtg\alpha_a + L) \tag{2.470}$$

$$F_{bn} = F_{abn} + F_{bbn} = D_1 \eta'_E \dot{x}(2xtg\alpha_a + L) \tag{2.471}$$

$$F_{c\tau y} = F_{bcy} + F_{xcy} = D_1 G_1 y(L + x) \tag{2.472}$$

$$F_{b\tau y} = F_{bby} + F_{xby} = D_1 \eta'_G \dot{y}(L + x) \tag{2.473}$$

Finally, the general viscoelastic forces can be expressed as

$$F_n = F_{cn} + F_{bn} = D_1 \{E' x(xtg\alpha_a + L) + \eta'_E \dot{x}(2xtg\alpha_a + L)\} \tag{2.474}$$

$$F_\tau = F_{b\tau y} + F_{c\tau y} = D_1 (L + x)(G' y + \eta'_G \dot{y}) \tag{2.475}$$

2.5.8 VISCOELASTIC CONTACT BETWEEN AN ELLIPTICAL CYLINDER AND THE FLAT SURFACE OF A SEMI-SPACE

The contact area between an elliptical cylinder and the flat surface of a semi-space has an elliptical form, see Figure 2.11.

According to Equations 2.94–2.101 and 2.106–2.113

$$dF_{acn} = D_1 E' dxda \tag{2.476}$$

$$dF_{bcn} = D_1 E' dx da \tag{2.477}$$

$$dF_{abn} = D_1 d\dot{x} da \tag{2.478}$$

$$dF_{bbn} = D_1 d\dot{x} db \tag{2.479}$$

$$dF_{bcy} = D_1 G' dy db \tag{2.480}$$

$$dF_{bby} = D_1 \eta'_G d\dot{y} db \tag{2.481}$$

$$dF_{xcy} = G' dy dx \tag{2.482}$$

$$dF_{xby} = \eta'_G d\dot{y} dx \tag{2.483}$$

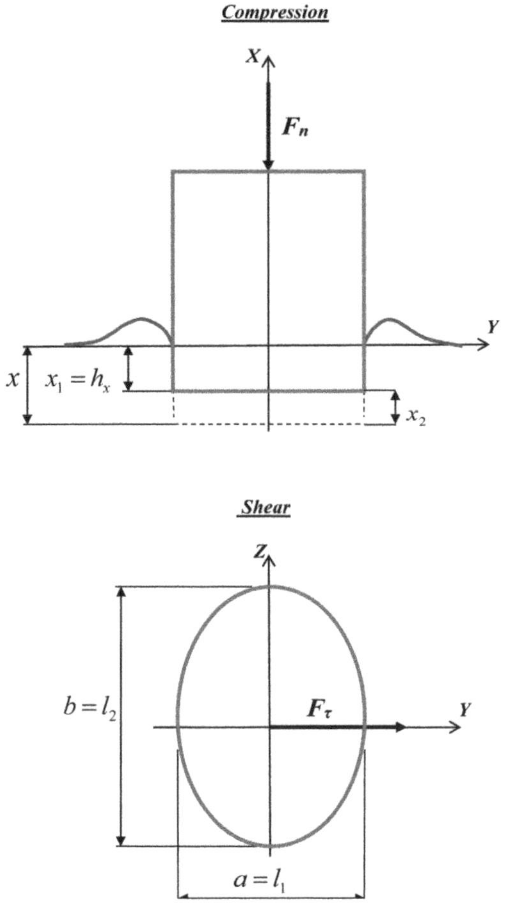

FIGURE 2.11 (See color insert.) Illustrations of the contact between an elliptical cylinder and the flat surface of a semi-space.

After integration of Equations 2.476–2.483, we get the following:

$$F_{acn} = D_1 E' \int dx \int da = D_1 E' xa \quad (2.484)$$

$$F_{bcn} = 2tg\alpha_b D_1 E' \int dx \int db = D_1 E' xb \quad (2.485)$$

$$F_{abn} = D_1 \eta'_E \int d\dot{x} \int da = D_1 \eta'_E \dot{x} a \quad (2.486)$$

$$F_{bbn} = D_1 \eta'_E \int d\dot{x} \int db = D_1 \eta'_E \dot{x} b \quad (2.487)$$

$$F_{bcy} = D_1 G' \int dy \int db = D_1 G' yb \quad (2.488)$$

$$F_{bby} = D_1 \eta'_G \int d\dot{y} \int db = D_1 \eta'_G \dot{y} x \quad (2.489)$$

$$F_{xcy} = G' \int dy \int^{h_x} dx = G' D_1 yx \quad (2.490)$$

$$F_{xby} = \eta'_G \int d\dot{y} \int^{h_x} dx = \eta'_G D_1 \dot{y} x \quad (2.491)$$

The viscous and elastic forces can be found after the summation of all partial elastic and viscous forces as follows:

$$F_{cn} = F_{acn} + F_{bcn} = D_1 E' x(a+b) \quad (2.492)$$

$$F_{bn} = F_{abn} + F_{bbn} = D_1 \eta'_E \dot{x}(a+b) \quad (2.492^*)$$

$$F_{c\tau y} = F_{bcy} + F_{xcy} = D_1 G' y(b+x) \quad (2.493)$$

$$F_{b\tau y} = F_{bby} + F_{xby} = D_1 \eta'_G \dot{y}(b+x) \quad (2.494)$$

Thus finally, according to Equations 1.1–1.3 the next system of equations for general viscoelastic forces can be expressed as follows:

$$F_n = F_{cn} + F_{bn} = D_1 (a+b)(E' x + \eta'_E \dot{x}) \quad (2.495)$$

$$F_\tau = F_{b\tau y} + F_{c\tau y} = D_1 (b+x)(\eta'_G \dot{y} + G' y) \quad (2.496)$$

In this case if the contact area is a circle, $a = b = d$, where d is the diameter, it follows

$$F_n = F_{cn} + F_{bn} = 2D_1 d(E' x + \eta'_E \dot{x}) \quad (2.495a)$$

$$F_\tau = F_{b\tau y} + F_{c\tau y} = D_1 (d+x)(\eta'_G \dot{y} + G' y) \quad (2.496b)$$

2.5.9 VISCOELASTIC CONTACT BETWEEN A WHEEL AND THE FLAT SURFACE OF A SEMI-SPACE

2.5.9.1 CYLINDRICAL SURFACE OF A WHEEL

In case of a cylindrical surface of a wheel and the flat semi-space, since $k_h = D_1$ and if take that L_w is the width of the wheel, $R = R_w$ is the radius of a wheel, $R_p = \infty$, see Figure 2.12, Equations 2.369–2.376 can be rewritten as

$$F_{acn} = \frac{2}{3} k_p E' R_w^{1/2} x^{3/2} \tag{2.497}$$

$$F_{bcn} = E' x L_w \tag{2.498}$$

$$F_{abn} = 2 k_p \eta'_E R_w^{1/2} \dot{x} x^{1/2} \tag{2.499}$$

$$F_{bbn} = 2\eta'_E \dot{x} L_w \tag{2.500}$$

$$F_{bcy} = G' L y \tag{2.501}$$

$$F_{bby} = \eta'_G L \dot{y} \tag{2.502}$$

$$F_{xcy} = D_1 G' x y \tag{2.503}$$

$$F_{xby} = D_1 \eta'_G x \dot{y} \tag{2.504}$$

Then, after their summation, it follows

$$F_{cn} = F_{acn} + F_{bcn} = \frac{2}{3} k_p E' R_w^{1/2} x^{3/2} + E' x L_w \tag{2.505}$$

$$F_{bn} = F_{abn} + F_{bbn} = 2 k_p \eta'_E R_w^{1/2} \dot{x} x^{1/2} + 2\eta'_E \dot{x} L_w \tag{2.506}$$

$$F_{c\tau y} = F_{bcy} + F_{xcy} = G' L_w y + D_1 G' x y \tag{2.507}$$

$$F_{b\tau y} = F_{bby} + F_{xby} = G' L_w \dot{y} + D_1 G' x \dot{y} \tag{2.508}$$

Thus, finally

$$F_n = F_{bn} + F_{cn} = 2\eta'_E (k_p R_w^{1/2} x^{1/2} + L_w)\dot{x} + E'\left(\frac{2}{3} k_p R_w^{1/2} x^{1/2} + L_w\right) x \tag{2.509}$$

$$F_\tau = F_{b\tau y} + F_{c\tau y} = \eta'_G P_{xw} \dot{y} + G' P_{xw} y \tag{2.510}$$

where

$$P_{xw} = D_1 x + L_w \tag{2.511}$$

2.5.9.2 TOROIDAL SURFACE OF A WHEEL

The contact area between a toroidal surface of a wheel and the flat surface of a semi-space has an elliptical form, see Figure 2.12. In this case, since we can take that $R_a = R_w$, $r_a = r_y$, $R_b = R_p$ and $r_b = r_z$, Equations 2.170, 2.171 and 2.174, 2.175 for the normal forces can be rewritten as:

$$F_{acn} = \frac{2}{3} k_{pa} E' R_w^{1/2} x^{3/2} \tag{2.512}$$

$$F_{bcn} = \frac{2}{3} k_{pb} E' R_p^{1/2} x^{3/2} \tag{2.513}$$

$$F_{abn} = 2 k_{pa} \eta_E' R_w^{1/2} \dot{x} x^{1/2} \tag{2.514}$$

$$F_{bbn} = 2 k_{pb} \eta_E' R_p^{1/2} \dot{x} x^{1/2} \tag{2.515}$$

where R_w is the radius of a wheel, R_p is the radius of wheel profile, and respectively $k_p = \sqrt{2 - \frac{x}{R_w}}$, $k_{pb} = \sqrt{2 - \frac{x}{R_p}}$. Since $F_{cn} = F_{acn} + F_{bcn}$ and $F_{bn} = F_{abn} + F_{bbn}$, it follows that:

$$F_{cn} = \frac{2}{3} E' x^{3/2} (k_{pa} R_w^{1/2} + k_{pb} R_p^{1/2}) = \frac{2}{3} E' \psi_w x^{3/2} \tag{2.516}$$

$$F_{bn} = 2 \eta_E' \dot{x} x^{1/2} (k_{pa} R_w^{1/2} + k_{pb} R_p^{1/2}) = 2 \eta_E' \psi_w \dot{x} x^{1/2} \tag{2.517}$$

Where, in this case $\psi_w = (k_{pa} R_w^{1/2} + k_{pb} R_p^{1/2})$.

Also, since in this case $h_{xa} = D_1$ Equations 2.178, 2.179 2.180 and 2.181 for the tangential forces can be rewritten as:

$$F_{bcy} = 2 k_{pb} G' R_p^{1/2} x^{1/2} y \tag{2.518}$$

$$F_{bby} = 2 k_{pb} \eta_G' R_p^{1/2} x^{1/2} \dot{y} \tag{2.519}$$

$$F_{xcy} = D_1 G' x y \tag{2.520}$$

$$F_{xby} = D_1 \eta_G' x \dot{y} \tag{2.521}$$

The equations for the effective tangential viscoelastic forces can be finally written as the sum of the elastic and the viscous tangential forces from series of Equations 2.518–2.521:

$$F_{c\tau y} = F_{xcy} + F_{bcy} = G' P_{bx} y \tag{2.522}$$

$$F_{b\tau y} = F_{xby} + F_{bby} = \eta_G' P_{bx} \dot{y} \tag{2.523}$$

where

$$P_{bx} = D_1 x + 2k_{pb} R_p^{1/2} x^{1/2} \tag{2.524}$$

Finally, using Equations 2.516, 2.517, 2.522 and 2.523 the next equations for the general viscoelastic forces can be written respectively as:

$$F_{cn} = F_{bn} + F_{cn} = 2\eta_E' \psi_w \dot{x} x^{1/2} + \frac{2}{3} E' \psi_w x^{3/2} \tag{2.525}$$

$$F_{by} = F_{b\tau y} + F_{c\tau y} = \eta_G' P_{bx} \dot{y} + G' P_{bx} y \tag{2.526}$$

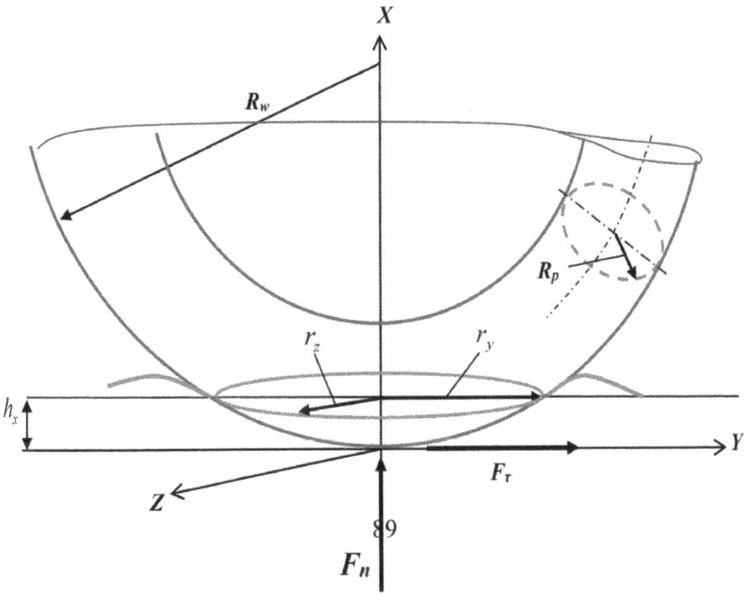

FIGURE 2.12 (See color insert.) Illustration of the contact a toroidal surface of a wheel and the flat surface of a semi-space.

2.5.10 VISCOELASTIC CONTACT BETWEEN SURFACE OF SLED AND THE FLAT SURFACE OF A SEMI-SPACE

2.5.10.1 CIRCULAR CYLINDRICAL FORWARD SURFACE OF A SLED

The contact area between a circular cylindrical forward surface of a sled and the flat surface of a semi-space has a form of rectangle, see Figure 2.13. Let the width of a sled be equal to the constant size b. (It is not shown in the Figure 2.13). Also, it is obvious that, in this case, the normal forces can be found as the sum of forces acting on the length L_s and on the radius $r = r_a$ of the contacting surface. Since $dr_a = \frac{1}{2} k_{pa} R_a^{1/2} \frac{dx}{x^{1/2}}$ and b=L=constant, we can write simple differential equations for all viscous and elastic differential forces, according to the 'MDSF', as follows:

$$dF_{acn} = E' dx dL_s + \frac{1}{2} k_{pa} R_a^{1/2} E' dx \frac{dx}{x^{1/2}} \tag{2.527}$$

$$dF_{abn} = \eta_E' d\dot{x} dL_s + \frac{1}{2} k_{pa} R^{1/2} \eta_E' d\dot{x} \frac{dx}{x^{1/2}} \tag{2.528}$$

$$dF_{bcn} = E' dx db \tag{2.529}$$

$$dF_{bbn} = \eta_E' d\dot{x} db \tag{2.530}$$

$$dF_{bcy} = G' dy db \tag{2.531}$$

$$dF_{bby} = \eta_G' d\dot{y} db \tag{2.532}$$

$$dF_{xcy} = G' dy dx \tag{2.533}$$

$$dF_{xby} = \eta_G' d\dot{y} dx \tag{2.534}$$

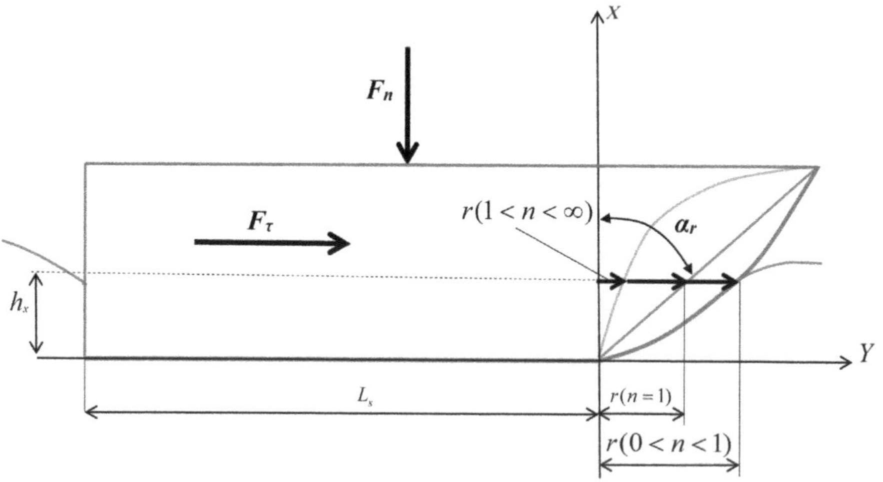

FIGURE 2.13 (See color insert.) Illustration of the contact between the surface of sled and the flat surface of a semi-space.

Then after integration, we get:

$$F_{acn} = E'D_1xL_s + \frac{1}{3}k_{pa}R_a^{1/2}E'x^{3/2} \tag{2.535}$$

$$F_{abn} = \eta'_E D_1\dot{x}L_s + k_{pa}R^{1/2}\eta'_E d\dot{x}x^{1/2} \tag{2.536}$$

$$F_{bcn} = D_1E'xL \tag{2.537}$$

$$F_{bbn} = D_1\eta'_E\dot{x}L \tag{2.538}$$

$$F_{bcy} = G'yL \tag{2.539}$$

$$F_{bby} = \eta'_G\dot{y}L \tag{2.540}$$

$$F_{xcy} = D_1G'yx \tag{2.541}$$

$$F_{xby} = D_1\eta'_G\dot{y}x \tag{2.542}$$

The equations for the effective normal viscoelastic forces can be written finally as sums:

$$F_{cn} = F_{acn} + F_{bcn} = E'(L_s + L)D_1x + \frac{1}{3}k_{pa}R_a^{1/2}E'x^{3/2} \tag{2.543}$$

$$F_{bn} = F_{abn} + F_{bbn} = \eta'_E(L_s + L)D_1\,\dot{x} + k_{pa}R^{1/2}\eta'_E d\dot{x}x^{1/2} \tag{2.544}$$

Also, the equations for the effective tangential viscoelastic forces can be finally written as sums:

$$F_{c\tau y} = F_{bcy} + F_{xcy} = G'y(L + D_1x) \tag{2.545}$$

$$F_{b\tau y} = F_{bby} + F_{xby} = \eta'_G\dot{y}(L + D_1x) \tag{2.546}$$

2.5.10.2 ARBITRARY CURVILINEAR CYLINDRICAL FORWARD SURFACE OF A SLED

The contact area between a sled, having an arbitrary cylindrical forward surface, and the flat semi-space has a form of rectangle, too, see Figure 2.13. As we know, in a general case, the radius of generatrix of the curvilinear surface can be expressed as function $r = C_rx^{1/n}$, see Equation 2.386. If $0 < n < 1$, we have the paraboloidal or hyperboloidal shapes of the cylinder of the forward surface of a sled. On the other hand, if $1 < n < \infty$, Equation 2.386

describes the inverse paraboloidal or hyperboloidal shapes of the cylinder of the forward surface of a body of sled. In the case, when $n = 1$ we get the flat shape of the forward surface, where $C_r = tg\alpha_r$, α_r denotes the angle between the flat forward surface of a sled and axis X, see Figure 2.13. As $h_x = k_h x$ and according to Equation 2.368 $h_x = D_1 x$ and

$$r = C_r D_1^{1/n} x^{1/n} \tag{2.547}$$

Taking the derivatives *da/dx* gives us

$$dr = \frac{C_r D_1^{1/n}}{n} x^{\frac{1-n}{n}} dx \tag{2.548}$$

Also, since b=L=constant, we can write according to the 'MDSF' simple differential equations for all viscous and elastic differential forces respectively as:

$$dF_{acn} = E' dx dL_s + E' dx \frac{C_r D_1^{1/n}}{n} x^{\frac{1-n}{n}} dx \tag{2.549}$$

$$dF_{abn} = \eta_E' d\dot{x} dL_s + \eta_E' d\dot{x} \frac{C_r D_1^{1/n}}{n} x^{\frac{1-n}{n}} dx \tag{2.550}$$

$$dF_{bcn} = E' dx db \tag{2.551}$$

$$dF_{bbn} = \eta_E' d\dot{x} db \tag{2.552}$$

$$dF_{bcy} = G' dy db \tag{2.553}$$

$$dF_{bby} = \eta_G' d\dot{y} db \tag{2.554}$$

$$dF_{xcy} = G' dy dx \tag{2.555}$$

$$dF_{xby} = \eta_G' d\dot{y} dx \tag{2.556}$$

Then, taking into account that integration of Equations 2.551–2.556 is same as in Section 2.5.10.1, we get:

$$F_{acn} = D_1 E' x L_s + E' \frac{C_r D_1^{1/n}}{n+1} x^{\frac{1+n}{n}} \tag{2.557}$$

$$F_{abn} = D_1 \eta_E' \dot{x} L_s + \eta_E' C_r D_1^{1/n} \dot{x} x^{1/n} \tag{2.558}$$

Taking into account that integration of Equations 2.551–2.556 is same as in paragraph before, the equations for the effective normal viscoelastic forces can be written finally as sums:

$$F_{cn} = F_{acn} + F_{bcn} = E'(L_s + L)D_1 x + E'\frac{C_r D_1^{1/n}}{n+1} x^{\frac{n+1}{n}} \tag{2.559}$$

$$F_{bn} = F_{abn} + F_{bbn} = \eta_E'(L_s + L)D_1 \dot{x} + \eta_E' C_r D_1^{1/n} \dot{x} x^{1/n} \tag{2.560}$$

In this case, the equations for the effective tangential viscoelastic forces are same as in Equations 2.545 and 2.546.

In the case of the flat shape of the forward surface, when $n = 1$ and where $C_r = tg\alpha$, the effective normal forces can be calculated by formulas:

$$F_{cn} = F_{acn} + F_{bcn} = E'(L_s + L)D_1 x + E'\frac{tg\alpha_r D_1}{n+1} x^2 \tag{2.561}$$

$$F_{bn} = F_{abn} + F_{bbn} = \eta_E'(L_s + L)D_1 \dot{x} + \eta_E' tg\alpha_r D_1 \dot{x} x \tag{2.562}$$

2.5.11 VISCOELASTIC CONTACT BETWEEN A BODY OF VESSEL AND A FLAT SEMI-SPACE

In the common case, a body of vessel can be got by moving an axis-symmetrical curvilinear surface, see Figure 2.8, along axis Y at a distance L_s, see Figure 2.14. Therefore, the contact area between curvilinear surface of a body of vessel and a flat surface of a semi-space can be presented by joining two half circles and a rectangle, as it is shown in Figure 2.13.

Also, as we can get the normal viscoelastic forces simply by adding normal force acting along this distance L_s to already found normal forces in the Section 2.5.4, see Equations 2.396 and 2.397, respectively as:

$$F_{acn} = D_1 E' x L_s + 2E'\frac{C_r D_1^{1/n}}{n+1} x^{\frac{1+n}{n}} \tag{2.563}$$

$$F_{abn} = D_1 \eta_E' \dot{x} L_s + 2\eta_E' C_r D_1^{1/n} \dot{x} x^{1/n} \tag{2.564}$$

and also

$$F_{bcn} = 2E'\frac{C_r D_1^{1/n}}{n+1} x^{\frac{1+n}{n}} \tag{2.565}$$

$$F_{bbn} = 2\eta_E' C_r D_1^{1/n} \dot{x} x^{1/n} \tag{2.566}$$

Thus, the equations for the effective normal viscoelastic forces can be written finally as sums:

$$F_{cn} = F_{acn} + F_{bcn} = E'L_s D_1 x + 4E' \frac{C_r D_1^{1/n}}{n+1} x^{\frac{n+1}{n}} \tag{2.567}$$

$$F_{bn} = F_{abn} + F_{bbn} = \eta_E' L_s D_1 \dot{x} + 4\eta_E' C_r D_1^{1/n} \dot{x} x^{1/n} \tag{2.568}$$

The tangential forces should be satisfied to Equations 2.404, 2.405 and 2.408.

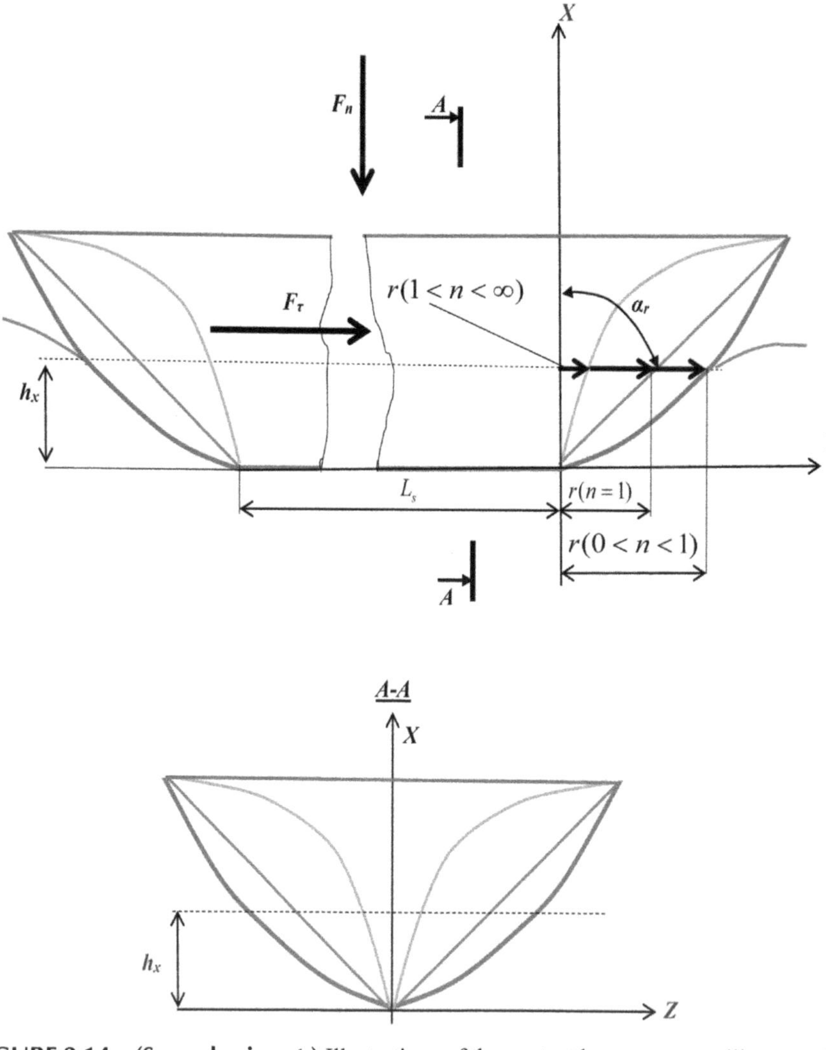

FIGURE 2.14 (See color insert.) Illustrations of the contact between a curvilinear surface of a body of vessel and the flat surface of a semi-space.

2.5.12 VISCOELASTIC CONTACT BETWEEN AN ARBITRARY SHAPE OF DISK AND THE FLAT SURFACE OF A SEMI-SPACE

The contact area between an arbitrary shape of disk and the flat surface of a semi-space is depicted in Figure 2.15. Since $a = 2r_a$, $b = 2r_b$, see Figure 2.15, and according to Equation 2.129, $r_a^2 = k_{pa}^2 R_a x$, and according to Equation 2.388, in this case, $b = 2r = 2C_r D_1^{1/n} x^{1/n}$, we can write that

$$da = \frac{k_{pa} R_a^{1/2}}{x^{1/2}} dx \quad (2.569)$$

$$db = \frac{2C_r D_1^{1/n}}{n} x^{\frac{1-n}{n}} dx \quad (2.570)$$

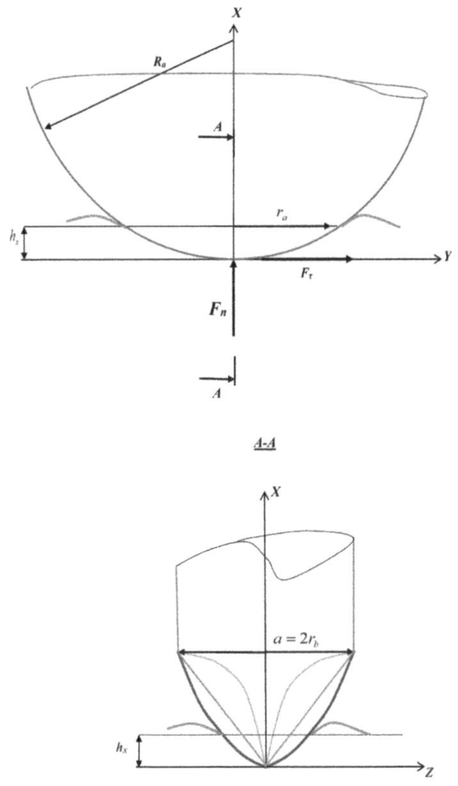

FIGURE 2.15 (See color insert.) Illustrations of the contact between an arbitrary shape of disk and the flat surface of a semi-space.

Now, since the expressions for *da* and *db* are already known, we can write, according to series of Equations 2.106–2.113, differential equations for viscous and elastic forces as:

$$dF_{acn} = k_{pa} R_a^{1/2} E' dx \frac{dx}{x^{1/2}} \tag{2.571}$$

$$dF_{abn} = k_{pa} R_a^{1/2} \eta_E' d\dot{x} \frac{dx}{x^{1/2}} \tag{2.572}$$

$$dF_{bcn} = \frac{2C_r D_1^{1/n}}{n} E' dx x^{\frac{1-n}{n}} dx \tag{2.573}$$

$$dF_{bbn} = \frac{2C_r D_1^{1/n} \eta_E'}{n} d\dot{x} x^{\frac{1-n}{n}} dx \tag{2.574}$$

$$dF_{bcy} = G' \frac{2C_r D_1^{1/n}}{n} dy x^{\frac{1-n}{n}} dx \tag{2.575}$$

$$dF_{bby} = \eta_G' \frac{2C_r D_1^{1/n}}{n} d\dot{y} x^{\frac{1-n}{n}} dx \tag{2.576}$$

$$dF_{xcy} = G' dy dx \tag{2.577}$$

$$dF_{xby} = \eta_G' d\dot{y} dx \tag{2.578}$$

After integration, we get:

$$F_{acn} = k_{pa} E' R_a^{1/2} \int\int dx \frac{dx}{x^{1/2}} = \frac{2}{3} k_{pa} E' R_a^{1/2} x^{3/2} \tag{2.579}$$

$$F_{abn} = k_{pa} \eta_E' \int d\dot{x} \int \frac{R_a^{1/2}}{x^{1/2}} dx = 2k_{pa} \eta_E' R_a^{1/2} \dot{x}(t) x^{1/2} \tag{2.580}$$

$$F_{bcn} = \frac{2C_r D_1^{1/n} E'}{n} \iint dx x^{\frac{1-n}{n}} dx = \frac{2C_r D_1^{1/n}}{n+1} E' x^{\frac{n+1}{n}} \tag{2.581}$$

$$F_{bbn} = \frac{2C_r D_1^{1/n} \eta_E'}{n} \int d\dot{x} \int x^{\frac{1-n}{n}} dx = 2C_r D_1^{1/n} \eta_E' \dot{x} x^{1/n} \tag{2.582}$$

$$F_{bcy} = G' \frac{2C_r D_1^{1/n}}{n} \int dy \int x^{\frac{1-n}{n}} dx = 2C_r D_1^{1/n} G' y x^{1/n} \tag{2.583}$$

$$F_{bby} = \eta_G' \frac{2C_r D_1^{1?n}}{n} \int d\dot{y} \int x^{\frac{1-n}{n}} dx = 2C_r D_1^{1/n} \eta_G' \dot{y} x^{1/n} \tag{2.584}$$

$$F_{xcy} = G'\int dy \int^{h_x} dx = G'yD_1x \tag{2.585}$$

$$F_{xby} = \eta'_G \int d\dot{y} \int^{h_x} dx = \eta'_G \dot{y} D_1 x \tag{2.586}$$

The viscous and elastic forces can be found after the summation of all partial elastic and viscous forces as follows:

$$F_{cn} = F_{acn} + F_{bcn} = \frac{2}{3}k_{pa}E'R_a^{1/2}x^{3/2} + \frac{2C_rD_1^{1/n}}{n+1}E'x^{\frac{n+1}{n}} \tag{2.587}$$

$$F_{bn} = F_{abn} + F_{bbn} = 2k_{pa}\eta'_E R_a^{1/2}\dot{x}(t)x^{1/2} + 2C_rD_1^{1/n}\eta'_E\dot{x}x^{1/n} \tag{2.588}$$

$$F_{c\tau y} = F_{bcy} + F_{xcy} = G'P_{xn}y \tag{2.589}$$

$$F_{b\tau y} = F_{bby} + F_{xby} = \eta'_G P_{xn}\dot{y} \tag{2.590}$$

where

$$P_{xn} = 2C_rD_1^{1/n}x^{1/n} + D_1x \tag{2.591}$$

KEYWORDS

- **differential specific forces**
- **normal deformation**
- **viscoelastic forces**
- **Kelvin–Voigt model**
- **Hertz theory**

CHAPTER 3

DYNAMICS OF VISCOELASTIC CONTACTS, TRIBOCYCLICITY, AND VISCOELASTIC LUBRICATION

ABSTRACT

Dynamics of contact between smooth curvilinear surfaces of two spherical bodies have been considered in this chapter. The differential equations of displacement at the compression, work and energy in the phases of compression and restitution, and the approximate solution to differential equations of the normal displacement by using the method of equivalent works have been developed.

A solution of the boundary value problem at impact and friction at impact between two spherical bodies have been considered and also the equations for contact stresses have been derived. It was proposed to consider the friction as the tribocyclic process of elementary contacts. Techniques of finding the normal and tangential viscoelastic constants in cases of indentation of a spherical and axis-symmetrical indenter or asperity into a semi-space during sliding and rolling motion have been given. Sliding and rolling friction at the tangential impact and also a viscoelastic sliding and rolling motion between a smooth surface of the body and the flat semi-space with a constant velocity have been examined here. Examples of sliding and rolling motion between bodies having different shapes and the flat semi-space have been proposed for consideration here in this chapter. Also, specificity of high-viscoelastic contact during sliding or rolling motion and viscoelastic lubrication have been examined.

3.1 DYNAMICS OF VISCOELASTIC COLLISION BETWEEN TWO SPHERICAL BODIES

3.1.1 INTRODUCTION

Let the two spherical bodies having the masses m_1 and m_2, the radii R_1 and R_2, and the initial relative velocities between centres of mass of the contacting bodies $\dot{x}_d$ $(t = 0) = V_x$ and $\dot{x}_d$ $(t = 0) = V_x$, $(V_z=0)$ come into an arbitrary viscoelastic contact at impact at the initial instance of time $t=0$ and at the initial point of contact A, see Figure 3.1. It is obvious that the viscoelastic forces F_n and F_τ are acting in the contact area between the surfaces of the contact and according to Newton's Second Law we can write:

$$F_n = -m\ddot{x}_d \tag{3.1}$$

$$F_n = -m\ddot{x}_d \tag{3.2}$$

$$M = -J_z\ddot{\varphi} \tag{3.3}$$

where m is the effective mass or reduced mass of the effective third body, $\ddot{y}_d, \ddot{x}_d$—the relative accelerations between the centres of mass of the bodies, J_z is the effective moment of inertia, φ is the relative angle of rotation of the bodies, $\ddot{\varphi}$ is the relative angular acceleration between bodies, $M = F_\tau l$ is the reactive moment, and where l is the effective radius of the effective third body or it is the contacting bodies and; therefore, it can be taken as the shoulder of tangential force, see Figure 3.1.

Remark: The term 'effective mass' has already been used by Stronge (2000); Dintwa (2006); Bordbar and Hyppänen (2007); Antypov et al. (2011) and by many others authors. Also, mass m was called the reduced mass by Landau and Lifshitz (1944, 1959) and Brilliantov et al. (1996).

We can see here that $x = x_d = x_1 + x_2$ is the distance of the mutual approach or the total deformation of compression between surfaces of the colliding bodies, and as well, at the same instance of time, it is the relative displacement between the centres of mass bodies relative to the initial point of contact A by axis X. At the impact of two bodies, we can consider the relative movement of the centre masses of colliding bodies with mass m_1 and m_2 similar as the movement of this third body having the effective mass m and the coordinates $x=0$, $y=0$ and $z=0$ in the initial moment of time $t=0$, and at the initial point A, see Figure 3.1.Further, in this book, this effective third body will be called simply the body. Thus, we have to understand that the all kinematic and dynamic parameters of the relative movement between centre masses of the contacting bodies are equal to the kinematic and dynamic parameters of the movement of the centre mass of this effective body having the effective mass m.

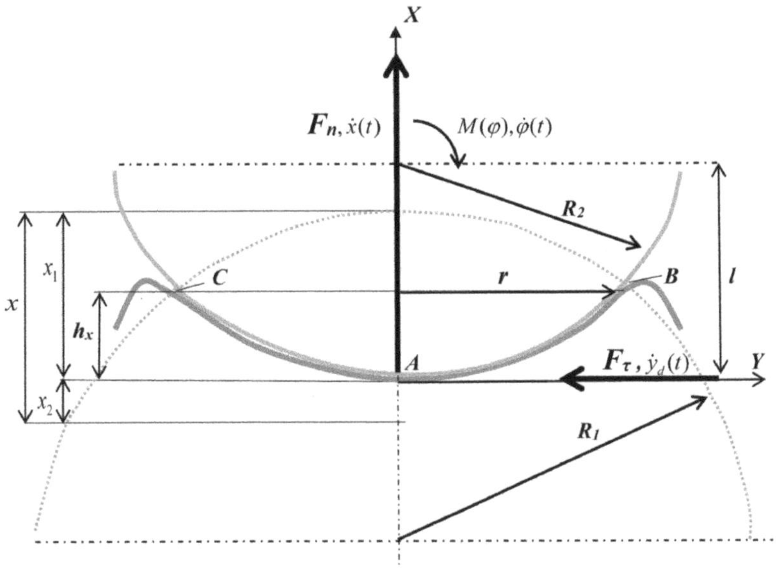

FIGURE 3.1 (See color insert.) Schematic illustration of the contact at impact between two spherical solid bodies.

Also, we can write that according to the second law of Newton

$$F_n = F_{n1} = F_{n2} = -m\frac{d\dot{x}_d}{dt} = -m_1\frac{d\dot{x}_{d1}}{dt} = -m_2\frac{dx_{d2}}{dt} \tag{3.4}$$

$$F_\tau = F_{\tau 1} = F_{\tau 2} = -m\frac{d\dot{y}_d}{dt} = -m_1\frac{d\dot{y}_{d1}}{dt} = -m_2\frac{d\dot{y}_{d2}}{dt} \tag{3.5}$$

and

$$M = M_1 = M_2 = -J_z\frac{d\omega}{dt} = -J_{z1}\frac{d\omega_1}{dt} = -J_{z2}\frac{d\omega_2}{dt} \tag{3.6}$$

where $\dot{x}_d = \dot{x}_{d1} + \dot{x}_{d2}$; $\dot{x}_{d1}$ is the relative velocity of displacement of the centre mass of a softer body by axis X, $\dot{x}_{d2}$ is the relative velocity of displacement of the centre mass of a harder body by axis X, $\dot{y}_d = \dot{y}_{d1} + \dot{y}_{d2}$; $\dot{y}_{d1}$ is the relative velocity of displacement of the centre mass of a softer body by axis Y, $\dot{y}_{d2}$ is the relative velocity of displacement of the centre mass of a harder body by axis Y, $\bar{\omega} = \bar{\omega}_1 + \bar{\omega}_2$ is the relative angular velocity between bodies; $\bar{\omega}_1$ and $\bar{\omega}_2$ are the angular velocities of the contacting bodies; J_z is the effective moment inertia; J_{z1} and J_{z2} are the moments of inertia of the contacting bodies. Since $\dot{x}_d = \dot{x}_{d1} + \dot{x}_{d2}$, Equation 3.4 can be written as follows

$$m = \frac{m_1 m_2}{m_1 + m_{z2}} \tag{3.7}$$

and as $J_z = J_{z1} + J_{z2}$, according to Equation 3.6

$$J_z = \frac{J_{z1} J_{z2}}{J_{z1} + J_{z2}} \tag{3.8}$$

The equations 3.7 and 3.8 are valid only for the movement of the centre mass of the body (third body). Since the moment of inertia of a spherical body is known as $J_z = \frac{2}{5} m l^2$, by taking Equation 3.8 into account, we get

$$l = \sqrt{\frac{5 J_{z1} J_{z2}}{2m(J_{z1} + J_{z2})}} \tag{3.9}$$

All authors, who use these equations, for example Stronge (2000); Dintwa (2006); Bordbar and Hyppänen (2007); Antypov et al. (2011); Landau and Lifshitz (1944, 1959), Brilliantov et al. (1996), have considered that x_d is the displacement between the centres mass of the bodies and x is the mutual approach (a compression, an overlapping) between contacting surfaces of these two bodies.

It is obvious that

$$x = x_d = x_{d1} + x_{d2} = x_1 + x_2 \tag{3.10}$$

where x_{d1} is the relative displacement of the centre mass of a softer body by axis X, x_{d2} is the relative displacement of the centre mass of a harder body by X, but we have to be aware that $x_{d1} \neq x_1$ and $x_{d2} \neq x_2$. Therefore, according to Equation 3.4

$$mx = m_1 x_{d1} = m_2 x_{d2} \tag{3.11}$$

Using Equations 3.10 and 3.11, we can express the relative displacements of the centres of masses x_{d2} and x_{d1} relative to axis X as follows

$$x_{d1} = K_{1m} x \tag{3.12}$$

$$x_{d2} = K_{2m} x \tag{3.13}$$

where K_{1m} is the coefficient of the mass of softer body, K_{2m} is the coefficient of the mass of harder body, which can be expressed as

$$K_{1m} = \frac{m_2}{m_1 + m_2} \tag{3.14}$$

$$K_{2m} = \frac{m_1}{m_1 + m_2} \tag{3.15}$$

According to Equations 2.84, 2.85, 3.12 and 3.13, it follows that

$$x = \frac{x_1}{D_1} = \frac{x_2}{D_2} = \frac{x_{d1}}{K_{m1}} = \frac{x_{d2}}{K_{m2}} \tag{3.16}$$

As well, in the case of a sliding contact between surfaces of two bodies, it is obvious that

$$y = y_d = y_{d1} + y_{d2} = y_1 + y_2 \tag{3.17}$$

where y_{d1} is the displacement of the centre mass of a softer body relative to the centre of oscillation by Y, y_{d2} is the displacement of the centre mass of a harder body relative to the centre of oscillation by Y, but $y_{d1} \neq y_1$ and $y_{d2} \neq y_2$.

Obviously, according to Equation 3.5, it follows that

$$my = m_1 y_{d1} = m_2 y_{d2} \tag{3.18}$$

And analogically we again get

$$y_{d1} = K_{1m} y \tag{3.19}$$

$$y_{d2} = K_{2m} y \tag{3.20}$$

Using Equations 2.88, 2.90, 3.19 and 3.20, we get

$$y = \frac{y_1}{D_{1\tau}} = \frac{y_2}{D_{2\tau}} = \frac{y_{d1}}{K_{m1}} = \frac{y_{d2}}{K_{m2}} \tag{3.21}$$

3.1.2 NORMAL DISPLACEMENT

3.1.2.1 DIFFERENTIAL EQUATIONS OF DISPLACEMENT AT THE COMPRESSION

First of all, let us consider the normal displacement at impact by axis X. Since $x = x_d$ and according to Equations 2.355 and 2.356, $F_{cn} = \frac{4}{3} E' R^{1/2} x^{3/2}$,

$F_{bn} = 4k_p\eta_E' R^{1/2}\dot{x}x^{1/2}$ by taking into account Equation 2.360, the equation for the general normal viscoelastic forces can be written as

$$F_n = F_{bn} + F_{cn} = b_x\dot{x} + c_x x = 4k_p\eta_E' R^{1/2}\dot{x}x^{1/2} + \frac{4}{3}k_p E' R^{1/2}x^{3/2} \tag{3.22}$$

where the expressions for the variable viscoelasticity parameters can be written as

$$b_x = 4\eta_E' R^{1/2}x^{1/2} \tag{3.23}$$

$$c_x = \frac{4}{3}E' R^{1/2}x^{1/2} \tag{3.24}$$

If the dynamic viscosities are replaced by the dynamic viscosity modulus according to the known expressions for Kelvin–Voigt model, see Equations 2.293 and 2.294, $\frac{E''}{\omega_x} = \eta_E'$, we get the next equation for the normal viscoelastic force

$$F_n = 4k_p\frac{E''}{\omega_x}R^{1/2}\dot{x}x^{1/2} + \frac{4}{3}k_p E' R^{1/2}x^{3/2} \tag{3.25}$$

Taking into account that $\frac{E''}{E'} = tg\beta_E$, we get

$$F_n = \frac{4}{3\omega_x}k_p E' R^{1/2}x^{1/2}(3tg\beta_E\dot{x} + \omega_x x) \tag{3.26}$$

Let us notice that the dynamic modulus of elasticity is also called as accumulation (and storage) modulus, and the dynamic modulus of viscosity is also called as the loss modulus.

According to Newton's Second Law, see Equation 3.1, and according to Equation 3.25, the differential equation of the movement (displacement) of the centre of mass of a body by axis X can be expressed as

$$m\ddot{x} + 4k_p\frac{E''}{\omega_x}R^{1/2}\dot{x}x^{1/2} + \frac{4}{3}k_p E' R^{1/2}x^{3/2} = 0 \tag{3.27}$$

or it can be also written in the canonical form as

$$m\ddot{x} + b_x\dot{x} + c_x x = 0 \tag{3.28}$$

Also, since $\frac{E''}{\omega_x} = \eta_E'$, the formula for the variable viscosity parameter can be written as

$$b_x = \frac{4E''R^{1/2}}{\omega_x} x^{1/2} \tag{3.29}$$

3.1.2.2 WORK AND ENERGY IN THE PHASES OF COMPRESSION AND RESTITUTION

The normal initial kinetic energy can be expressed as $W_x = \frac{mV_x^2}{2}$. On the other hand, the kinetic energy at the instance of rebound can be expressed as $W_{tx} = \frac{mV_{tx}^2}{2}$, where $V_{tx} = \dot{x}(\tau_x)$ is the normal relative velocity between the centres of mass of the bodies at the instance of the rebound $t = \tau_x$ (τ_x) is the time period of the contact). The graphical illustration of the functional dependences between the normal viscoelastic forces and the displacement of the centre of mass of a body is depicted in Figure 3.2a. Also, the 'rheological model of Kelvin–Voigt', which usually is used for the viscoelastic contact, is represented in Figure 3.2b.

As we know the period of time at impact includes two principally different phases such as the phases of compression and restitution; and it is obvious that the normal initial kinetic energy W_x is spent for the work A_{xm} of the normal viscoelastic force F_n in the compression phase. But on the other hand, A_{xm} can be found as the sum of works A_{xcm} and A_{xbm}, where A_{xcm} is the work of the normal elastic force F_{cn} and A_{xbm} is the work of the normal viscous force F_{bn} in the compression phase. Also, we can say that during the time of the phase of compression, the part of the kinetic energy W_x is transformed into the potential energy of an elastic deformation of the nonlinear elastic element (spring), see Figure 3.2b, and the other part of this kinetic energy is dissipated during the time of deformation at the compression of the nonlinear viscous element (dashpot). However, on the other hand, the work A_{xt} of the normal viscoelastic force F_n in the restitution phase equals to the normal energy of a body W_{tx} at the instance of rebound, and also A_{xt} can be found as the difference between A_{xct} and A_{xbt}, where A_{xct} is the work of the normal elastic force F_{cn} and A_{xbt} is the work of the normal viscous force F_{bn} in the restitution phase. Consequently, we can write that

$$A_{xm} = A_{xcm} + A_{xbm} = \frac{mV_x^2}{2} \tag{3.30}$$

$$A_{xt} = A_{xct} - A_{xbt} = \frac{mV_{tx}^2}{2} \tag{3.31}$$

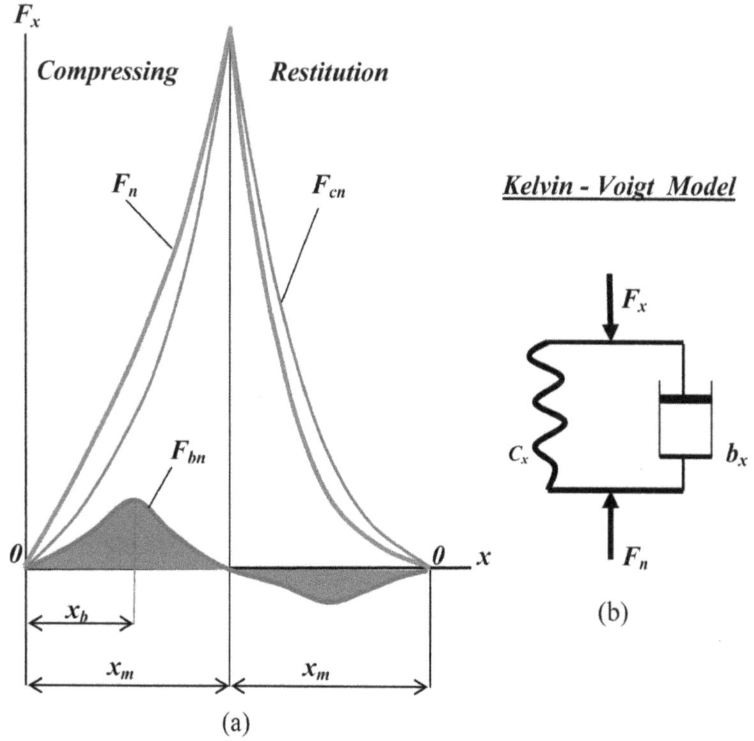

FIGURE 3.2 **(See color insert.)** (a) The graphical illustration of the functional dependences between the normal viscoelastic forces and the displacement *x(t)* of the centre of mass of a body, (b) The 'nonlinear rheological model of Kelvin–Voigt', where c_x and b_x magnitudes are not constant.

Source: Reprinted from Goloshchapov, 2015a, with permission from SAGE Publications. http://journals.sagepub.com/doi/abs/10.1177/1056789514560912).

It is obvious that $A_{xcm} = A_{xct}$ and; hence, the potential energy which has been accumulated inside of the elastic element (spring) fully returns back to the body at the instance of rebound. The works A_{xcm} and A_{xbm} at the compression can be found by integration:

$$A_{xcm} = \int_0^{x_m} F_{cn} dx = \frac{4}{3} k_p E' R^{1/2} \int_0^{x_m} x^{3/2} dx = \frac{8}{15} k_p E' R^{1/2} x_m^{5/2} \tag{3.32}$$

and

$$A_{xbm} = \int_0^{x_m} F_{bn} dx = \int_0^{x_m} \frac{4k_p E'' R^{1/2}}{\omega_x} \dot{x} x^{1/2} dx = \left(\frac{4k_p E'' R^{1/2}}{\omega_x} \right) \frac{\int_0^{x_m} \int dx x^{1/2} dx}{\int_0^{\tau_1} dt} = \frac{8k_p E'' R^{1/2} x_m^{5/2}}{5\omega_x \tau_1} \tag{3.33}$$

Analogically, the works A_{xct} and A_{xbt} in the restitution phase can be found as follows:

$$A_{xct} = -\int_{x_m}^{0} F_{cn} dx = -\int_{x_m}^{0} \frac{4}{3} k_p E' R^{1/2} x^{3/2} dx = \frac{8}{15} k_p E' R^{1/2} x_m^{5/2} \tag{3.34}$$

and

$$A_{xbt} = -\int_{x_m}^{0} F_{bn} dx = -\int_{x_m}^{0} \frac{4k_p E'' R^{1/2}}{\omega_x} \dot{x} x^{1/2} dx = -\left(\frac{4k_p E'' R^{1/2}}{\omega_x} \right) \frac{\int_{x_m}^{0} x^{1/2} dx \int dx}{\int_{\tau_1}^{\tau_x} dt} = \frac{8k_p E'' R^{1/2} x_m^{5/2}}{5\omega_x \tau_2} \tag{3.35}$$

where $\tau_x = \tau_1 + \tau_2$ is the time period of the contact, τ_1 is the time period of the compression, τ_2 is the time period of the restitution, x_m is the maximum magnitude of the compression between surfaces of the contacting bodies (also, as we already know, it is the maximum displacement of the centre of mass of a body, which is equal to the maximum of mutual approach between bodies).

According to Equations 3.30–3.35, the equations for the work of the compression and the restitution can be written as follows:

$$A_{xm} = A_{xcm} + A_{xbm} = \frac{8}{15} k_p R^{1/2} x_m^{5/2} \left(E' + \frac{3E''}{\omega_x \tau_1} \right) \tag{3.36}$$

$$A_{xt} = A_{xct} - A_{xbt} = \frac{8}{15} k_p R^{1/2} x_m^{5/2} \left(E' - \frac{3E''}{\omega_x \tau_2} \right) \tag{3.37}$$

and since $tg\beta_E = \frac{E''}{E'}$, it follows

$$A_{xm} = A_{xcm} + A_{xbm} = \frac{8}{15} k_p R^{1/2} x_m^{5/2} E' \left(1 + \frac{3tg\beta_E}{\omega_x \tau_1} \right) \tag{3.38}$$

$$A_{xt} = A_{xct} - A_{xbt} = \frac{4}{15} k_p R^{1/2} x_m^{5/2} E' \left(1 - \frac{3tg\beta_E}{\omega_x \tau_2} \right) \tag{3.39}$$

Since $A_{xm} = W_x = \frac{mV_x^2}{2}$ and $A_{xt} = W_{tx} = \frac{mV_{tx}^2}{2}$, by using Equation 3.38 we get the formula for x_m as follows

$$x_m = \left[\frac{15 m \omega_x \tau_1 V_x^2}{16(3tg\beta_E + \omega_x \tau_1) k_p E' R^{1/2}} \right]^{2/5} \tag{3.40}$$

Also, we can define the energetic coefficient of restitution e_x, which equals to the square of the kinematic coefficient of restitution k_x (further, it will be named simply as the coefficient of restitution), the ratio between W_{tx} and W_x is as follows:

$$e_x = k_x^2 = \frac{V_{tx}^2}{V_x^2} = \left(\frac{\omega_x \tau_2 - 3tg\beta}{\omega_x \tau_1 + 3tg\beta} \right) \frac{\tau_1}{\tau_2} \tag{3.41}$$

Since we can take that

$$x_m \approx \frac{/V_x/}{2} \tau_1 = \frac{/V_{tx}/}{2} \tau_2 \tag{3.42}$$

we get

$$k_x = \frac{\tau_1}{\tau_2} \tag{3.43}$$

Then, using Equations 3.41 and 3.43 gives us

$$tg\beta_E = \frac{\omega_x \tau_1}{3} \times \frac{1 - k_x}{k_x} \tag{3.44}$$

Thus, we have got the equation, which binds the coefficient of restitution and the tangent of the angle of mechanical losses. So, if $k_x = 1$, $tg\beta \rightarrow 0$, we get the totally elastic impact, but if $k_x = 0$, $tg\beta \rightarrow \infty$, then we can get the totally viscous impact. Using Equation 3.44 we can write the formula for the restitution coefficient as

$$k_x = \left[\frac{\omega_x \tau_1}{(3tg\beta_E + \omega_x \tau_1)} \right] \tag{3.45}$$

By comparing Equations 3.40 and 3.45, we can finally get the expression for the maximum magnitude of the compression between a body and semi-space as

$$x_m = \left[\frac{15 m V_x^2}{16 k_p E' R^{1/2}} k_x \right]^{2/5} \tag{3.46}$$

In the case of a totally elastic impact, when $k_x = 1$ and $k_p = 1$ we get the same result, as it has been obtained by L. Landau and Lifshitz (1944, 1959) according to the Hertz theory for absolute elastic contact.

3.1.2.3 APPROXIMATE SOLUTION TO DIFFERENTIAL EQUATIONS OF THE NORMAL DISPLACEMENT BY USING THE METHOD OF EQUIVALENT WORKS

For practical application of the differential Equation 3.28 with the variable viscoelasticity parameters, we can find their approximate solutions in the same manner as for the equations with the equivalent constant viscoelasticity parameters, if we choose the equivalent constant parameters B_x, C_x so that the works A_{xcm} and A_{xbm} with the variable viscoelasticity parameters c_x, b_x will be equal to the work with the constant viscoelasticity parameters. Thus, according to this statement and according to the boundary conditions $t = \tau_1$, $x = x_m$ and using the known expressions for works A_{xcm} and A_{xbm} from Equations 3.32 and 3.33, we can write next equations as follows

$$A_{xcm} = C_x \int_0^{x_m} x dx = \frac{1}{2} C_x x_m^2 = \frac{8}{15} k_p E' R^{1/2} x_m^{5/2} \tag{3.47}$$

and

$$A_{xbm} = B_x \int_0^{x_m} \dot{x} dx = B_x \frac{\int_0^{x_m} x dx}{\int_0^{\tau_1} dt} = B_x \frac{x_m^2}{2\tau_1} = \frac{8 k_p E'' R^{1/2} x_m^{5/2}}{5 \omega_x \tau_1} \tag{3.48}$$

Hence, according to the results obtained in Equations 3.47 and 3.48, we can write the expressions for the equivalent constant viscoelasticity parameters, respectively as:

$$C_x = \frac{16}{15} k_p E' R^{1/2} x_m^{1/2} \tag{3.49}$$

$$B_x = \frac{16 E'' k_p R^{1/2}}{5 \omega_x} x_m^{1/2} \tag{3.50}$$

Thus, Equation 3.28 with variable parameters can be rewritten as the equations with constant parameters as

$$m\ddot{x} + B_x \dot{x} + C_x x = 0 \tag{3.51}$$

Equation 3.51 is the equation of damped oscillations and the solution to this equation is known as

$$x = C_1 e^{-\delta_x t} \sin(\omega_x t) + C_2 e^{-\delta_x t} \cos(\omega_x t) \tag{3.52}$$

where $\omega_x = \sqrt{\omega_{0x}^2 - \delta_x^2}$, $\delta_x = \frac{B_x}{2m}$ is the normal damping factor, $\omega_{0x} = \sqrt{\frac{C_x}{m}}$ is the angular frequency of free harmonic oscillations by axis X. The constants of integration C_1 and C_2, which can be found from the initial conditions, if $t = 0$ follows $C_2 = 0$ and

$$x = C_1 e^{-\delta_x t} \sin(\omega_x t) \tag{3.53}$$

After integration we get

$$\dot{x} = C_1 e^{-\delta_x t}[\omega_x \cos(\omega_x t) - \delta_x \sin(\omega_x t)] \tag{3.54}$$

According to the initial condition, $\dot{x} = V_x$ follows that $C_1 = \frac{V_x}{\omega_x}$, and after substitution follows

$$x = \frac{V_x}{\omega_x} e^{-\delta_x t} \sin(\omega_x t) \tag{3.55}$$

and we get the equation for the normal relative velocity between centres of mass of the contacting bodies as well

$$\dot{x} = \frac{V_x}{\omega_x} e^{-\delta_x t}[\omega_x \cos(\omega_x t) - \delta_x \sin(\omega_x t)] \tag{3.56}$$

It is obvious that the period of time of the contact τ_x is equal to the semi-period of damped oscillations $T_x/2$ by axis X. Hence,

$$\tau_x = \frac{T_x}{2} = \frac{\pi}{\omega_x} \tag{3.57}$$

Since $\tau_x = \tau_1 + \tau_2$ and also by using Equations 3.41, 3.43, 3.44 and 3.57, we get

$$tg\beta_E = \frac{\pi}{3} \times \frac{(1 - k_x)}{(1 + k_x)} \tag{3.58}$$

We can write the equation for the restitution coefficient as follows

$$k_x = \frac{(\pi - 3tg\beta_E)}{(\pi + 3tg\beta_E)} \tag{3.59}$$

If $tg\beta_E = 0$; hence, $k_x = 1$, it is a totally elastic impact, but if $tg\beta_E = \pi/2$; hence, $k_x = 0$ and $x = 0$, it is absolutely plastic impact. Both of these two cases are not possible in nature.

Finally, from Equations 3.46 and 3.59, it follows that

$$x_m = \left[\frac{15mV_x^2}{16k_p E' R^{1/2}} \times \frac{(\pi - 3tg\beta_E)}{(\pi + 3tg\beta_E)} \right]^{2/5} \quad (3.60)$$

Thus, we have a very simple way to calculate x_m, if we know the value of $tg\beta$. For example, according to Equations 2.17, 2.74 and 2.75, the effective $tg\beta$ can be calculated also by formula

$$tg\beta_E = \frac{E''}{E'} = \frac{E_1'' E_2'' (E_1' + E_2')}{E_1' E_2' (E_1'' + E_2'')} \quad (3.61)$$

Using Equation 3.56 for the velocity, the duration of the time of the impact equals to the period of the time of the contact, and it can be found now from the conditions $\dot{x} = V_{tx}$ and $t = \tau_x$ as

$$\tau_x = -\frac{\ln k_x}{\delta_x} \quad (3.62)$$

where

$$\delta_x = \frac{B_x}{2m} = \frac{8k_p E'' R^{1/2}}{5m\omega_x} x_m^{1/2} = \frac{8k_p E' tg\beta_E}{5\pi m} \tau_x R^{1/2} x_m^{1/2} \quad (3.63)$$

and since $tg\beta$ is known from Equation 3.58, by using Equations 3.46, 3.62 and 3.63 we get

$$\tau_x^2 = -\frac{2(1+k_x)\ln k_x}{V_{0x}^{2/5}(1-k_x)k_x^{1/5}} \times \left(\frac{5m}{8k_p E' R^{1/2}} \right)^{4/5} \quad (3.64)$$

3.1.3 TANGENTIAL DISPLACEMENT

3.1.3.1 DIFFERENTIAL EQUATIONS OF DISPLACEMENT AT THE ROLLING SHEAR

As it is known that in the case of rolling between the contacting surfaces, the instantaneous centre of rotation—the centre of velocities always is placed at the point A, see Figures 3.1 and 1.2. Therefore, it is obvious that the velocity of shear at point A always equals to zero, $\dot{y}_A = 0$, and as well the velocity of the body, in case of the rolling at point A, always equals zero, too, $\dot{y}_{dA} = 0$. As well, since the point B is placed on the border of the contact area and since x is the depth of indentation of the effective third body, having the mass m and the effective radius of rotation l, we can write for the angular velocity of the relative rotation between the colliding bodies that, $\omega = \frac{\dot{y}_r}{x} = \frac{\dot{y}}{l}$ and the velocity of the tangential deformation at shear in the case of rolling contact can be found as

$$\dot{y}_r = \frac{x}{l}\dot{y} \tag{3.65}$$

Also, you can look at Equations 1.68 and 1.70, where it already obtained the similar result. We can rewrite Equation 3.65 as $dy_r = \frac{x}{l}dy$, but since y_d and x are linearly independent, the tangential deformation at shear in the case of rolling contact can be expressed as

$$y_r = \frac{x}{l}y \tag{3.66}$$

Now, according to Equations 2.357 and 2.358, the equations for the tangential viscoelastic forces for the rolling contact can be written as

$$F_{c\tau r} = G'P_s y \tag{3.67}$$

$$F_{b\tau r} = \eta'_G P_s \dot{y} \tag{3.68}$$

where, $P_x = k_h x + 2k_p R^{1/2} x^{1/2}$, therefore,

$$P_s = \frac{x}{l}P_x = \frac{x^{3/2}}{l}(k_h x^{1/2} + 2k_p R^{1/2}) \tag{3.69}$$

and $k_h = \left(\frac{R - D_2 R_2}{R_2}\right)$.

By taking into account that $F_{\tau r} = F_{b\tau r} + F_{c\tau r}$ and $\frac{G''}{\omega_{yr}} = \eta'_G$, the equation for the general tangential viscoelastic force can be expressed as

$$F_{\tau r} = \frac{G''}{\omega_{yr}}P_s\dot{y} + G'P_s y \tag{3.70}$$

where the expressions for the variable viscoelasticity parameters can be written as follows

$$b_y = \frac{G''}{\omega_{yr}}P_s \tag{3.71}$$

$$c_y = G'P_s \tag{3.72}$$

And since $\frac{G''}{G'} = tg\beta_G$, we can write that

$$F_{\tau r} = P_s G'(\frac{tg\beta_G}{\omega_{yr}}\dot{y} + y) \tag{3.73}$$

If we take $M = F_\tau \times l$, the equation for the reactive moment can be written as follows

$$M = P_\varphi \left(\frac{G''}{\omega_{yr}}\dot{y} + G'y \right) \tag{3.74}$$

where $P_\varphi = P_s l = x^{3/2}(k_h x^{1/2} + 2k_p R^{1/2})$

And now, according to Newton's Second Law, see Equations 3.2, 3.3, 3.70 and 3.74 the differential equation of the movement (displacement) of the centre of mass of a body by axis Y can be expressed as

$$m\ddot{y} + \frac{G''}{\omega_{yr}} P_s \dot{y} + G'P_s y = 0 \tag{3.75}$$

$$J_z \ddot{\varphi} + P_\varphi \left(\frac{G''}{\omega_{yr}}\dot{y} + G'y \right) = 0 \tag{3.76}$$

or it can be also written in the canonical form as

$$m\ddot{y} + b_y \dot{y} + c_y y = 0 \tag{3.77}$$

$$J_z \ddot{\varphi} + P_\phi \left(b_y \dot{y} + c_y y \right) = 0 \tag{3.78}$$

3.1.3.2 WORK AND ENERGY AT THE ROLLING SHEAR

In duration of the time of the collision, the tangential initial kinetic energy can be expressed as $W_y = \frac{mV_y^2}{2}$. On the other hand, the tangential kinetic energy at the instance of rebound can be expressed as $W_{ty} = \frac{mV_{ty}^2}{2}$; where $V_{ty} = \dot{y}_t$ is the tangential relative velocity between the centres of mass of the bodies at the instance of rebound. It is obvious that during the time of the displacement and of the rolling shear along axis Y, the initial tangential kinetic energy of the bodies (the third body) W_y is spent for the work A_y of the tangential viscoelastic force F_τ. The work A_y can be found as the sum of the works A_{yb} and A_{yc}, where A_{yb} is the work of the tangential viscous force $F_{b\tau}$; and A_{yc} is the work of the tangential elastic force. But on the other hand, it is obvious as well that the work A_{yb} is transformed into the dissipative energy Q_ω and the work A_{yc} is transformed into the work A_ω of the relative rotation between bodies. Thus, according to the 'law of preservation of energy for a nonconservative (dissipative) mechanical systems', we can write the equations for the relative displacement of the centres of mass of the bodies and for the relative rotation of the bodies, as follows:

$$\frac{m}{2}\left(\frac{dy}{dt}\right)^2 + A_y = \frac{mV_y^2}{2} \tag{3.79}$$

$$\frac{J_z}{2}\left(\frac{d\varphi}{dt}\right)^2 + A_\omega + Q_\omega = \frac{J_z\omega_0^2}{2} \tag{3.80}$$

where $A_y = \int F_\tau dy_d$, $A_\omega = -\int M d\varphi$; $Q_\omega = \int F_{b\tau} dy_d$, ω_0 is the initial angular velocity and where $M = F_\tau l$. Since $F_\tau = F_{c\tau} + F_{b\tau}$ and since we can take that $d\varphi = dy_d/l$; hence,

$$A_\omega = -\int M d\varphi = -\int F_\tau dy \tag{3.81}$$

Also, if the initial angular velocity ω_0 equals zero, we can write Equations 3.79 and 3.80 for the boundary conditions at the instance of time $t = \tau_x$, as follows:

$$\frac{mV_{ty}^2}{2} + \int_0^{y_t} F_\tau dy = \frac{mV_y^2}{2} \tag{3.82}$$

$$\frac{J_z\omega_t^2}{2} - \int_0^{y_t}(F_{c\tau} + F_{b\tau})dy + \int_0^{y_t} F_{b\tau} dy = 0 \tag{3.83}$$

where $V_{ty} = \dot{y}_t$ is the velocity at the instance of time $t = \tau_x$, ω_t is the relative angular velocity between bodies at the instant of time $t = \tau_x$, y_t is displacement of the centres of mass of the bodies along axis Y at the instance of time $t = \tau_x$. Equations 3.82 and 3.83 can be rewritten as

$$A_y = \int_0^{y_t}(F_{c\tau} + F_{b\tau})dy = \frac{mV_y^2}{2} - \frac{mV_{ty}^2}{2} \tag{3.84}$$

$$\frac{J_z\omega_t^2}{2} = \int_0^{y_t} F_{c\tau} dy = A_{yc} \tag{3.85}$$

Now, we can find the works of the rolling shear in the moment of time $t = \tau_1$ and $x = x_m$, respectively as

$$A_{ycm} = \int_0^{y_m} F_{c\tau y} dy = G' P_s \int_0^{y_m} y dy = \frac{G'}{2} P_m y_m^2 \tag{3.86}$$

$$A_{ybm} = \int_0^{y_m} F_{b\tau y} dy = \frac{G''}{\omega_y} \int_0^{P_m} dP_x \int_0^{y_m} \dot{y} dy = \frac{G''}{\omega_{yr}} P_m \frac{\int_0^{y_m} \int dy dy}{\int_0^{\tau_1} dt} = \frac{G''}{2\omega_{yr}} P_m \frac{y_m^2}{\tau_1} \tag{3.87}$$

where

$$P_m = \frac{x_m^{3/2}}{l} (k_h x_m^{1/2} + 2k_p R^{1/2}) \tag{3.88}$$

and where y_m is the deformation of the shear between the surfaces of the contacting bodies along axis Y at the instance of time.

We can find the approximate solutions for equations of motion in the same manner as for the equations with the equivalent constant viscoelasticity parameters, if we choose the equivalent constant parameters B_y, C_y so that the works A_{ycm} and A_{ybm} with the variable viscoelasticity parameters c_y, b_y will be equal to the work with the constant viscoelasticity parameters. Thus, according to this statement and boundary conditions $t = \tau_1$, $x = x_m$ and $y = y_m$, and also by using the expressions for works A_{ycm} and A_{ybm} from Equations 3.86 and 3.87 in the phase of the rolling shear, we get

$$A_{ycm} = C_{yr} \int_0^{y_m} y dy = \frac{1}{2} C_{yr} y_m^2 = \frac{1}{2} G' P_m y_m^2 \tag{3.89}$$

$$A_{ybm} = B_{yr} \int_0^{y_m} \dot{y} dy = B_{yr} \frac{\int_0^{y_m} \int dy dy}{\int_0^{\tau_1} dt} = B_{yr} \frac{y_m^2}{2\tau_1} = \frac{G''}{2\omega_{yr}\tau_1} P_m y_m^2 \tag{3.90}$$

Hence, we can write the expressions for the equivalent constant viscoelasticity parameters as:

$$C_{yr} = G' P_m \tag{3.91}$$

$$B_{yr} = \frac{G''}{\omega_{yr}} P_m \tag{3.92}$$

Thus, Equation 3.77 for displacement between the centre masses can be written as the equation with constant parameters as follows

$$m\ddot{y} + B_{yr}\dot{y} + C_{yr}y = 0 \tag{3.93}$$

Equation 3.93 is the equation of the damped oscillations and according to the initial condition $y = 0$ and $\dot{y} = V_y$, the solutions to this equation is known as

$$y = \frac{V_y}{\omega_{yr}} e^{-\delta_{yr}t} \sin(\omega_{yr}t) \tag{3.94}$$

where $\omega_{yr} = \sqrt{\omega_{0yr}^2 - \delta_{yr}^2}$, $\delta_{yr} = \dfrac{B_y}{2m}$ is the tangential damping factor, $\omega_{0yr} = \sqrt{\dfrac{C_{yr}}{m}}$ is the angular frequency of the harmonic oscillations by axis Y.

The equation for the tangential relative velocity between centres of mass of the contacting bodies can be received as well by differentiation of Equation 3.94 as

$$\dot{y} = \frac{V_y}{\omega_{yr}} e^{-\delta_{yr}t} [\omega_{yr}\cos(\omega_{yr}t) - \delta_{yr}\sin(\omega_{yr}t)] \tag{3.95}$$

The full changing of the energy of dissipative system at the rolling shear can be found as the difference between A_y and A_{yc} as follows

$$\Delta W_y = A_y - A_{yc} = \frac{mV_y^2}{2} - \frac{mV_{ty}^2}{2} - \frac{J_z\omega_t^2}{2} = A_{yb} \tag{3.96}$$

Now, let us find the work A_{yc} and A_{yb} for the full period time of contact:

$$A_{yc} = C_{yr}\int_0^{y_t} y\,dy = \frac{1}{2}C_{yr}y_t^2 \tag{3.97}$$

$$A_{yb} = Q_\omega = B_{yr}\int_0^{y_t} \dot{y}\,dy = B_y \frac{\int_0^{y_t}\int dy\,dy}{\int_0^{\tau_x} dt} = B_{yr}\frac{y_t^2}{2\tau_x} \tag{3.98}$$

According to Equation 3.85, a conclusion can be drawn that the work A_{yc} is transformed into the kinetic energy of the relative rotation between the bodies, but on the other hand the work A_{yb} is transformed into dissipative energy Q_ω in the process of the internal friction. According to Equations 3.85, 3.91 and 3.97

$$A_{yc} = \frac{J_z\omega_t^2}{2} = \frac{1}{2}G'P_m y_t^2 \tag{3.99}$$

Hence, the equation for the relative angular velocity at the instance time of rebound can be written as follows

$$\omega_t = \left(\frac{G'P_m}{J_z}\right)^{\frac{1}{2}} y_t \qquad (3.100)$$

3.1.4 BOUNDARY VALUE PROBLEM AND MAIN RELATIONS IN THE KELVIN–VOIGT MODEL AT IMPACT

It is obvious from Equation 3.46 that in the case of $k_x = 1$ and $k_p = 1$ we have the same result as was obtained by Landau and Lifshitz (1944, 1965) for a totally elastic impact by using the Hertz Theory. It proves the correctness of the way of finding solutions for viscoelastic contact. But we have to understand that the application of the obtained equations for the viscoelastic contact has the boundaries of application which can be found by solving the next equation given below:

$$\omega_x = \sqrt{\omega_{0x}^2 - \delta_x^2} \qquad (3.101)$$

First of all, since $\omega_{0x}^2 = \frac{C_x}{m}$ and $\delta_x = \frac{B_x}{2m}$, it follows $\delta_x = \frac{B_x \omega_{0x}^2}{2C_x}$, then, taking into account Equations 3.49 and 3.50, and since $tg\beta_E = \frac{E''}{E'}$, as well follows that $\delta_x = \frac{3\omega_{0x}^2}{2\omega_x} tg\beta_E$. After substituting this ratio into Equation 3.101, we get the next algebraic equation

$$\omega_x^4 - \omega_{0x}^2 \omega_x^2 + \frac{9}{4}\omega_{0x}^4 tg^2\beta_E = 0 \qquad (3.102)$$

This equation has only the one valid solution

$$\omega_x^2 = \frac{\omega_{0x}^2}{2}(1 + \sqrt{1 - 9tg^2\beta_E}) \qquad (3.103)$$

and it has the valid root only when $1 - 9tg^2\beta_E \geq 0$; therefore, for the compression we get

$$tg\beta_E = \frac{E''}{E'} \leq \frac{1}{3} \qquad (3.104)$$

and according to Equation 3.58, we get the following for a viscoelastic contact:

$$k_x \leq \frac{\pi - 1}{\pi + 1} \tag{3.105}$$

In the case when $k_x \vartriangle \frac{\pi - 1}{\pi + 1}$, the plastic deformations of a compression will be having place in the zone of the contact.

Also, we can find the relations between the retardation times and the time periods of contact and shear. It is known from Equations 2.293 and 2.294 that $\frac{E''}{\omega_x} = \eta_E'$, $\frac{G''}{\omega_y} = \eta_G'$ and also already known from Equations 2.291 and 2.292 that $tg\beta_E = \frac{E''}{E'}$, $tg\beta_G = \frac{G''}{G'}$, and also it is known (Lee, 1962; Flügge, 1975; McCrum et al., 1997; Lakes, 1998; Menard and Kevin, 1999; Hosford, 2005; Roylance, 2001) that

$$\tau_{KE} = \frac{\eta_E'}{E} \tag{3.106}$$

$$\tau_{KG} = \frac{\eta_G'}{G} \tag{3.107}$$

where τ_{KE}, τ_{KG} are the relaxation times for deformations of the compression and the shear, respectively, which also are named as the retardation times in the Kelvin–Voigt model: E, G are modules elasticity of tension/compression and shear (or they are also named as the JUNG modules). Since in the Kelvin–Voigt model $E(t) = E = E'(\omega_x)$, $G(t) = G = G'(\omega_y)$ and $\frac{E''}{\omega_x} = \eta_E'$, $\frac{G''}{\omega_y} = \eta_G'$ and as

$$\tau_{KE} = \frac{\eta_E'}{E'} \tag{3.108}$$

$$\tau_{KG} = \frac{\eta_G'}{G'} \tag{3.109}$$

we get

$$tg\beta_E = \omega_x \tau_{KE} \tag{3.109*}$$

$$tg\beta_G = \omega_{yr} \tau_{KG} \tag{3.110}$$

Or in the case of the sliding contact, it follows $tg\beta_G = \omega_y \tau_{KG}$, where ω_y the frequency of oscillation in the case of a slide. Thus, it follows that

$$\frac{tg\beta_E}{tg\beta_G} = \frac{\tau_{KE}}{\tau_{KG}}\frac{\omega_x}{\omega_{yr}} \tag{3.111}$$

Thus, since $\tau_{KE} = \frac{\eta'_E}{E'}$, $\tau_{KG} = \frac{\eta'_G}{G'}$ and $\frac{E'}{G'} = 2(1+\nu)$ we can write one important ratio as

$$\frac{\eta'_E}{\eta'_G} = 2(1+\nu)\frac{tg\beta_E}{tg\beta_G}\frac{\omega_{yr}}{\omega_x} \tag{3.112}$$

where v is the effective Poisson's coefficient. Hence, it can be calculated as

$$\nu = \frac{1}{2}\frac{\eta'_E\omega_{yr}tg\beta_E}{\eta'_G\omega_x tg\beta_G} - 1 \tag{3.113}$$

On the other hand, since $\omega_{yr} = \sqrt{\omega_{0yr}^2 - \delta_{yr}^2}$, $\delta_{yr} = \frac{B_{yr}}{2m}$, $\omega_{0yr}^2 = \frac{C_{yr}}{m}$ and $tg\beta_G = \frac{G''}{G'}$, taking into account Equations 3.91 and 3.92, it follows as $\delta_{yr} = \frac{\omega_{0yr}^2}{2\omega_{yr}}tg\beta_G$, and we get the next algebraic equation

$$\omega_{yr}^4 - \omega_{0yr}^2\omega_{yr}^2 + \frac{1}{4}\omega_{0yr}^4 tg^2\beta_G = 0 \tag{3.114}$$

This equation has only the one valid solution

$$\omega_{yr}^2 = \frac{\omega_{0yr}^2}{2}(1+\sqrt{1-tg^2\beta_G}) \tag{3.115}$$

and it has the valid root only when $1 - tg^2\beta_G \geq 0$; therefore, for a rolling shear we get

$$tg\beta_G = \frac{G''}{G'} \leq 1 \tag{3.116}$$

Now, we can find the attitude $\frac{\omega_x}{\omega_{yr}}$ as follows:

$$\frac{\omega_x}{\omega_{yr}} = \frac{\omega_{0x}}{\omega_{0yr}}\left(\frac{1+\sqrt{1-9tg^2\beta_E}}{1+\sqrt{1-tg^2\beta_G}}\right)^{1/2} \tag{3.117}$$

Since $\omega_{0x} = \sqrt{\frac{C_x}{m}}$, according to Equation 3.49, it follows as

$$\omega_{0x} = \sqrt{\frac{16}{15m}k_p E' R^{1/2} x_m^{1/2}} \tag{3.118}$$

And since $\omega_{0yr} = \sqrt{\frac{C_{yr}}{m}}$, according to Equations 3.91 and 3.88, it follows that

$$\omega_{0yr} = \sqrt{\frac{G'}{m}\frac{x_m^{3/2}}{l}(k_h x_m^{1/2} + 2k_p R^{1/2})} \tag{3.119}$$

Also, it is obvious that $2R^{1/2}k_p >> k_h x^{1/2}$, we can write

$$\omega_{0yr} = \sqrt{\frac{G'}{m}\frac{x_m^{3/2}}{l}2k_p R^{1/2}} \tag{3.120}$$

Substitution of ω_{0yr} from Equation 3.120 and $\omega_0 x$ from Equation 3.118 into 3.117 gives:

$$\frac{\omega_x}{\omega_{yr}} = \sqrt{\frac{8}{15}\frac{E'}{G'}\frac{l}{x_m}} \times \left(\frac{1+\sqrt{1-9tg^2\beta_E}}{1+\sqrt{1-tg^2\beta_G}} \right)^{1/2} \tag{3.121}$$

Since $\frac{E'}{G'} = 2(1+\nu)$, finally

$$\frac{\omega_x}{\omega_{yr}} = \sqrt{\frac{16}{15}\frac{l}{x_m}(1+\nu)} \times \left(\frac{1+\sqrt{1-9tg^2\beta_E}}{1+\sqrt{1-tg^2\beta_G}} \right)^{1/2} \tag{3.122}$$

Let us consider the example of the collision between a hard spherical body and a high-elastic semi-space. For example, $l = R$ is the radius of a body, $x_m = 0.1R$, $\nu = 0.5$, $tg\beta_G = tg\beta_E = 0.1$. Then, after calculation we get that $\frac{\omega_x}{\omega_{yr}} = 4 \times (0.99) \approx 4$. As we can see, in the can see of the case of a small damping, the frequency of damped oscillation at the compression is approximately four times more than under a rolling shear.

3.1.5 FRICTION AT IMPACT BETWEEN TWO SPHERICAL BODIES

Taking into account Equations 3.26 and 3.74, the expression for calculating the coefficient of friction for the viscoelastic dynamic contact at impact can be written now simply as

$$f_r = \frac{F_{\tau r}}{F_n} = \frac{3G'P_S\omega_x(\dot{y}tg\beta_G + \omega_{yr}y)}{4k_pE'R^{1/2}x^{1/2}\omega_{yr}(3\dot{x}tg\beta_E + \omega_x x)} \tag{3.123}$$

Substituting P_s from Equation 3.69 into 3.123 we get

$$f_r = \frac{3G'x\omega_x(\dot{y}tg\beta_G + \omega_{yr}y)(k_h x^{1/2} + 2k_p R^{1/2})}{4lk_p E'R^{1/2}\omega_{yr}(3\dot{x}tg\beta_E + \omega_x x)} \tag{3.124}$$

In the case of viscoelastic contact $R >> x_m$, consequently we can write that

$$f_r = \frac{3G'x\omega_x(\dot{y}tg\beta_G + \omega_{yr}y)}{2lE'\omega_{yr}(3\dot{x}tg\beta_E + \omega_x x)} \tag{3.125}$$

The maximum value of the coefficient of friction will be in the moment $t = \tau_1$, $\dot{x} = \dot{x}_m = 0$, $x = x_m$ and $y = y_m$, $\dot{y} = \dot{y}_m$, and we get

$$f_{rm} = \frac{3G'\omega_{yr}(\dot{y}_m tg\beta_G + \omega_{yr}y_m)}{2lE'} \tag{3.126}$$

Since $\frac{E'}{G'} = 2(1+\nu)$ it follows that

$$f_{rm} = \frac{F_\tau}{F_n} = \frac{3\omega_{yr}}{4l}\frac{(\dot{y}_m tg\beta_G + \omega_{yr}y_m)}{(1+\nu)} \tag{3.127}$$

For the moment of the time of maximum compression and shear, $t = \tau_1$, according to Equations 3.94 and 3.95 follow

$$y_m = \frac{V_y}{\omega_{yr}}e^{-\delta_{yr}\tau_1}\sin(\omega_{yr}\tau_1) \tag{3.128}$$

$$\dot{y}_m = \frac{V_y}{\omega_{yr}}e^{-\delta_{yr}\tau_1}[\omega_{yr}\cos(\omega_{yr}\tau_1) - \delta_{yr}\sin(\omega_{yr}\tau_1)] \tag{3.129}$$

Substituting Equations 3.128 and 3.129 into Equation 3.127 we finally get

$$f_{rm} = \frac{3}{4}\frac{V_y}{l(1+\nu)}e^{-\delta_{yr}\tau_1}[\omega_{yr}tg\beta_G\cos(\omega_{yr}\tau_1) + (\omega_{yr} - \delta_y tg\beta_G)\sin(\omega_{yr}\tau_1)] \tag{3.130}$$

As we can see, the coefficient of friction changes in the process of collision between bodies; and it reaches maximum value in the moment of the maximum of indentation. The coefficient of friction is not a constant value. It changes cyclically in the period of time of contact. Slipping is not observed here between colliding bodies as some theories teach us. Curvilinear surfaces only roll relatively on each other, and work of elastic deformation completely turns into the kinetic energy of mutual rotation.

3.1.6 CONTACT STRESSES

3.1.6.1 AVERAGE CONTACT STRESSES

As it is known, the basic condition of workability of a contact surface is that the maximum stress in the centre of contact area should not be higher than the endurance limit σ_{lim}. Also, it is obvious that dangerous values of stresses reach maximum, when the normal viscous force reaches a maximum magnitude, too, at the instance of time $t=\tau_b$, $x=x_b$, $y_d=y_b$ and $t=\tau_1$, $x=x_m$, $y_d=y_{dm}$, see Figure 3.2a. The problem is finding the contact maximal stress in these moments of the time.

First of all, we should find the average normal contact stresses of viscoelasticity in the centre of the contact as

$$\tilde{\sigma}_n = \frac{F_n}{S_x} \tag{3.131}$$

where $S_x = \pi r^2 = \pi k_p^2 Rx$ is the square of the cyclic contact area. Since $\frac{E''}{\omega_x} = \eta'_E$ and $tg\beta_E = \frac{E''}{E'}$, the equation for the normal viscoelastic force from Equation 3.25 can be expressed as

$$F_n = \frac{4}{3} k_p E' R^{1/2} x^{1/2} (x + 3\frac{tg\beta_E}{\omega_x}\dot{x}) \tag{3.132}$$

Thus, the expression for general normal viscoelastic stress can be written as

$$\tilde{\sigma}_n = \frac{4E'}{3\pi k_p R^{1/2} x^{1/2}} (x + 3\frac{tg\beta_E}{\omega_x}\dot{x}) \tag{3.133}$$

It is obvious that the tangential contact stresses of viscoelasticity arise in the tangential section S_x (plane *YAZ*), in the tangential section S_y (plane *XAY*), see Figure 2.8. Since $\dot{y}_r = \frac{x}{l}\dot{y}$, $y_r = \frac{x}{l} y$ and $\frac{G''}{\omega_{yr}} = \eta'_G$, Equations 2.178–2.181 for the case of the rolling contact between the surfaces of two spherical bodies at impact can be rewritten as

$$F_{bcyr} = 2k_p G' \frac{R^{1/2} x^{3/2}}{l} y \tag{3.134}$$

$$F_{bbyr} = 2k_p \frac{G''}{\omega_{yr}} \frac{R^{1/2} x^{3/2}}{l} \dot{y} \tag{3.135}$$

$$F_{xcyr} = k_h G' \frac{x^2}{l} y \tag{3.136}$$

$$F_{xbyr} = k_h \frac{G'' x^2}{\omega_{yr} l} \dot{y} \tag{3.137}$$

As we know that $tg\beta = \frac{E''}{E'}$, the equation for the average tangential viscoelastic stresses in the section S_x, which acts along axis Y can be expressed as

$$\tilde{\sigma}_{\tau by} = \frac{F_{byr}}{S_x} = \frac{2G' x^{1/2}}{\pi k_p R^{1/2} l}(y + \frac{tg\beta_G}{\omega_{yr}}\dot{y}) \tag{3.138}$$

whereas, according to Equations 3.134 and 3.135,

$$F_{byr} = F_{bcyr} + F_{bbyr} = 2k_p G' \frac{R^{1/2} x^{3/2}}{l}(y + \frac{tg\beta_G}{\omega_{yr}}\dot{y}) \tag{3.139}$$

On the other hand, the equation for the average tangential viscoelastic stresses in the section S_y, acting along axis Y can be written as

$$\tilde{\sigma}_{\tau xy} = \frac{F_{xy}}{S_y} = G' \frac{3x^{1/2}}{4k_h^{1/2} R^{1/2} l}(y + \frac{tg\beta_G}{\omega_{yr}}\dot{y}) \tag{3.140}$$

whereas according to Equations 3.136 and 3.137,

$$F_{xyr} = F_{xcyr} + F_{xbyr} = k_h G' \frac{x^2}{l}(y + \frac{tg\beta_G}{\omega_{yr}}\dot{y}) \tag{3.141}$$

where S_y can be found by integration $dS_y = 2rdx$, see Figures 2.1 and 2.2,

$$S_y = 2\int_0^{h_{xa}} r dx = 2R^{1/2} \int_0^{h_{xa}} x^{1/2} dx = \frac{4}{3} R^{1/2} h_x^{3/2} \tag{3.142}$$

and since $h_x = k_h x$, the equation can be also written as

$$S_y = \frac{4}{3} k_h^{3/2} R^{1/2} x^{3/2} \tag{3.143}$$

Now, the average tangential viscoelastic stresses acting along the axis Y can be expressed as follows

$$\tilde{\sigma}_\tau = \tilde{\sigma}_{\tau by} + \tilde{\sigma}_{\tau xy} = \frac{G' x^{1/2}}{R^{1/2} l} \varsigma_y (y + \frac{tg\beta_G}{\omega_{yr}}\dot{y}) \tag{3.144}$$

where

$$\varsigma_y = \left(\frac{2}{\pi k_p} + \frac{3}{4k_h^{1/2}} \right) \tag{3.145}$$

According to the hypothesis of maximum tangential stresses, the equation for the average contact stresses of viscoelasticity in the contact area can be expressed as

$$\tilde{\sigma} = \sqrt{\tilde{\sigma}_n^2 + 4\tilde{\sigma}_\tau^2} \tag{3.146}$$

Finally, by substituting σ_n and $\tilde{\sigma}_\tau$ into Equation 3.146, we get the equation for the average contact stresses of viscoelasticity as given below:

$$\tilde{\sigma} = \sqrt{\left(\frac{4E'}{3\pi k_p R^{1/2} x^{1/2}} (x + 3\frac{tg\beta_E}{\omega_x}\dot{x}) \right)^2 + 4\left(\varsigma_y \frac{G' x^{1/2}}{R^{1/2} l} (y + \frac{tg\beta_G}{\omega_y}\dot{y}) \right)^2} \tag{3.147}$$

Finally, according to the conditions $t=\tau_1$, $x=x_m$, $y=y_m$, the equation for the maximum of average contact stress can be written as

$$\tilde{\sigma}_m = \sqrt{\left(\frac{4E'}{3\pi k_p R^{1/2} x_m^{1/2}} (x_m + 3\frac{tg\beta_E}{\omega_x}\dot{x}_m) \right)^2 + 4\left(\varsigma_y \frac{G' x_m^{1/2}}{R^{1/2} l} (y_m + \frac{tg\beta_E}{\omega_{yr}}\dot{y}_m) \right)^2} \tag{3.148}$$

The basic condition of workability of a contact surface is that the average stress should not be higher than the endurance limit $\sigma_{\lim}$.

Also, it is obvious that dangerous stress values have taken place when the normal viscous force reaches a maximum at the instance of time $t=\tau_b$, see Figure 3.2a, and it is known that the maximum value of continuous function reaches the point when its first derivative equals zero:

$$(F_{bn})' = \frac{4E'' R^{1/2}}{\omega_x} (x^{1/2}\dot{x})' = 0 \tag{3.149}$$

So, we receive the differential equation as follows

$$2\ddot{x}x + \dot{x}\dot{x} = 0 \tag{3.150}$$

After the substitution of the functions $x(\tau_b)$, $\dot{x}(\tau_b)$ and $\ddot{x}(\tau_b)$ into Equation 3.150 and then a simplification, we get the next equation for the calculation of the period of time τ_b as

$$(3\delta_x^2 - 2\omega_x^2) tg^2(\omega_x \tau_b) - 6\omega_x \delta_x tg(\omega_x \tau_b) + \omega_x^2 = 0 \tag{3.151}$$

This very simple algebraic equation can be solved by relating to the trigonometric function $tg(\omega_x \tau_b)$, and then the period of time τ_b can be calculated as

$$\tau_b = \frac{1}{\omega_x} \operatorname{arctg}\left(\frac{6\omega_x \delta_x \pm \omega_x \sqrt{6\delta_x^2 + \omega_x^2}}{2(3\delta_x^2 - 2\omega_x^2)} \right) \quad (3.152)$$

We should choose the valid positive root from the two roots calculated according to Equation 3.152. Then, we can find the values of the displacements x_b and y_b by using Equations 3.55 and 3.94, respectively as

$$x_b = \frac{V_x}{\omega_x} e^{-\delta_x \tau_b} \sin(\omega_x \tau_b) \quad (3.153)$$

$$y_b = \frac{V_y}{\omega_{yr}} e^{-\delta_{yr} \tau_b} \sin(\omega_{yr} \tau_b) \quad (3.154)$$

Then, the velocities of displacements $\dot{x}_b$ and $\dot{y}_b$ can be found by using Equations 3.56 and 3.95, respectively

$$\dot{x}_b = \frac{V_x}{\omega_x} e^{-\delta_x \tau_b} [\omega_x \cos(\omega_x \tau_b) - \delta_x \sin(\omega_x \tau_b)] \quad (3.155)$$

$$\dot{y}_b = \frac{V_y}{\omega_{yr}} e^{-\delta_{yr} \tau_b} [\omega_{yr} \cos(\omega_{yr} \tau_b) - \delta_{yr} \sin(\omega_{yr} \tau_b)] \quad (3.156)$$

Substituting the calculated values of x_b, $\dot{x}_b$ and y_b, $\dot{y}_b$ into Equation 3.147, we get the second condition of the workability of a contact surface

$$\tilde{\sigma}_b = \sqrt{\left(\frac{2E'}{\pi k_p R^{1/2} x_b^{1/2}} (x_b + 3\frac{tg\beta_E}{\omega_x} \dot{x}_b) \right)^2 + 4\left(\varsigma_y \frac{G' x_b^{1/2}}{R^{1/2} l} (y_b + \frac{tg\beta_G}{\omega_y} \dot{y}_b) \right)^2} \quad (3.157)$$

We can calculate the effective radius l of the third (effective) body having the effective mass $m = \frac{m_1 m_2}{m_1 + m_2}$ using Equation 2.127; where $l = \sqrt{\frac{5J_{z1} J_{z2}}{2m(J_{z1} + J_{z2})}}$.

Since $J_{z1} = \frac{2}{5} m_1 R_1^2$ and $J_{z2} = \frac{2}{5} m_2 R_2^2$, we get

$$l^2 = \frac{R_1^2 R_2^2 (m_1 + m_2)}{m_1 R_1^2 + m_2 R_2^2} \quad (3.158)$$

Finally, the formula for calculation of the radius of the third effective body can be expressed as

$$l = R_1 R_2 \sqrt{\frac{m_1 + m_2}{m_1 R_1^2 + m_2 R_2^2}} \tag{3.159}$$

Also, we can calculate the approximate value of l if the deformations between contacting surfaces are very small, when $R >> x$. Since $\bar{\omega} = \bar{\omega}_1 + \bar{\omega}_2$, see Figure 2.8, we can write that

$$\frac{\dot{y}}{l} = \frac{\dot{y}_2}{R_2 - x_2} + \frac{\dot{y}_1}{R_1 - x_1}\Big|_{t=0} = \frac{\bar{V}_y}{l} = \frac{\bar{V}_{2y}}{R_2 - x_2} + \frac{\bar{V}_{1y}}{R_1 - x_1} \tag{3.160}$$

where $\dot{y}_1$, $\dot{y}_2$ are the velocities of displacements of centres of mass of the contacting bodies by axis Y, $\bar{V}_{1y}$ and $\bar{V}_{2y}$ are the initial velocities of displacements of centres of mass of the contacting bodies by axis Y in the initial moment of time $t=0$. If we take that in case of contact between two rigid surfaces where $R_1 >> x_1$ and $R_2 >> x_2$, we can write that

$$\frac{\bar{V}_y}{l} \approx \frac{\bar{V}_{2y}}{R_2} + \frac{\bar{V}_{1y}}{R_1} \tag{3.161}$$

Thus,

$$l \approx \frac{\bar{V}_y R_1 R_2}{R_1 \bar{V}_{2y} + R_2 \bar{V}_{1y}} \tag{3.162}$$

In the case of contact between a spherical body and a flat semi-space, when $R_1 = \infty$ and $\bar{V}_y = \bar{V}_{2y}$, it follows that $l = R_2 = R$.

3.1.6.2 MAXIMUM CONTACT STRESSES

The normal contact stresses of viscoelasticity in the centre of the contact area at the point A should be equal to the maximum of pressure $\sigma_n = P_c = \frac{3}{2} P_a$, see Equation 2.141, and since $P_a = \tilde{\sigma}_n = \frac{F_n}{S_x}$, we get the following

$$\sigma_n = \frac{3}{2}\tilde{\sigma}_n = \frac{3F_n}{2S_x} \tag{3.163}$$

Substitution of $\tilde{\sigma}_n$ from Equation 3.133 in Equation 3.163 gives

$$\sigma_n = \frac{2E'}{\pi k_p R^{1/2} x^{1/2}} (x + 3\frac{tg\beta_E}{\omega_x}\dot{x}) \tag{3.164}$$

Also, since the surface of the contact has approximately the parabolic shape and the magnitude of shear on the border of contact area equals zero, $y = 0$;

therefore, the shear elastic stresses are also equal to zero, $\sigma_{\tau c} = 0$. Let us take that the radial distribution of the tangential elastic stress inside of this area changes like the parabolic function analogically as we have taken for the normal stress distribution

$$\tilde{\sigma}_{\tau c} = \sigma_{Ac\tau}\left(1-\frac{r_y^2}{r^2}\right) \tag{3.165}$$

where r_y is a current radius of the contact area; r is the radius of the contact area by axis Y, $\sigma_{A\tau c}$ is the magnitude of the stress in the centre of the contact area.

Further since the square under the function in Equation 3.165 and the square under the linear function of the mean tangential elastic stress in the contact area are equal, we can write that

$$\sigma_{Ac\tau}\int_0^r\left(1-\frac{r_y^2}{r^2}\right)dr_y = \tilde{\sigma}_{c\tau}r \tag{3.166}$$

where $\tilde{\sigma}_{c\tau}$ is the mean tangential elastic stress. Then, after integration, it follows

$$\sigma_{Ac\tau}\left(r-\frac{1}{3}r\right) = \tilde{\sigma}_{c\tau}r \tag{3.167}$$

and, finally, the ratio between maximum and the mean tangential elastic stress in the contact zone, see Figure 3.3, can be found as

$$\sigma_{Ac\tau} = \frac{3}{2}\tilde{\sigma}_{c\tau} \tag{3.168}$$

Where the average elastic tangential stress can be found as

$$\tilde{\sigma}_{c\tau} = \tilde{\sigma}_{bcy} + \tilde{\sigma}_{xcy} \tag{3.169}$$

since F_{bcyr} is known from Equation 3.134,

$$\tilde{\sigma}_{bcy} = \frac{F_{bcy}}{S_x} = \frac{2G'x^{1/2}}{\pi k_p l R^{1/2}}y \tag{3.170}$$

and since F_{xcyr} is known from Equation 3.136,

$$\tilde{\sigma}_{xcy} = \frac{F_{xcy}}{S_y} = \frac{3G'x^{1/2}}{4k_h^{1/2} l R^{1/2}}y \tag{3.171}$$

Thus,

$$\tilde{\sigma}_{c\tau} = \frac{G'x^{1/2}}{lR^{1/2}} \varsigma_y y \tag{3.172}$$

Substituting Equation 3.172 in 3.168 we get

$$\sigma_{Ac\tau} = \frac{3G'x^{1/2}}{2lR^{1/2}} \varsigma_y y \tag{3.173}$$

According to the hypothesis of maximum tangential stresses, the equation for the general contact stresses of viscoelasticity in the centre of the contact area can be expressed as

$$\sigma_A = \sqrt{\sigma_n^2 + 4\sigma_{Ac\tau}^2} \tag{3.174}$$

Substituting expressions from Equations 3.164 and 3.173 into 3.174 we get the following

$$\sigma_A = \sqrt{\left(\frac{2E'}{\pi k_p R^{1/2} x^{1/2}} (x + 3\frac{tg\beta_E}{\omega_x} \dot{x})\right)^2 + 4\left(\frac{3G'x^{1/2}}{2lR^{1/2}} \varsigma_y y\right)^2} \tag{3.175}$$

It is obvious that the stresses at the point A reach the maximum magnitude in the moment $t = \tau_1$, $x = x_m$, $y_d = y_{dm}$ and when the velocity of compression reaches the minimum $\dot{x} = 0$; therefore, the formula for calculation of the maximum stresses can be expressed as

$$\sigma_{Am} = \sqrt{\left(\frac{2E'x_m^{1/2}}{\pi k_p R^{1/2}}\right)^2 + 4\left(\frac{3G'x_m^{1/2}}{2lR^{1/2}} \varsigma_y y_m\right)^2} \tag{3.176}$$

Also, since the magnitude of the velocity of the shear on the border of contact area at point B and the velocity of restitution at point C are equal maximum magnitudes, $\dot{y} = \dot{y}_B$ and $\dot{y} = \dot{y}_C$; therefore, the tangential viscous stresses are equal maximum magnitudes too, $\sigma_{b\tau} = \sigma_{Bb\tau}$ and $\sigma_{b\tau} = \sigma_{Cb\tau}$, but at point A the velocity of the shear is equal to zero and, thus, the tangential viscous stresses are equal to zero, too, see Figure 3.3. Let us take that the radial distribution of the tangential viscous stress inside of this area changes like the parabolic function

$$\tilde{\sigma}_{b\tau} = \sigma_{Bb\tau}\left(\frac{r_y^2}{r^2}\right) \tag{3.177}$$

where r_y is a current radius of the contact area r is the radius of the contact area by the axis, $\sigma_{Bb\tau}$ is the magnitude of the stress on the border of the contact area at the point *B*, see Figure 3.3.

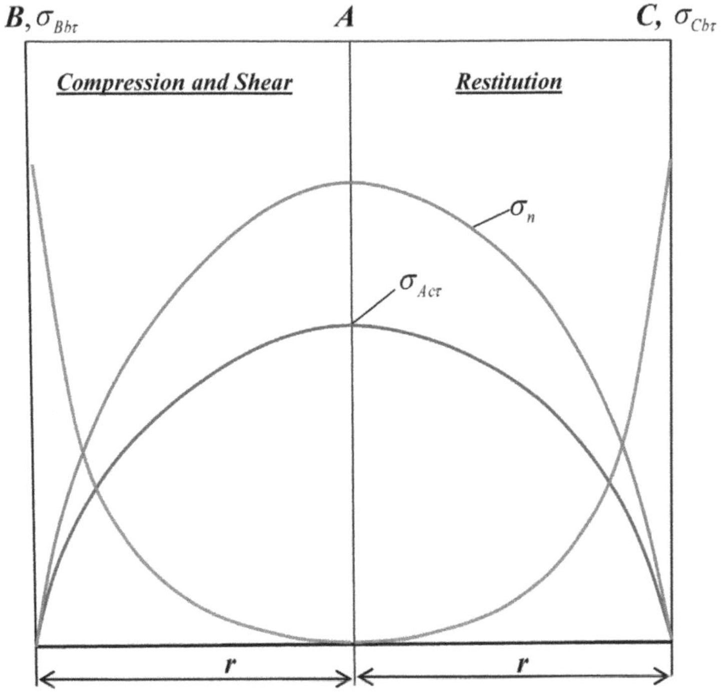

FIGURE 3.3 (See color insert.) Distribution of the normal and the tangential stresses in the area of the contact.

Further, since the square under the function in Equation 3.177 and the square under the linear function of the mean tangential viscous stress in the contact area are equal, we get

$$\sigma_{Bb\tau}\int_0^r \left(\frac{r_y^2}{r^2}\right) dr_y = \tilde{\sigma}_{b\tau} r \tag{3.178}$$

where $\tilde{\sigma}_{b\tau}$ is the mean tangential viscous stress. Then, after integration it follows

$$\sigma_{Bb\tau}\left(\frac{1}{3}r\right) = \tilde{\sigma}_{b\tau} r \tag{3.179}$$

and, finally, the ratio between maximum and the mean tangential viscous stress in the contact zone, see Figure 3.3, can be found as

$$\sigma_{Bb\tau} = 3\tilde{\sigma}_{b\tau} \tag{3.180}$$

where the average viscous tangential stress can be found as

$$\tilde{\sigma}_{b\tau} = \tilde{\sigma}_{bby} + \tilde{\sigma}_{xby} \tag{3.181}$$

where

$$\tilde{\sigma}_{bby} = \frac{F_{bby_r}}{S_x} = \frac{2G''x^{1/2}}{\pi k_p \omega_y l R^{1/2}}\dot{y} \tag{3.182}$$

and

$$\tilde{\sigma}_{xby} = \frac{F_{xbyr}}{S_y} = \frac{3G''x^{1/2}}{4\omega_y l k_h^{1/2} R^{1/2}}\dot{y} \tag{3.183}$$

Thus, in this case, we get

$$\tilde{\sigma}_{b\tau} = \frac{G''x^{1/2}}{\omega_y l R^{1/2}}\varsigma_y \dot{y} \tag{3.184}$$

By substituting Equation 3.184 in 3.180 we get

$$\sigma_{Bb\tau} = \frac{3G''x^{1/2}}{\omega_y l R^{1/2}}\varsigma_y \dot{y} \tag{3.185}$$

The stresses in Equation 3.185 take the maximum magnitude at the moment $t=\tau_1$, $x=x_m$, $y=y_m$, and; therefore, the maximum of these stresses can be expressed as

$$\sigma_{Bb\tau} = \frac{3G''x_m^{1/2}}{\omega_y l R^{1/2}}\varsigma_y \dot{y}_m \tag{3.186}$$

3.2 TRIBOCYCLICITY

3.2.1 FREE DAMPED OSCILLATION OF A BODY OF INDENTER OR ASPERITY ON THE FLAT HORIZONTAL SURFACE OF SEMI-SPACE

Let a rigid body, indenter or asperity having the weight $P_i = m_i g$ be put very slowly on the flat viscoelastic surface of a semi-space or let the body be lifted up by somebody from the equilibrium static size $x_t = x_0$ of indentation of a body on the surface of a semi-space and in the initial moment of time $t=0$, the initial deformation of the compression $x = 0$. It is obvious in this case that the volume of deformation begins to oscillate together with a body of indenter along the axis X. Thus, we get free damped oscillations under an action of the normal reactive force as follows

$$F_n(x,\dot{x}) = C_x x + B_x \dot{x} \tag{3.187}$$

which very soon will go down, and in some moment of time $t \to \infty$ a body will stop oscillating and its vertical velocity will be equal to zero, $\dot{x}(t_0) = 0$, and also obviously the normal force will reach the constant equilibrium magnitude $F_{n0} = C_x x_0 = m_i g$; where x_0 is the equilibrium static size of indentation of a body into a semi-space.

Also from Equation 3.4, it follows that $\frac{1}{m_i} = \frac{1}{m_{2i}} + \frac{1}{m_1} = 0$ and as in the case of the contact between a body and a semi-space $m_1 = \infty$, and hence, $m_{2i} = m_i$. Therefore, it is proved that in the case of contact between a body and a semi-space, the mass of a body of indenter or asperity is equal to the effective mass m_i.

Further, according to the Kelvin–Voigt model, since the force of compression between a body and a semi-space $F_d = P_i + F_n$, according to the second law of Newton

$$m_i \frac{d^2 x}{dt^2} = P_i - F_d = -F_n \tag{3.188}$$

where $F_n = C_x x_0 + B_x \dot{x}$, x is the displacement between the centres of mass of a body and semi-space which, as we remember, is the mutual approach or indentation x between bodies, too.

It is better to consider the motion of the body relative to $x_t = x_0 - x$, where x_t is the displacement of the centre mass of a body relative to the centre of oscillation x_0, which is the initial constant, which is equal to the indentation of a body at time $t \to \infty$, and since $\dot{x} = -\dot{x}_t$ and $\ddot{x} = -\ddot{x}_t$, we can write the differential equation of oscillator as follows

$$m_i \frac{d^2 x_t}{dt^2} + B_x \dot{x}_t + C_x x_t = 0 \tag{3.189}$$

It is obvious that x_m can be replaced by x_0 in Equations 3.49 and 3.50; the normal viscoelastic constants in the case of indentation of a spherical indenter can be expressed, respectively as

$$C_x = \frac{16}{15} k_p E' R^{1/2} x_0^{1/2} \tag{3.190}$$

$$B_x = \frac{16E'' k_p R^{1/2}}{5\omega_x} x_0^{1/2} \tag{3.191}$$

Equation 3.189 is the equation of the damped oscillations, and the solution to this equation is known (Biryukov et al., 2011) as

$$x_t = C_1 e^{-\delta_x t} \sin(\omega_x t) + C_2 e^{-\delta_x t} \cos(\omega_x t) \tag{3.192}$$

According to initial conditions $t = 0$, $x = 0$ and $x_t = x_0$, it follows that $C_2 = x_0$ and we can write

$$x_t = C_1 e^{-\delta_x t} \sin(\omega_x t) + x_0 e^{-\delta_x t} \cos(\omega_x t) \tag{3.193}$$

After the integration we get

$$\dot{x}_t = C_1 e^{-\delta_x t} \{\omega_x \cos(\omega_x t) - \delta_x \sin(\omega_x t)\} - x_0 e^{-\delta_x t} \{\omega_x \sin(\omega_x t) + \delta_x \cos(\omega_x t)\} \tag{3.194}$$

According to initial conditions $t = 0$ and $\dot{x}_i = 0$, it follows, and we can write the solution as

$$x_t = e^{-\delta_x t} \{\frac{\delta_x}{\omega_x} x_0 \sin(\omega_x t) + x_0 \cos(\omega_x t)\} \tag{3.195}$$

where $x_0 = \frac{m_i g}{C_x} = \frac{g}{\omega_{0x}^2}$, $\omega_x = \sqrt{\omega_{0x}^2 - \delta_x^2}$, $\delta_x = \frac{B_x}{2m_i}$ is the normal damping factor; $\omega_{0x} = \sqrt{\frac{C_x}{m_i}}$ is the angular frequency of free harmonic oscillations.

Let $A \sin(\omega_x t) + B \cos(\omega_x t) = \xi_t$, where $A = C_1 = \frac{\delta_x}{\omega_x} x_0$ and $B = C_2 = x_0$. Then, let us divide and multiply the right part of equality by expression $\sqrt{A^2 + B^2}$ as

$$\xi_t = \left(\sqrt{A^2 + B^2}\right) \times \left(\frac{A \sin(\omega_x t)}{\sqrt{A^2 + B^2}} + \frac{B \cos(\omega_x t)}{\sqrt{A^2 + B^2}}\right) \tag{3.196}$$

At the same time ratios are carried out as follows

$$\frac{A}{\sqrt{A^2 + B^2}} < 1 \tag{3.197}$$

$$\frac{B}{\sqrt{A^2+B^2}}<1 \tag{3.198}$$

$$\left(\frac{A}{\sqrt{A^2+B^2}}\right)^2+\left(\frac{B}{\sqrt{A^2+B^2}}\right)^2=1 \tag{3.199}$$

Therefore, we can write that:

$$\frac{A}{\sqrt{A^2+B^2}}=\cos\varphi_0 \tag{3.200}$$

$$\frac{B}{\sqrt{A^2+B^2}}=\sin\varphi_0 \tag{3.201}$$

Therefore, we can rewrite Equation 3.196 as

$\xi_t=\sqrt{A^2+B^2}\{\sin(\omega_x t)\cos\varphi_0+\cos(\omega_x t)\sin\varphi_0\}$ or

$\xi_t=\sqrt{A^2+B^2}\sin(\omega_x t+\varphi_0)$. Hence, the solution finally can be expressed as

$$x_t=x_0\sqrt{(\delta_x/\omega_x)^2+1}\times e^{-\delta_x t}\sin(\omega_x t+\varphi_0) \tag{3.202}$$

It is better to present it as

$$x_t=A_{0x}e^{-\delta_x t}\sin(\omega_x t+\varphi_0) \tag{3.203}$$

where $A_{0x}=x_0\sqrt{(\delta_x/\omega_x)^2+1}$ and designates the initial amplitude, $\varphi_0=\arcsin\dfrac{1}{\sqrt{(\delta_x/\omega_x)^2+1}}$

3.2.1.1 FINDING THE NORMAL VISCOELASTIC CONSTANTS IN CASE OF INDENTATION OF SPHERICAL INDENTER OR ASPERITY INTO A SEMI-SPACE

We have to be aware that the depth of indentation $x = x_0 - x_t$, because when $t = 0$, $x_t = x_0$ and $x = 0$. When a body will stop oscillating and its vertical velocity will be equal to zero, $\dot{x}(t_0) = 0$, the magnitude of x_0 and the normal force will reach the constant equilibrium magnitude $F_{n0} = m_i g$. According to Equation 2.355, we can write that $F_{cn}=F_{n0}=\frac{4}{3}E'R^{1/2}x_0^{3/2}$, thus it is followed by

$$x_0 = \left(\frac{4m_i g}{3k_p E' R^{1/2}} \right)^{2/3} \tag{3.204}$$

Thus, we obtain

$$C_x = \left(\frac{4}{3} k_p R^{1/2} E' \right)^{2/3} (m_i g)^{1/3} \tag{3.205}$$

and

$$B_x = \frac{3E''}{\omega_x} \left(\frac{4}{3} k_p R^{1/2} \right)^{2/3} \left(\frac{m_i g}{E'} \right)^{1/3} \tag{3.206}$$

Also, in the case of indentation of a spherical indenter, it follows

$$\omega_{0x}^2 = \frac{C_x}{m_i} = g^{1/3} \left(\frac{4}{3m_i} k_p R^{1/2} E' \right)^{2/3} \tag{3.207}$$

and since according to Equation 3.115 $\omega_x^2 = \frac{\omega_{0x}^2}{2}(1+\sqrt{1-9tg^2\beta_E})$, it follows

$$\omega_x^2 = \frac{g^{1/3}}{2}\left(1+\sqrt{1-9tg^2\beta_E}\right) \cdot \left(\frac{4}{3m_i} k_p R^{1/2} E' \right)^{2/3} \tag{3.208}$$

Remark: In the case of viscoelastic contact for finding ω_x we should use Equation 3.115.

Also, according to Equations 3.206 and 3.208, we obtain that

$$\delta_x = \frac{B_x}{2m_i} = \frac{3E''}{\sqrt{2}\left(1+\sqrt{1-9tg^2\beta_E}\right)^{1/2}} \left(\frac{4k_p R^{1/2} g^{1/2}}{3m_i (E')^2} \right)^{1/3} \tag{3.209}$$

3.2.1.2 FINDING THE NORMAL VISCOELASTIC CONSTANTS IN CASES OF INDENTATION OF AXIS-SYMMETRICAL INDENTER OR ASPERITY INTO SEMI-SPACE

In the case of indentation of an axis-symmetrical curvilinear surface of indenter or asperity into the flat surface of a semi-space, see Figure 2.8. Since general equations for the viscoelastic forces are known from Equations 2.402 and 2.403, $F_{cn} = \frac{4C_r D_1^{1/n}}{n+1} E' x^{\frac{n+1}{n}}$ and $F_{bn} = 4C_r D_1^{1/n} \eta_E' \dot{x} x^{1/n}$, we can find the work for these forces as follows

$$A_{xc0} = \int_0^{x_0} F_{cn} dx = \frac{4C_r D_1^{1/n}}{n+1} E' \int_0^{x_m} x^{\frac{n+1}{n}} dx = \frac{4nC_r D_1^{1/n}}{(n+1)(2n+1)} E' x_0^{\frac{2n+1}{n}} \tag{3.210}$$

and

$$A_{xb0} = \int_0^{x_0} F_{bn} dx = 4C_r D_1^{1/n} \eta_E' \frac{\int_0^{x_0} x^{1/n} \int dx dx}{\int_0^{\tau_1} dt} = \frac{4nC_r D_1^{1/n}}{(2n+1)\tau_1} \eta_E' x_0^{\frac{2n+1}{n}} \tag{3.211}$$

Now again, by using the method of equivalent works, it follows

$$A_{xc0} = C_x \int_0^{x_0} x dx = \frac{1}{2} C_x x_0^2 = \frac{4nC_r D_1^{1/n}}{(n+1)(2n+1)} E' x_0^{\frac{2n+1}{n}} \tag{3.212}$$

and

$$A_{xb0} = \int_0^{x_0} \dot{x} dx = B_x \frac{\int_0^{x_0} x dx}{\int_0^{\tau_1} dt} = B_x \frac{x_0^2}{2\tau_1} = \frac{4nC_r D_1^{1/n}}{(2n+1)\tau_1} \eta_E' x_0^{\frac{2n+1}{n}} \tag{3.213}$$

Thus, taking into account that $\eta_E' = \frac{E''}{\omega_x}$, we get the next general expressions for the normal viscoelastic constants as

$$C_x = \frac{8nC_r D_1^{1/n}}{(n+1)(2n+1)} E' x_0^{\frac{1}{n}} \tag{3.214}$$

$$B_x = \frac{8nC_r D_1^{1/n}}{(2n+1)\omega_x} E'' x_0^{\frac{1}{n}} \tag{3.215}$$

Since in the general case $F_{n0} = C_x x_0 = m_i g$ and $m_i g = \frac{4C_r D_1^{1/n}}{n+1} E' x_0^{\frac{n+1}{n}}$, the magnitude of x_0 can be found as

$$x_0 = \left(\frac{m_i g(n+1)}{4C_r D_1^{1/n} E'} \right)^{\frac{n}{n+1}} \tag{3.216}$$

Thus, the expression for the angular frequency of free harmonic oscillations by axis X can be expressed in the common case as

$$\omega_{0x}^2 = \frac{C_x}{m_i} = \frac{2nC_r D_1^{\left(\frac{1}{n^2+n}\right)}}{(2n+1)} \left(\frac{4g^n}{(n+1)m_i} E' \right)^{\frac{1}{n+1}} \tag{3.217}$$

and since according to Equation 3.115, $\omega_x^2 = \frac{\omega_{0x}^2}{2}(1 + \sqrt{1 - 9tg^2 \beta_E})$, it follows as

$$\omega_x^2 = \frac{C_x}{m_i} = \frac{nC_r D_1^{\left(\frac{1}{n^2+n}\right)}}{(2n+1)}\left(1+\sqrt{1-9tg^2\beta_E}\right)\cdot\left(\frac{4g^n}{(n+1)m_i}E'\right)^{\frac{1}{n+1}} \tag{3.218}$$

According to Equations 3.215 and 3.216, we can obtain that

$$\delta_x = \frac{B_x}{2m_i} = \frac{4^{\frac{2n-1}{2(n+1)}} n^{1/2} (n+1)^{\frac{3}{2(n+1)}} C_r^{\frac{2n-1}{2(n+1)}} D_1^{\frac{2n^2(n+1)-(n^2+1)}{2n(n^2+1)(n+1)}} g^{\frac{1}{n+1}} (E'')^{\frac{1}{n+1}}}{\sqrt{2n+1}\left(1+\sqrt{1-9tg^2\beta_E}\right)^{1/2} m_i^{\frac{1}{2(n+1)}} (E')^{\frac{1}{n+1}}} \tag{3.219}$$

Thus, for example, in the case of the conical shape of indenter or asperity, when $n = 1$ and $C_r = tg\alpha_r$, see Figure 2.8, we get

$$C_x = \frac{4D_1 tg\alpha_r}{3} E' x_0 \tag{3.220}$$

$$B_x = \frac{8D_1 tg\alpha_r}{3\omega_x} E'' x_0^{1/2} \tag{3.221}$$

$$\omega_{0x}^2 = \frac{C_x}{m_i} = \frac{2tg\alpha_r D_1^{\left(\frac{1}{2}\right)}}{3}\left(\frac{2g^n}{m_i}E'\right)^{\frac{1}{2}} \tag{3.222}$$

$$\delta_x = \frac{B_x}{2m_i} = \frac{4^{\frac{5}{4}} (tg\alpha_r)^{\frac{1}{4}} D_1^{\frac{1}{4}} g^{\frac{1}{2}} (E'')^{\frac{1}{2}}}{\sqrt{3}\left(1+\sqrt{1-9tg^2\beta_E}\right)^{1/2} m_i^{\frac{1}{4}} (E')^{\frac{1}{2}}} \tag{3.223}$$

But on the other hand, in the case of parabolic shape of indenter or asperity, $n = 2$, we can write that

$$C_x = \frac{16C_r D_1^{1/2}}{15} E' x_0^{\frac{1}{2}} \tag{3.224}$$

$$B_x = \frac{16C_r D_1^{1/2}}{5\omega_x} E'' x_0^{\frac{1}{2}} \tag{3.225}$$

$$\omega_{0x}^2 = \frac{C_x}{m_i} = \frac{4C_r D_1^{\left(\frac{1}{6}\right)}}{5}\left(\frac{4g^2}{3m_i}E'\right)^{\frac{1}{3}} \tag{3.226}$$

$$\delta_x = \frac{B_x}{2m_i} = \frac{2\sqrt{6} C_r^{\frac{1}{2}} D_1^{\frac{19}{60}} g^{\frac{1}{3}} (E'')^{\frac{1}{3}}}{\sqrt{5}\left(1+\sqrt{1-9tg^2\beta_E}\right)^{1/2} m_i^{\frac{1}{6}} (E')^{\frac{1}{3}}} \tag{3.227}$$

3.2.2 SLIDING BETWEEN AN ARBITRARY SHAPE OF INDENTER OR ASPERITY AND THE FLAT SURFACE OF SEMI-SPACE

3.2.2.1 EXAMPLE 1

Let a rigid body or indenter go in contact with the flat viscoelastic surface of a semi-space with the initial velocity $\dot{y}(t = 0) = V_y$, see Figure 2.8. In this case, we will get a sliding at a tangential impact. It is obvious that the volume of deformation of a semi-space, in this case, begins to oscillate again together with a body of indenter along the axes X and Y, too. Thus, we get a nonequilibrial process under actions of the reactive normal force and the tangential force. In this case, we have got the damping oscillations, and it is obvious that the relative displacement of the centre mass of a body x_t to the centre of oscillation x_0 should satisfy the known Equation 3.203. But, on the other hand, the displacement by axis Y should satisfy the equation

$$m_i \frac{d^2 y}{dt^2} + B_y \dot{y} + C_y y = 0 \tag{3.228}$$

Equation 3.228 is the equation of the damped oscillations and the solution to this equation can be written as

$$y = \frac{V_y}{\omega_y} e^{-\delta_y t} \sin(\omega_y t) \tag{3.229}$$

where $\omega_y = \sqrt{\omega_{0y}^2 - \delta_y^2}$, $\delta_y = \dfrac{B_y}{2m_i}$ is the tangential damping factor, $\omega_{0y} = \sqrt{\dfrac{C_y}{m_i}}$ is the angular frequency of free harmonic oscillations by axis Y.

As well, the equation for the tangential relative velocity between centres of mass of the body and the effective volume of deformation V_d by axis X can be received by differentiation of Equation 3.229 as

$$\dot{y} = \frac{V_y}{\omega_y} e^{-\delta_y t} [\omega_y \cos(\omega_y t) - \delta_y \sin(\omega_y t)] \tag{3.230}$$

3.2.2.2 FINDING THE TANGENTIAL VISCOELASTIC CONSTANTS IN THE CASE OF INDENTATION OF A BODY OF SPHERICAL INDENTER OR ASPERITY INTO SEMI-SPACE

For finding the tangential viscoelastic constant B_y and C_y in Equation 3.228, we should find, first of all, the works of the tangential forces. In the case of indentation of a spherical indenter, these works can be found by integration in analogy with Equations 3.89 and 3.90 as follows:

$$A_{yc0} = \int_0^{y_0} F_{c\tau y} dy = G' \int_0^{P_0} dP_x \int_0^{y_0} y dy = \frac{G'}{2} P_0 y_0^2 \tag{3.231}$$

$$A_{yb0} = \int_0^{y_0} F_{b\tau y} dy = \frac{G''}{\omega_y} \int_0^{P_0} dP_x \int_0^{y_0} \dot{y} dy = \frac{G''}{\omega_y} P_0 \frac{\int_0^{y_0} \int dy dy}{\int_0^{\tau_1} dt} = \frac{G''}{2\omega_y} P_0 \frac{y_0^2}{\tau_1} \tag{3.232}$$

where initial tangential amplitude $y_0 = \frac{V_y}{\omega_y}$ and $P_x = k_h x + 2k_p R^{1/2} x^{1/2}$, it follows as

$$P_0 = x_0^{1/2}(k_h x_0^{1/2} + 2k_p R^{1/2}) \tag{3.233}$$

Also, by using already known method of the equivalent work in the phase of the shear, we get

$$A_{yc0} = C_y \int_0^{y_0} y dy = \frac{1}{2} C_y y_0^2 = \frac{1}{2} G' P_0 y_0^2 \tag{3.234}$$

$$A_{yb0} = B_y \int_0^{y_0} \dot{y} dy = B_y \frac{\int_0^{y_0} \int dy dy}{\int_0^{\tau_1} dt} = B_y \frac{y_0^2}{2\tau_1} = \frac{G''}{2\omega_y \tau_1} P_0 y_0^2 \tag{3.235}$$

and

$$C_y = G' P_0 \tag{3.236}$$

$$B_y = \frac{G''}{\omega_y} P_0 \tag{3.237}$$

Further, we can find

$$\delta_y = \frac{B_y}{2m_i} = \frac{G''}{2m_i \omega_y} P_0 \tag{3.238}$$

$$\omega_{0y}^2 = \frac{G'}{m_i} P_0 \tag{3.239}$$

The problem includes the calculation of these parameters. First of all, we need to find P_0 by substituting x_0 from Equation 3.204 into 3.233 as follows:

$$P_0 = \left(\frac{15 m_i g}{16 k_p E' R^{1/2}} \right)^{1/3} \left(k_h \left(\frac{15 m_i g}{16 k_p E' R^{1/2}} \right)^{1/3} + 2k_p R^{1/2} \right) \tag{3.240}$$

On the other hand, since $\omega_y = \sqrt{\omega_{0y}^2 - \delta_y^2}$, $\delta_y = \frac{B_{yr}}{2m_i}$, $\omega_{0y}^2 = \frac{C_y}{m_i}$, $\omega_y = \omega_x$, and $tg\beta_G = \frac{G''}{G'}$ and by taking into account Equations 3.238 and 3.239, it follows as $\delta_{yr} = \frac{\omega_{0y}^2}{2\omega_y} tg\beta_G$, and ultimately we get the next algebraic equation as

$$\omega_y^4 - \omega_{0y}^2\omega_y^2 + \frac{1}{4}\omega_{0y}^4 tg^2\beta_G = 0 \tag{3.241}$$

This equation has only one valid solution

$$\omega_y^2 = \frac{\omega_{0y}^2}{2}(1 + \sqrt{1 - tg^2\beta_G}) \tag{3.242}$$

Substitution of Equation 3.239 in 3.242 gives

$$\omega_y^2 = \frac{G'}{2m_i} P_0(1 + \sqrt{1 - tg^2\beta_G}) \tag{3.243}$$

Thus, it follows as

$$\omega_y = \left(\frac{G'}{2m_i}\right)^{1/2} P_0^{1/2}\left(1 + \sqrt{1 - tg^2\beta_G}\right)^{1/2} \tag{3.244}$$

Also, according to Equations 3.240 and 3.242, we can write that

$$\omega_{0y}^2 = \frac{G'}{m_i}\left(\frac{15m_i g}{16k_p E'R^{1/2}}\right)^{1/3}\left(k_h\left(\frac{15m_i g}{16k_p E'R^{1/2}}\right)^{1/3} + 2k_p R^{1/2}\right) \tag{3.245}$$

By substituting this result into Equation 3.245 we get

$$\omega_y^2 = \frac{G'}{2m_i}\left(\frac{15m_i g}{16k_p E'R^{1/2}}\right)^{1/3}\left(k_h\left(\frac{15m_i g}{16k_p E'R^{1/2}}\right)^{1/3} + 2k_p R^{1/2}\right)\left(1 - \sqrt{1 - tg^2\beta_G}\right) \tag{3.246}$$

Also using Equations 3.244, 3.237 and 3.238 we can write that:

$$B_y = \frac{G'' P_0^{1/2}}{\left(\frac{G'}{2m_i}\right)^{1/2}\left(1 + \sqrt{1 - tg^2\beta_G}\right)^{1/2}} \tag{3.247}$$

$$\delta_y = \frac{G'' P_0^{1/2}}{(2m_i G')^{1/2}\left(1 + \sqrt{1 - tg^2\beta_G}\right)^{1/2}} \tag{3.248}$$

3.2.2.3 FINDING THE TANGENTIAL VISCOELASTIC CONSTANTS IN THE CASES OF INDENTATION OF AXIS-SYMMETRICAL BODY OF INDENTER OR ASPERITY INTO SEMI-SPACE

For finding the tangential viscoelastic constants B_y and C_y first of all let us find the works of the tangential forces $F_{c\tau y} = G'P_{xn}y$ and $F_{b\tau y} = \eta'_G P_{xn}\dot{y}$, where $P_{xn} = 2C_r D_1^{1/n} x^{1/n} + D_1 x$, see Equations 2.404–2.406. In the case of indentation of an axis-symmetrical curvilinear surface of indenter or asperity into the flat surface of a semi-space, these works can be found by integration as follows:

$$A_{yc0} = \int_0^{y_0} F_{c\tau y} dy = G' \int_0^{P_{n0}} dP_{xn} \int_0^{y_0} y dy = \frac{G'}{2} P_{n0} y_0^2 \tag{3.249}$$

$$A_{yb0} = \int_0^{y_0} F_{b\tau y} dy = \frac{G''}{\omega_y} \int_0^{P_{n0}} dP_{xn} \int_0^{y_0} \dot{y} dy = \frac{G''}{\omega_y} P_{n0} \frac{\int_0^{y_0} \int dy dy}{\int_0^{\tau_1} dt} = \frac{G''}{2\omega_y} P_{0n} \frac{y_0^2}{\tau_1} \tag{3.250}$$

where

$$P_{n0} = 2C_r D_1^{1/n} x_0^{1/n} + D_1 x_0 \tag{3.251}$$

Also, by using already known method of the equivalent work in the phase of the shear, we get

$$A_{yc0} = C_y \int_0^{y_0} y dy = \frac{1}{2} C_y y_m^2 = \frac{1}{2} G' P_{n0} y_0^2 \tag{3.252}$$

$$A_{yb0} = B_y \int_0^{y_0} \dot{y} dy = B_y \frac{\int_0^{y_0} \int dy dy}{\int_0^{\tau_1} dt} = B_y \frac{y_0^2}{2\tau_1} = \frac{G''}{2\omega_y \tau_1} P_{n0} y_0^2 \tag{3.253}$$

and

$$C_y = G' P_{n0} \tag{3.254}$$

$$B_y = \frac{G''}{\omega_y} P_{n0} \tag{3.255}$$

Further, we can find that

$$\delta_y = \frac{B_y}{2m_i} = \frac{G''}{2m_i \omega_y} P_{n0} \tag{3.256}$$

$$\omega_{0y}^2 = \frac{G'}{m_i} P_{n0} \tag{3.257}$$

First of all again, we need to find P_{n0} by substituting x_0 from Equation 3.216 into 3.251 as follows

$$P_{n0} = 2C_r D_1^{1/n} \left(\frac{m_i g(n+1)}{4C_r D_1^{1/n} E'} \right)^{\frac{1}{n+1}} + D_1 \left(\frac{m_i g(n+1)}{4C_r D_1^{1/n} E'} \right)^{\frac{n}{n+1}} \tag{3.258}$$

Thus, substituting P_{n0} in Equations 3.254–3.257 gives:

$$C_y = 2G' C_r D_1^{1/n} \left(\frac{m_i g(n+1)}{4C_r D_1^{1/n} E'} \right)^{\frac{1}{n+1}} + D_1 \left(\frac{m_i g(n+1)}{4C_r D_1^{1/n} E'} \right)^{\frac{n}{n+1}} \tag{3.259}$$

$$B_y = \frac{G''}{\omega_y} P_{n0} \tag{3.260}$$

Further, we can find

$$\delta_y = \frac{B_y}{2m_i} = \frac{G''}{2m_i \omega_y} P_{n0} \tag{3.261}$$

$$\omega_{0y}^2 = \frac{G'}{m_i} P_{n0} \tag{3.262}$$

3.2.2.4 EXAMPLE 2

Then, when a body of indenter or asperity will stop moving and oscillate, it will be in a steady calm state, $x_t = 0$. Let us try to push a body along the tangential axis Y with the initial velocity V_y. It is obvious that, in this case, the volume of deformation of a semi-space begins to oscillate again together with a body of indenter along the axes X and Y. Thus, we get again a nonequilibrial process under actions of the reactive normal force and the tangential force, too. Also, as we can see the motion of the centre of oscillation is parallel to the surface on the distance x_0 without rebound is possible only, if the initial normal velocity $V_x \leq x_0 \omega_x$. Therefore, let us consider the variant when $V_x = x_0 \omega_x$. In this case, the relative displacement of the centre mass of a body x_t to the centre of oscillation x_0 should be satisfied to the solution

$$x_t = \frac{V_x}{\omega_x} e^{-\delta_x t} \sin(\omega_x t) = x_0 e^{-\delta_x t} \sin(\omega_x t) \tag{3.263}$$

But, if the initial normal velocity $V_x = 0$, it follows that $x_t = 0$ and that the indentation of a body keeps the constant value $x = x_0$ during the time of impact. The solution for tangential displacement satisfies already obtained Equation 3.229.

3.2.2.5 SLIDING FRICTION AT THE TANGENTIAL IMPACT

It is obvious that in common form the friction force equals to the tangential force

$$F_\tau = C_y y + B_y \dot{y} \tag{3.264}$$

and since $F_n = C_x x + B_x \dot{x}$, the coefficient of sliding friction, as we know, can be expressed as

$$f_s = \frac{F_\tau}{F_n} = \frac{C_y y + B_y \dot{y}}{C_x x + B_x \dot{x}} \tag{3.265}$$

where $x = x_0 - x_t$, the solution for x according to Equation 3.203 can be written as

$$x = x_0 - A_{0x} e^{-\delta_x t} \sin(\omega_x t + \varphi_0) \tag{3.266}$$

where, as we know, $A_{0x} = x_0 \sqrt{(\delta_x / \omega_x)^2 + 1}$ is the initial amplitude and $\varphi_0 = \arcsin \frac{1}{\sqrt{(\delta_x / \omega_x)^2 + 1}}$ is the initial phase angle.

The examples for calculation of the parameters of viscoelasticity B_x, C_x and B_y, C_y already have been discussed in this topic. As we can see again that, in both considered examples, the friction coefficient is not the constant value because the tangential and the normal forces fluctuate in process of damping oscillation. The maximum of the elastic tangential force will be at the time $t = t_e$, see Figures 3.4 and 3.5. Also, the equation for the current amplitude of the velocity of sliding can be expressed as

$$\dot{y}_t = V_y e^{-\delta_y t} \tag{3.267}$$

Then, after integration, we obtain the equation for the motion of centre of oscillation, see Figure 3.4, as follows

$$y_t = \frac{V_y}{\delta_y}\left(1 - e^{-\delta_y t}\right) \tag{3.268}$$

It is obvious that the common tangential displacement of a centre of mass of a body by axis Y, see Figure 3.4, respectively can be found as

$$y_d = y_t + y = \frac{V_y}{\delta_y}\left(1 - e^{-\delta_y t}\right) + \frac{V_y}{\omega_y} e^{-\delta_y t} \sin(\omega_y t) \tag{3.269}$$

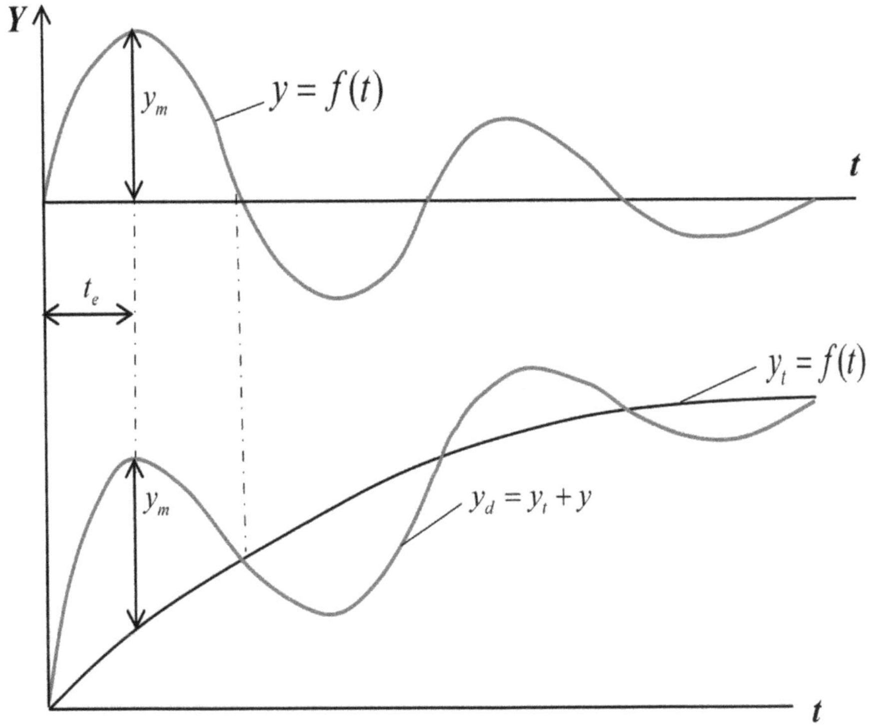

FIGURE 3.4 (See color insert.) The graphical illustrations of a free damped oscillation, graphs $y = f(t)$, $y_t = f(t)$ and $y_d = f(t)$, at a sliding or rolling impact under initial conditions $t = 0$, $\dot{y} = V_y$, $y = 0$.

3.2.3 VISCOELASTIC SLIDING BETWEEN SMOOTH SURFACES OF SOLID BODY AND THE FLAT SEMI-SPACE WITH A CONSTANT VELOCITY

As you remember, the simplified model of friction already has been considered in Chapter 1, see Figure 1.2. Indeed, how in reality does the friction between smooth contact surfaces happen? Let us consider the friction between two smooth surfaces as the dynamic impulse cyclical process of the shear and restitution of elementary volumes of deformation V_{di}. We can take that during the time of sliding or rolling between points *B* and *C*, the elementary volumes of contacting surfaces move to each other by elementary jumps of deformation. All these elementary jumps will happen at the same time on all length of a body at once as soon as a body and volume of deformation will start to go out of balance. Thus, the elastic frictional force can be expressed as $F_{cs} = c_y \sum_{i=1}^{n} (y_{is} - y_{ir})$; where $n = \frac{t_{BC}}{\tau_G}$ is the quantity of elementary impulses of shear, τ_G is the time of impulse cycle, t_{BC} is the time of motion between points *B* and *C*, y_{is} denotes the elementary tangential deformation of a shear, y_{ir} denotes the elementary tangential deformation of restitution. Also, it is obvious that, in reality, in the process of sliding or rolling, the full work of the elastic force for each cycle of deformation equals zero, because the energy of the tangential elastic forces in the shear (compression) phases will be returned fully to a body in the restitution phase. Thus, the sum of action of the tangential elastic force in process of stationary motion with a constant velocity equals zero, and we can neglect this force in this case. And also, according to the Kelvin–Voigt model $y_{is} = y_{ir}$, it follows that $F_{cs} = 0$, like it was obtained in Chapter 1. In the similar way, we can write for the viscous force of sliding friction that $F_{bs} = b_y \frac{1}{n} \sum_{i=1}^{n} \left(\frac{V_{is} + V_{ir}}{2} \right)$, where $V_{is} = V_y + \dot{y}_{is}$ is is the maximum of velocity of the elementary volume of deformation in the shear phase, and $V_{ir} = V_y + \dot{y}_{ir}$ is the minimum of velocity of the elementary volume of deformation in the restitution phase; $\dot{y}_{is}$ and $\dot{y}_{ir}$ are the amplitudes of the velocity in the shear and the restitution phases, respectively. Also, since $\dot{y}_{is} = -\dot{y}_{ir}$ we can write finally similar as in Equation 1.62 that $F_{\tau s} = F_{bs} = b_y V_y$

$$F_{\tau s} = F_{bs} = B_y V_y \tag{3.270}$$

As one can see, we have received the similar formula as Equation 1.63. But now, we have the method of how the parameters of viscosity B_y can be found.

Also, as we know, the motion of thc centre mass of a body can be expressed as $y_d = y_t + y$, see Equation 3.269, whereas in the case of motion with a constant velocity, $y_t = V_y t$, we can write that motion of centre mass of a body $y_d = V_y t + y$. Also, since the magnitude of the relative damped oscillation $y = \frac{V_y}{\omega_y} e^{-\delta_y t} \sin(\omega_y t)$ decreases eventually and it can be neglected after time $t = t_{sR}$, where t_{sR} is the relaxation of damped oscillations at sliding. Therefore, in the stationary conditions of motion of the centre mass of a body with a constant velocity relative to fixed coordinates, it can be considered that $y_d = V_y t$. It is obvious that we have to apply the real constant driving force for knowing at what constant velocity V_y a body could be moved. Also, in case of $V_y = const$, the acceleration of oscillator $\frac{d^2y}{dt^2} = 0$, because the motion with a constant velocity is equivalent to a steady equilibrium state. Thus, in the process of sliding or rolling with a constant velocity, the centre mass of a body keeps an equilibrium state, and it is obvious that, in this case, the motion with a constant velocity is possible only if we apply the tangential constant driving force $F_{0y} = a_{0y}m_i$ to a body, which should be equal to the frictional force $F_{\tau s}$ and; therefore, we can write that

$$F_{0y} = F_{\tau s} = B_y V_y = a_{0y} m_i \tag{3.271}$$

where a_{0y} denotes the acceleration of the driving force.

Also, since the normal velocity of the centre of mass of a body $\dot{x} = 0$, it follows that $x_t = 0$ and the indentation of a body keeps the constant value $x = x_0$ during the time of sliding with the stationary velocity V_y, because, as we know, the motion with a constant velocity equivalents to a steady equilibrium state and; therefore, the normal force will be equal to the constant magnitude, as already was proposed in Equation 1.62. But usually in the case of viscoelastic sliding or rolling motion with a constant velocity $F_{cn} >> F_{bn}$ and if to neglected by the normal viscous force, for example in Equation 1.76, we can take that

$$F_n = C_x x_0 \tag{3.272}$$

Or if the motion is placed along the horizontal axis,

$$F_n = m_i g \tag{3.273}$$

The coefficient of sliding friction under a constant velocity can be expressed in the same manner as in Equation 1.58

$$f_s = \frac{F_{\tau s}}{F_n} = \frac{B_y V_y}{C_x x_0} \tag{3.274}$$

As we can see, indeed, that in conditions of sliding with a constant velocity between two smooth contacting surfaces, the coefficient of friction keeps a constant magnitude!

Also, in case of contact with the horizontal surface of a semi-space, since $F_n = C_x x_0 = m_i g$ and $B_y = 2\delta_y m_i$, the expression for the coefficient of sliding friction can be written as

$$f_s = \frac{2\delta_y V_y}{g} \tag{3.275}$$

or since $\delta_y = \frac{1}{t_{sR}}$, where t_{sR} is the time of relaxation under sliding, we get

$$f_s = \frac{2V_y}{g t_{sR}} \tag{3.276}$$

At first glance, the obtained expression seems very simple but as we know and have to understand that the relaxation time is the function of velocity, temperature and physical–mechanical properties of materials of the contacting surface. If the velocity will change, the time of relaxation also will change automatically. Thus, if we are able to find the time of relaxation for conditions of sliding with the velocity V_y, can find the friction coefficient, too.

3.2.4 TANGENTIAL CONTACTS BETWEEN ROLLING BODY AND THE FLAT SURFACE OF A SEMI-SPACE

3.2.4.1 EXAMPLE 1

Let a rigid rolling body (a ball, cylinder, wheel or disk) having the mass m_i enter in contact with the flat viscoelastic surface of a semi-space with the initial velocity $\dot{y}(t = 0) = V_y$, see Figure 2.12. In this case, we will get a rolling motion at a tangential impact. In this case, we have got the damping oscillations and it is obvious that the relative displacement of the centre mass of a body x_t to the centre of oscillation x_0 should satisfy known Equation 3.203. But, on the other hand, the displacement by axis Y should satisfy to the equation

$$m_i \ddot{y} + B_{yr} \dot{y} + C_{yr} y = 0 \tag{3.277}$$

Equation 3.277 is the equation of the damped oscillations and it is similar to Equation 3.93, and the solutions to this equation are known as

$$y = \frac{V_y}{\omega_{yr}} e^{-\delta_{yr} t} \sin(\omega_{yr} t) \tag{3.278}$$

where $\omega_{yr} = \sqrt{\omega_{0yr}^2 - \delta_{yr}^2}$, $\delta_{yr} = \frac{B_{yr}}{2m_i}$ is the tangential damping factor, $\omega_{0yr} = \sqrt{\frac{C_{yr}}{m_i}}$ is the angular frequency of the harmonic oscillations by axis Y.

As well the equation for the tangential relative velocity between centres of mass of the contacting bodies can be received by differentiation of Equation 3.278 as

$$\dot{y} = \frac{V_y}{\omega_{yr}} e^{-\delta_{yr} t} [\omega_{yr} \cos(\omega_{yr} t) - \delta_{yr} \sin(\omega_{yr} t)] \tag{3.279}$$

3.2.4.2 FINDING THE TANGENTIAL VISCOELASTIC CONSTANTS IN CASE OF ROLLING CONTACT BETWEEN SPHERICAL BODY AND SEMI-SPACE

The tangential viscoelastic forces for the rolling contact of a spherical body are known from Equations 3.67 and 3.68 as $F_{c\tau r} = G'P_s y$ and $F_{b\tau r} = \eta'_G P_s \dot{y}$, respectively, where $P_s = \frac{x}{l} P_x$, $P_x = k_h x + 2k_p R^{1/2} x^{1/2}$ and $k_h = \left(\frac{R - D_2 R_2}{R_2} \right)$. But, in the case of contact with a semi-space, it is obvious that since $R = R_2 = l$; where R_2 is the radius of the spherical body or a ball, $k_h = D_1$ and $P_x = D_1 x + 2k_p R^{1/2} x^{1/2}$

$$P_s = \frac{x}{R}(D_1 x + 2k_p R^{1/2} x^{1/2}) \tag{3.280}$$

In the case of indentation of a spherical body, we can find works by integration in analogy with Equations 3.89 and 3.90 as follows:

$$A_{yc0} = \int_0^{y_0} F_{c\tau y} dy = G' \int_0^{P_{0r}} dP_s \int_0^{y_0} y dy = \frac{G'}{2} P_{0r} y_0^2 \tag{3.281}$$

$$A_{yb0} = \int_0^{y_0} F_{b\tau y} dy = \frac{G''}{\omega_{yr}} \int_0^{P_{0r}} dP_s \int_0^{y_0} \dot{y} dy = \frac{G''}{\omega_{yr}} P_{0r} \frac{\int_0^{y_0} \int dy dy}{\int_0^{\tau_1} dt} = \frac{G''}{2\omega_{yr}} P_{0r} \frac{y_0^2}{\tau_1} \tag{3.282}$$

where initial tangential amplitude $y_0 = \frac{V_y}{\omega_{yr}}$, and where

$$P_{0r} = \frac{x_0}{R}(D_1 x_0 + 2k_p R^{1/2} x_0^{1/2}) \tag{3.283}$$

Also using already known method of the equivalent work in the phase of the shear, we get

$$A_{yc0} = C_y \int_0^{y_0} y dy = \frac{1}{2} C_y y_0^2 = \frac{1}{2} G' P_{0r} y_0^2 \tag{3.284}$$

$$A_{yb0} = B_y \int_0^{y_0} \dot{y}dy = B_y \frac{\int_0^{y_0} \int dydy}{\int_0^{\tau_1} dt} = B_y \frac{y_0^2}{2\tau_1} = \frac{G''}{2\omega_{yr}\tau_1} P_{0r} y_0^2 \tag{3.285}$$

and

$$C_y = G' P_0 \tag{3.286}$$

$$B_{yr} = \frac{G''}{\omega_{yr}} P_{0r} \tag{3.287}$$

Further, we can find

$$\delta_{yr} = \frac{B_{yr}}{2m_i} = \frac{G''}{2m_i \omega_{yr}} P_{0r} \tag{3.288}$$

$$\omega_{0yr}^2 = \frac{G'}{m_i} P_{0r} \tag{3.289}$$

For finding ω_{yr}, we can use Equation 3.115, where $\omega_{yr}^2 = \frac{\omega_{0yr}^2}{2}(1+\sqrt{1-tg^2\beta_G})$.

3.2.4.3 FINDING THE VISCOELASTIC CONSTANTS IN CASE OF ROLLING CONTACT BETWEEN CYLINDRICAL WHEEL AND SEMI-SPACE

In the case of a rolling contact between a cylindrical wheel and a semi-space, see Figure 2.12, and since $\frac{G''}{\omega_{yr}} = \eta_G'$ and $\frac{E''}{\omega_x} = \eta_E'$ we can rewrite already obtained Equations 2.506–2.509 as follows:

$$F_{cn} = E'\left(\frac{2}{3}k_p R_w^{1/2} x^{3/2} + L_w x\right) \tag{3.290}$$

$$F_{bn} = 2\frac{E''}{\omega_x}(k_p R_w^{1/2} x^{1/2} + L_w)\dot{x} \tag{3.291}$$

$$F_{c\tau y} = G' P_{xw} y_r \tag{3.292}$$

$$F_{b\tau y} = \frac{G''}{\omega_{yr}} P_{xw} \dot{y}_r \tag{3.293}$$

where $P_{xw} = D_1 x + L_w$

The works for the normal forces for the equilibrium position $x = x_0$ can be found by integration as follows:

$$A_{xc0} = \int_0^{x_0} F_{cn} dx = E' \int_0^{x_0} \left(\frac{2}{3} k_p R_w^{1/2} x^{3/2} + L_w x \right) dx = E' \left(\frac{4}{15} k_p R_w^{1/2} x_0^{5/2} + \frac{1}{2} x_0^2 L_w \right) \quad (3.294)$$

and

$$A_{xb0} = \int_0^{x_0} F_{bn} dx = \frac{2E''}{\omega_x} \int_0^{x_0} \left(k_p R_w^{1/2} x^{1/2} + L_w \right) \dot{x} dx = \frac{2E''}{\omega_x} \frac{\int_0^{x_0} \int dx \left(k_p R_w^{1/2} x^{1/2} + L_w \right) dx}{\int_0^{\tau_1} dt}$$

$$= \frac{2E''}{\omega_x \tau_1} \left(\frac{2}{5} k_p R_w^{1/2} x_0^{5/2} + \frac{1}{2} x_0^2 L_w \right) \quad (3.295)$$

Now, by using the method of equivalent works, it follows as

$$A_{xc0} = C_x \int_0^{x_0} x dx = \frac{1}{2} C_x x_0^2 = E' x_0^2 \left(\frac{4}{15} k_p R_w^{1/2} x_0^{1/2} + \frac{1}{2} L_w \right) \quad (3.296)$$

and

$$A_{xb0} = \int_0^{x_0} \dot{x} dx = B_x \frac{\int_0^{x_0} x dx}{\int_0^{\tau_1} dt} = B_x \frac{x_0^2}{2\tau_1} = \frac{2E''}{\omega_x \tau_1} x_0^2 \left(\frac{2}{5} k_p R_w^{1/2} x_0^{1/2} + \frac{1}{2} L_w \right) \quad (3.297)$$

Thus, in the case of a rolling contact between a cylindrical wheel and a semi-space we get next expressions for the normal viscoelastic constants

$$C_x = 2E' \left(\frac{4}{15} k_p R_w^{1/2} x_0^{1/2} + \frac{1}{2} L_w \right) \quad (3.298)$$

$$B_x = \frac{4E''}{\omega_x} \left(\frac{2}{5} k_p R_w^{1/2} x_0^{1/2} + \frac{1}{2} L_w \right) \quad (3.299)$$

Further, since in the case of a rolling contact, according to Equations 3.65 and 3.66 $\dot{y}_r = \frac{x}{l} \dot{y}$, $y_r = \frac{x}{l} y$ and since in this case $l = R_w$, see Figure 2.12, the equations for the tangential viscoelastic forces for the rolling contact can be rewritten as

$$F_{c\tau y} = G' P_{rw} y \quad (3.300)$$

$$F_{b\tau y} = \frac{G''}{\omega_{yr}} P_{rw} \dot{y} \quad (3.301)$$

and for cylindrical wheel

$$P_{rw} = \frac{x}{R_w} (D_1 x + L_w) \quad (3.302)$$

The works for the tangential forces can be found by integration as follows:

$$A_{yc0} = \int_0^{y_0} F_{c\tau y} dy = G' \int_0^{P_{rw0}} dP_{rw} \int_0^{y_0} y dy = \frac{G'}{2} P_{rw0} y_0^2 \quad (3.303)$$

$$A_{yb0} = \int_0^{y_0} F_{b\tau y} dy = \frac{G''}{\omega_y} \int_0^{P_{rw0}} dP_{rw} \int_0^{y_0} \dot{y} dy = \frac{G''}{\omega_{yr}} P_{rw0} \frac{\int_0^{y_0} \int dy dy}{\int_0^{\tau_1} dt} = \frac{G''}{2\omega_{yr}} P_{rw0} \frac{y_0^2}{\tau_1} \quad (3.304)$$

where

$$P_{rw0} = \frac{x_0}{R_w}(D_1 x_0 + L_w) \quad (3.305)$$

Using the method of the equivalent work in the phase of the shear, we get

$$A_{yc0} = C_{yr} \int_0^{y_0} y dy = \frac{1}{2} C_{yr} y_m^2 = \frac{1}{2} G' P_{rw0} y_0^2 \quad (3.306)$$

$$A_{yb0} = B_{yr} \int_0^{y_0} \dot{y} dy = B_{yr} \frac{\int_0^{y_0} \int dy dy}{\int_0^{\tau_1} dt} = B_{yr} \frac{y_0^2}{2\tau_1} = \frac{G''}{2\omega_y \tau_1} P_{rw0} y_0^2 \quad (3.307)$$

and

$$C_{yr} = G' P_{rw0} \quad (3.308)$$

$$B_{yr} = \frac{G''}{\omega_{yr}} P_{rw0} \quad (3.309)$$

Also, we can find

$$\delta_{yr} = \frac{B_{yr}}{2m_i} = \frac{G''}{2m_i \omega_{yr}} P_{rw0} \quad (3.310)$$

$$\omega_{0yr}^2 = \frac{G'}{m_i} P_{rw0} \quad (3.311)$$

We can use Equation 3.115 to find ω_{yr}.

3.2.4.4 FINDING THE VISCOELASTIC CONSTANTS IN CASE OF ROLLING CONTACT BETWEEN TOROIDAL SURFACE OF WHEEL AND SEMI-SPACE

In the case of a rolling contact between a toroidal surface of wheel and a semi-space, see Figure 2.12, and since $\frac{G''}{\omega_{yr}} = \eta_G'$ and $\frac{E''}{\omega_x} = \eta_E'$ we can rewrite already obtained Equations 2.517, 2.518, 2.523 and 2.524 as

$$F_{cn} = \frac{2}{3} E' \psi_w x^{3/2} \tag{3.312}$$

$$F_{bn} = 2 \frac{E''}{\omega_x} \psi_w \dot{x} x^{1/2} \tag{3.313}$$

where $\psi_w = (k_{pa} R_w^{1/2} + k_{pb} R_p^{1/2})$ and

$$F_{c\tau y} = G' P_{bx} y_r \tag{3.314}$$

$$F_{b\tau y} = \frac{G''}{\omega_{yr}} P_{bx} \dot{y}_r \tag{3.315}$$

where $P_{bx} = D_1 x + 2k_{pb} R_p^{1/2} x^{1/2}$

The works for the normal forces for the equilibrium position $x = x_0$ can be found by integration as follows:

$$A_{xc0} = \int_0^{x_0} F_{cn} dx = \frac{2}{3} E' k_p \psi_w \int_0^{x_0} x^{3/2} dx = \frac{4}{15} E' \psi_w x_0^{5/2} \tag{3.316}$$

$$A_{xb0} = \int_0^{x_0} F_{bn} dx = \frac{2E''}{\omega_X} \psi_w \int_0^{x_0} \dot{x} x^{1/2} dx = \frac{2E''}{\omega_x} \psi_w \frac{\int_0^{x_0} \int dx x^{1/2} dx}{\int_0^{\tau_1} dt} = \frac{4E''}{5\omega_x \tau_1} \psi_w x_0^{5/2} \tag{3.317}$$

Now, by using the method of equivalent works, it follows as

$$A_{xc0} = C_x \int_0^{x_0} x dx = \frac{1}{2} C_x x_0^2 = \frac{4}{15} E' \psi_w x_0^{5/2} \tag{3.318}$$

and

$$A_{xb0} = \int_0^{x_0} \dot{x} dx = B_x \frac{\int_0^{x_0} x dx}{\int_0^{\tau_1} dt} = B_x \frac{x_0^2}{2\tau_1} = \frac{4E''}{5\omega_x \tau_1} \psi_w x_0^{5/2} \tag{3.319}$$

Thus, in the case of a rolling contact between a toroidal surface of wheel and a semi-space we get the next expressions for the normal viscoelastic constants as

$$C_x = \frac{8}{15} E' \psi_w x_0^{1/2} \tag{3.320}$$

$$B_x = \frac{8E''}{5\omega_x \tau_1} \psi_w x_0^{1/2} \tag{3.321}$$

Further, since in the case of a rolling contact, according to Equations 3.65 and 3.66, $\dot{y}_r = \frac{x}{l} \dot{y}$, $y_r = \frac{x}{l} y$, and since in this case $l = R_w$, see Figure 2.12,

Equations 3.314 and 3.315 for the tangential viscoelastic forces for the rolling contact can be rewritten as

$$F_{c\tau y} = G' P_{tw} y \tag{3.322}$$

$$F_{b\tau y} = \frac{G''}{\omega_{yr}} P_{tw} \dot{y} \tag{3.323}$$

whereas for a toroidal surface of wheel

$$P_{tw} = \frac{x}{R_w}(D_1 x + 2k_{pb} R_p^{1/2} x^{1/2}) \tag{3.324}$$

The works for the tangential forces can be found by integration as follows:

$$A_{yc0} = \int_0^{y_0} F_{c\tau y} dy = G' \int_0^{P_{tw0}} dP_{tw} \int_0^{y_0} y dy = \frac{G'}{2} P_{tw0} y_0^2 \tag{3.325}$$

$$A_{yb0} = \int_0^{y_0} F_{b\tau y} dy = \frac{G''}{\omega_{yr}} \int_0^{P_{rw0}} dP_{tw} \int_0^{y_0} \dot{y} dy = \frac{G''}{\omega_{yr}} P_{tw0} \frac{\int_0^{y_0} \int dy dy}{\int_0^{\tau_1} dt} = \frac{G''}{2\omega_{yr}} P_{tw0} \frac{y_0^2}{\tau_1} \tag{3.326}$$

where

$$P_{tw0} = \frac{x_0}{R_w}(D_1 x_0 + 2k_{pb} R_p^{1/2} x_0^{1/2}) \tag{3.327}$$

Using the method of the equivalent work in the phase of the shear, we get

$$A_{yc0} = C_y \int_0^{y_0} y dy = \frac{1}{2} C_{yr} y_m^2 = \frac{1}{2} G' P_{tw0} y_0^2 \tag{3.328}$$

$$A_{yb0} = B_y \int_0^{y_0} \dot{y} dy = B_y \frac{\int_0^{y_0} \int dy dy}{\int_0^{\tau_1} dt} = B_{yr} \frac{y_0^2}{2\tau_1} = \frac{G''}{2\omega_{yr}\tau_1} P_{tw0} y_0^2 \tag{3.329}$$

and

$$C_{yr} = G' P_{tw0} \tag{3.330}$$

$$B_{yr} = \frac{G''}{\omega_{yr}} P_{tw0} \tag{3.331}$$

Also, we can find

$$\delta_{yr} = \frac{B_{yr}}{2m_i} = \frac{G''}{2m_i \omega_{yr}} P_{tw0} \tag{3.332}$$

$$\omega_{0yr}^2 = \frac{G'}{m_i} P_{tw0} \tag{3.333}$$

For finding ω_{yr} we can use Equation 3.115.

3.2.4.5 FINDING THE VISCOELASTIC CONSTANTS IN CASE OF ROLLING CONTACT BETWEEN AN ARBITRARY SHAPE OF DISK AND SEMI-SPACE

In the case of a rolling contact between an arbitrary shape of disk and a semi-space, see Figure 2.15, and since $\frac{G''}{\omega_{yr}} = \eta_G', \frac{E''}{\omega_x} = \eta_E'$ we can rewrite Equations 2.584–2.587 as follows:

$$F_{cn} = 2E'(\frac{1}{3}k_{pa}R_a^{1/2}x^{3/2} + \frac{C_r D_1^{1/n}}{n+1}x^{\frac{n+1}{n}}) \tag{3.334}$$

$$F_{bn} = 2\frac{E''}{\omega_x}\dot{x}(k_{pa}R_a^{1/2}x^{1/2} + C_r D_1^{1/n}x^{1/n}) \tag{3.335}$$

$$F_{c\tau y} = G'P_{xn}y_r \tag{3.336}$$

$$F_{b\tau y} = \frac{G''}{\omega_{yr}}P_{xn}\dot{y}_r \tag{3.337}$$

where $P_{xn} = 2C_r D_1^{1/n}x^{1/n} + D_1 x$.

The works for the normal forces for the equilibrium position $x = x_0$ can be found by integration as follows:

$$\begin{aligned} A_{xc0} &= \int_0^{x_0} F_{cn}dx = 2E'\int_0^{x_0}\left(\frac{1}{3}k_{pa}R_a^{1/2}x^{3/2} + \frac{C_r D_1^{1/n}}{n+1}x^{\frac{n+1}{n}}\right)dx \\ &= 2E'\left(\frac{2}{15}k_{pa}R_a^{1/2}x_0^{5/2} + \frac{nC_r D_1^{1/n}}{(n+1)(2n+1)}x_0^{\frac{2n+1}{n}}\right) \end{aligned} \tag{3.338}$$

and

$$\begin{aligned} A_{xb0} &= \int_0^{x_0} F_{bn}dx = \frac{2E''}{\omega_x}\int_0^{x_0} \dot{x}\left(k_{pa}R_a^{1/2}x^{1/2} + C_r D_1^{1/n}x^{1/n}\right)dx \\ &= \frac{2E''}{\omega_x}\frac{\int_0^{x_0}\int dx\left(k_{pa}R_a^{1/2}x^{1/2} + C_r D_1^{1/n}x^{1/n}\right)dx}{\int_0^{\tau_1}dt} = \frac{2E''}{\omega_x\tau_1}\left(\frac{2}{5}k_{pa}R_a^{1/2}x_0^{5/2} + \frac{nC_r D_1^{1/n}}{(2n+1)}x_0^{\frac{2n+1}{n}}\right) \end{aligned} \tag{3.339}$$

Now, by using the method of equivalent works, it follows

$$A_{xc0} = C_x \int_0^{x_0} x dx = \frac{1}{2} C_x x_0^2 = 2E' \left(\frac{2}{15} k_{pa} R_a^{1/2} x_0^{5/2} + \frac{nC_r D_1^{1/n}}{(n+1)(2n+1)} x_0^{\frac{2n+1}{n}} \right) \quad (3.340)$$

and

$$A_{xb0} = \int_0^{x_0} \dot{x} dx = B_x \frac{\int_0^{x_0} x dx}{\int_0^{\tau_1} dt} = B_x \frac{x_0^2}{2\tau_1} = \frac{2E''}{\omega_x \tau_1} \left(\frac{2}{5} k_{pa} R_a^{1/2} x_0^{5/2} + \frac{nC_r D_1^{1/n}}{(2n+1)} x_0^{\frac{2n+1}{n}} \right) \quad (3.341)$$

Thus, in the case of a rolling contact between a toroidal surface of a wheel and a semi-space, we get next expressions for the normal viscoelastic constants as

$$C_x = 4E' \left(\frac{2}{15} k_{pa} R_a^{1/2} x_0^{1/2} + \frac{nC_r D_1^{1/n}}{(n+1)(2n+1)} x_0^{\frac{1}{n}} \right) \quad (3.342)$$

$$B_x = \frac{4E''}{\omega_x} \left(\frac{2}{5} k_{pa} R_a^{1/2} x_0^{1/2} + \frac{nC_r D_1^{1/n}}{(2n+1)} x_0^{\frac{1}{n}} \right) \quad (3.343)$$

Since in the case of a rolling contact, according to Equations 3.65 and 3.66, $\dot{y}_r = \frac{x}{l} \dot{y}$ and $y_r = \frac{x}{l} y$ and since in this case $l = R_a$, see Figure 2.15, Equations 3.336 and 3.337 for the tangential viscoelastic forces for the rolling contact can be rewritten as

$$F_{c\tau y} = G' P_{xn} y_r \quad (3.344)$$

$$F_{b\tau y} = \frac{G''}{\omega_y} P_{xn} \dot{y}_r \quad (3.345)$$

where

$$P_{xd} = \frac{x}{R_a} (2C_r D_1^{1/n} x^{1/n} + D_1 x) \quad (3.346)$$

The works for the tangential forces can be found by integration as follows:

$$A_{yc0} = \int_0^{y_0} F_{c\tau y} dy = G' \int_0^{P_{xd0}} dP_{xd} \int_0^{y_0} y dy = \frac{G'}{2} P_{xd0} y_0^2 \quad (3.347)$$

$$A_{yb0} = \int_0^{y_0} F_{b\tau y} dy = \frac{G''}{\omega_y} \int_0^{P_{dw0}} dP_{xd} \int_0^{y_0} \dot{y} dy = \frac{G''}{\omega_{yr}} P_{xd0} \frac{\int_0^{y_0} \int dy dy}{\int_0^{\tau_1} dt} = \frac{G''}{2\omega_y} P_{xd0} \frac{y_0^2}{\tau_1} \quad (3.348)$$

where

$$P_{xd0} = \frac{x_0}{R_a}(2C_r D_1^{1/n} x_0^{1/n} + D_1 x_0) \tag{3. 349}$$

Using the method of the equivalent work, we get

$$A_{yc0} = C_y \int_0^{y_0} y dy = \frac{1}{2} C_{yr} y_0^2 = \frac{1}{2} G' P_{xd0} y_0^2 \tag{3.350}$$

$$A_{yb0} = B_y \int_0^{y_0} \dot{y} dy = B_y \frac{\int_0^{y_0} \int dy dy}{\int_0^{\tau_1} dt} = B_{yr} \frac{y_0^2}{2\tau_1} = \frac{G''}{2\omega_{yr}\tau_1} P_{xd0} y_0^2 \tag{3.351}$$

and

$$C_{yr} = G' P_{xd0} \tag{3.352}$$

$$B_{yr} = \frac{G''}{\omega_{yr}} P_{xd0} \tag{3.353}$$

Also, we can find that

$$\delta_{yr} = \frac{B_{yr}}{2m_i} = \frac{G''}{2m_i \omega_{yr}} P_{xd0} \tag{3.354}$$

$$\omega_{0yr}^2 = \frac{G'}{m_i} P_{xd0} \tag{3.355}$$

As we already know, for finding ω_{yr} we can use Equation 3.115.

3.2.4.6 EXAMPLE 2

Let a rigid rolling body (a ball, wheel or disk), having the mass *m*, stop moving and it will be in a steady calm state, $x_t = 0$. Then, let this body be pushed along the tangential horizontal axis *Y* with the initial velocity V_y. In this case, we get the rolling contact. But, in the case of a sliding contact also, as we can see, the motion of the centre of oscillation parallel to the surface on the distance x_0 without rebound is only possible if the initial normal velocity $V_x \leq x_0\omega_x$. Here, let us consider variant, when $V_x = x_0\omega_x$. We take these conditions because they are satisfied with the motion with the stationary velocity, too. In this case, the relative displacement of the centre mass of a body x_t to the centre of oscillation x_0 should be satisfied to the solution in Equation 3.263 that we got in the case of sliding motion. But, if the initial normal velocity $V_x = 0$, it follows that $x_t = 0$ and the indentation of a body will keep the constant value $x = x_0$ during the time of impact. But on the other hand, in this case, the solution for rolling motion along the horizontal axis Y is satisfied and already obtained from Equation 3.278.

3.2.4.7 ROLLING FRICTION AT THE TANGENTIAL IMPACT

As we already know, in common form the friction force equals to the tangential force

$$F_{\tau r} = C_{yr} y + B_{yr} \dot{y} \tag{3.356}$$

and since $F_n = C_x x + B_x \dot{x}$, the coefficient of rolling friction can be expressed in this case as

$$f_r = \frac{F_{\tau r}}{F_n} = \frac{C_{yr} y + B_{yr} \dot{y}}{C_x x + B_x \dot{x}} \tag{3.357}$$

where the solution for x can be found according to Equation 3.266 as it was done for a sliding contact. The examples for calculation of the parameters of viscoelasticity $B_{x,}$ C_x and $B_{yr,}$ C_{yr} in cases of a rolling contact have already been given in this topic. As we can see again that, for example for a sliding contact, the friction coefficient in case of rolling contact is not the constant value, too, because the tangential and normal forces fluctuate in process of damping oscillation. The maximum of the elastic tangential force will be at the time $t = t_e$, see Figures 3.4.

Also, the equation for the current amplitude of the velocity of sliding can be expressed as

$$\dot{y}_t = V_y e^{-\delta_{yr} t} \tag{3.358}$$

Then, after integration, we obtain the equation for the motion of the centre of oscillation, see Figures 3.4, as follows

$$y_t = \frac{V_y}{\delta_{y_r}} \left(1 - e^{-\delta_{yr} t}\right) \tag{3.359}$$

It is obvious that the common tangential displacement of a centre of mass of a body by axis Y, see Figure 3.4, can be found as

$$y_d = y_t + y = \frac{V_y}{\delta_{y_r}} \left(1 - e^{-\delta_{yr} t}\right) + \frac{V_y}{\omega_{yr}} e^{-\delta_{yr} t} \sin(\omega_{yr} t) \tag{3.360}$$

3.2.5 VISCOELASTIC ROLLING MOTION BETWEEN SMOOTH SURFACE OF ROLLING BODY AND THE FLAT SEMI-SPACE WITH A CONSTANT VELOCITY

The finding solutions for a viscoelastic rolling motion between a smooth surface of a rolling body and the flat semi-space with a constant velocity can be done in same way as it was done in the case of the sliding contact. As we know, the rolling motion is possible only if we apply the tangential driving force $F_{yr} = F_{0yr} = a_{0yr}m_i$ to a body, which should be equal to the average frictional rolling force $F_{\tau r} = B_{yr}V_y$. Therefore, we can write that

$$F_{0yr} = F_{\tau r} = B_{yr}V_y = a_{yr}m_i \tag{3.361}$$

where a_{0yr} denotes the acceleration of the rolling driving force.

In the case of rolling motion, similarly as in the case of sliding, the normal force will be equal to the constant magnitude as proposed in Equation 1.76 as follows

$$F_n = C_x x_0 \tag{3.362}$$

Or if the rolling motion has been placed along the horizontal axis

$$F_n = m_i g \tag{3.363}$$

The coefficient of sliding friction under the stationary velocity can be expressed as

$$f_r = \frac{F_{\tau s}}{F_n} = \frac{B_{yr}V_y}{C_x x_0} \tag{3.364}$$

As we can see, indeed, that in conditions of sliding with the stationary velocity between two smooth contacting surfaces, the coefficient of friction keeps a constant magnitude!

Also, in case of contact with the horizontal surface of a semi-space, since $F_n = C_x x_0 = m_i g$ and $B_{yr} = 2\delta_{yr}m_i$, the expression for the coefficient of sliding friction can be written as

$$f_r = \frac{2\delta_{yr}V_y}{g} \tag{3.365}$$

or since $\delta_{yr} = \frac{1}{t_{rR}}$, where t_{rR} is the time of relaxation under rolling motion, we get

$$f_r = \frac{2V_y}{g t_{rR}} \tag{3.366}$$

3.3 EXAMPLES OF A VISCOELASTIC SLIDING AND ROLLING FRICTION BETWEEN A SOLID BODY AND THE FLAT SEMI-SPACE WITH A CONSTANT VELOCITY

3.3.1 VISCOELASTIC SLIDING AND ROLLING MOTION BETWEEN SPHERICAL BODY AND THE FLAT SEMI-SPACE

SLIDING MOTION:

According to Equation 3.237, $B_y = \frac{G''}{\omega_y} P_0$, and Equation 3.233, $P_0 = x_0^{1/2}(k_h x_0^{1/2} + 2k_p R^{1/2})$ and for contact body and a semi-space, it is known that $k_h = D_1$, the sliding friction force in Equation 3.271 can be expressed as

$$F_{\tau s} = V_y \frac{G''}{\omega_y} x_0^{1/2}(D_1 x_0^{1/2} + 2k_p R^{1/2}) \tag{3.367}$$

Also, according to Equation 3.190, $C_x = \frac{16}{15} k_p E' R^{1/2} x_0^{1/2}$, the coefficient of sliding friction in Equation 3.274 can be expressed now as

$$f_s = V_y \frac{15}{16\omega_y} \frac{G''}{E'} \frac{(D_1 x_0^{1/2} + 2k_p R^{1/2})}{k_p R^{1/2} x_0} \tag{3.368}$$

Usually, in the case of viscoelastic contact $R >> x$ we can approximately take that

$$f_s = V_y \frac{15}{8x_0\omega_y} \frac{G''}{E'} \tag{3.369}$$

Also, since $G'' = G' tg\beta_G$ and $\frac{E'}{G'} = 2(1+\nu)$, where ν is the effective Poisson's coefficient, see Equation 3.113, it follows

$$f_s = \frac{15V_y}{16x_0\omega_y} \frac{tg\beta_G}{(1+\nu)} \tag{3.370}$$

As we can see that, in case of a totally elastic contact, when $tg\beta_G = 0$, it follows that $f_s = 0$.

ROLLING MOTION:

According to Equation 3.304, $B_{yr} = \frac{G''}{\omega_{yr}} P_{0r}$ and Equation 3.283, $P_{0r} = \frac{x_0}{R}(D_1 x_0 + 2k_p R^{1/2} x_0^{1/2})$, the rolling friction force in Equation 3.361 can be expressed as

$$F_{\tau r} = V_y \frac{G''}{\omega_{yr}} \frac{x_0^{3/2}}{R} (D_1 x_0^{1/2} + 2k_p R^{1/2}) \qquad (3.371)$$

Also, according to Equation 3.190 $C_x = \frac{16}{15} k_p E' R^{1/2} x_0^{1/2}$, the coefficient of rolling friction in Equation 3.364 can be found now as

$$f_r = V_y \frac{15}{16\omega_{yr}} \frac{G''}{E'} \frac{(D_1 x_0^{1/2} + 2k_p R^{1/2})}{k_p R^{3/2}} \qquad (3.372)$$

Usually in the case of viscoelastic contact $R >> x$, we can approximately take that

$$f_r = V_y \frac{15}{8R\omega_{yr}} \frac{G''}{E'} \qquad (3.373)$$

And since $G'' = G' tg\beta_G$ and $\frac{E'}{G'} = 2(1+\nu)$, it follows that

$$f_r = \frac{15V_y}{16R\omega_{yr}} \frac{tg\beta_G}{(1+\nu)} \qquad (3.374)$$

Now, we can find the next ratio between sliding and rolling as

$$\frac{F_{\tau s}}{F_{\tau r}} = \frac{f_s}{f_r} = \frac{R}{x_0} \times \frac{\omega_{yr}}{\omega_y} \times \frac{G''(\omega_y, V_y)}{G''(\omega_{yr}, V_y)} \qquad (3.375)$$

where $G''(\omega_y, V_y)$ is the viscosity modulus in conditions of sliding and $G''(\omega_{yr}, V_{yr})$ is the viscosity modulus in conditions of rolling motion, since $\dot{y}_r = \frac{x}{l}\dot{y}$, see Equation 3.65, it follows that $V_{yr} = V_y \frac{x_0}{R}$. We should be aware that a viscosity modulus depends on a velocity, frequency and temperature, and; therefore, modules in process of sliding and rolling are completely different. Also, according to Equations 3.115 and 3.242, $\omega_{yr}^2 = \frac{\omega_{0yr}^2}{2}(1+\sqrt{1-tg^2\beta_G})$ and $\omega_y^2 = \frac{\omega_{0y}^2}{2}(1+\sqrt{1-tg^2\beta_G})$, and according to Equations 3.239 and 3.289, $\omega_{0y}^2 = \frac{G'}{m_i} P_0$ and $\omega_{0yr}^2 = \frac{G'}{m_i} P_{0r}$, it follows that $\frac{\omega_{yr}}{\omega_y} = \frac{\omega_{0yr}}{\omega_{0y}} = \sqrt{\frac{P_{0r}}{P_0}} = \sqrt{\frac{x_0}{R}}$. Thus, we can write that

$$\frac{F_{\tau s}}{F_{\tau r}} = \frac{f_s}{f_r} = \sqrt{\frac{R}{x_0}} \times \frac{G''(\omega_y, V_y)}{G''(\omega_{yr}, V_y)} \qquad (3.376)$$

But, also on the other hand, as we know, since in common case $\eta'_G = \frac{G''}{\omega_y}$, for dynamic viscosities we can write that $\eta'_G(V_{yr}) = \frac{G''(\omega_{yr}, V_{yr})}{\omega_{yr}}$ and $\eta'_G(V_y) = \frac{G''(\omega_y, V_y)}{\omega_{yr}}$

$$\frac{f_s}{f_r} = \frac{R}{x_0} \times \frac{\eta'_G(V_y)}{\eta'_G(V_{yr})} \quad (3.377)$$

It is usually, in reality, that $\frac{\eta'_G(V_y)}{\eta'_G(V_{yr})} \geq 1$ because if a velocity goes up, a viscosity goes up too, and; thus, we usually get that

$$\frac{f_s}{f_r} \geq \frac{R}{x_0} \quad (3.378)$$

3.3.2 VISCOELASTIC SLIDING BETWEEN AXIS-SYMMETRICAL BODY OF INDENTER, ASPERITY AND SEMI-SPACE

According to Equation 3.255, $B_y = \frac{G''}{\omega_y} P_{n0}$, and Equation 3.251, $P_{n0} = 2C_r D_1^{1/n} x_0^{1/n} + D_1 x_0$ and since for contact body and a semi-space it is known that $k_h = D_1$, the sliding friction force in Equation 3.271 can be expressed as

$$F_{\tau s} = V_y \frac{G''}{\omega_y} (2C_r D_1^{1/n} x_0^{1/n} + D_1 x_0) \quad (3.379)$$

Also, according to Equation 3.214 $C_x = \frac{8nC_r D_1^{1/n}}{(n+1)(2n+1)} E' x_0^{\frac{1}{n}}$, the coefficient of sliding friction in Equation 3.274 can be expressed now as

$$f_s = \frac{V_y}{\omega_y} \frac{G''}{E'} \frac{(n+1)(2n+1)(2C_r D_1^{1/n} x_0^{1/n} + D_1 x_0)}{8nC_r x_0^{\frac{1+n}{n}}} \quad (3.380)$$

Also, since $G'' = G' tg\beta_G$ and $\frac{E'}{G'} = 2(1+\nu)$, it follows that

$$f_s = \frac{V_y}{\omega_y} \frac{tg\beta_G}{16(1+\nu)} \frac{(n+1)(2n+1)(2C_r D_1^{1/n} x_0^{1/n} + D_1 x_0)}{nC_r x_0^{\frac{1+n}{n}}} \quad (3.381)$$

Thus, for example, in the case of the conical shape of indenter or asperity, when $n = 1$ and $C_r = tg\alpha_r$, see Figure 2.8, we get

$$F_{\tau s} = V_y \frac{G''}{\omega_y}(2tg\alpha_r + 1)D_1 x_0 \sin(\omega_{0y} t) \tag{3.382}$$

and

$$f_s = \frac{V_y}{\omega_y}\frac{tg\beta_G}{16(1+\nu)}\frac{6D_1(2tg\alpha_r + 1)}{x_0 tg\alpha_r} \tag{3.383}$$

But on the other hand, in the case of parabolic shape of indenter or asperity, when $n = 2$, see Figure 2.8, we can write that

$$F_{\tau s} = V_y \frac{G''}{\omega_y}(2C_r D_1^{1/2} x_0^{1/2} + D_1 x_0)\sin(\omega_{0y} t) \tag{3.384}$$

and

$$f_s = \frac{V_y}{\omega_y}\frac{15tg\beta_G}{32(1+\nu)}\frac{(2C_r D_1^{1/2} x_0^{1/2} + D_1 x_0)}{C_r x_0^{3/2}} \tag{3.385}$$

3.3.3 VISCOELASTIC SLIDING BETWEEN SURFACE OF SLED AND SEMI-SPACE

3.3.3.1 CIRCULAR CYLINDRICAL FORWARD SURFACE OF SLED

According to Equations 2.544 and 2.545, see also Figure 2.13, the equations for the normal viscoelastic forces can be written as

$$F_{cn} = E'(L_s + L)D_1 x + \frac{1}{3}k_{pa} R_a^{1/2} E' x^{3/2} \tag{3.386}$$

$$F_{bn} = \eta_E'(L_s + L)D_1 \dot{x} + k_{pa} R^{1/2} \eta_E' d\dot{x}x^{1/2} \tag{3.387}$$

The works for the normal forces for the equilibrium position $x = x_0$ can be found by integration as follows:

$$A_{xc0} = \int_0^{x_0} F_{cn} dx = E'(L_s + L)D_1 \int_0^{x_0} x dx + \frac{1}{3}k_{pa} R^{1/2} E' \int_0^{x_0} x^{3/2} dx = E' x_0^2 \left(\frac{1}{2}(L_s + L) + \frac{2}{15}k_{pa} R^{1/2} x_0^{1/2}\right) \tag{3.388}$$

$$A_{xb0}=\int_0^{x_0}F_{bn}dx=\eta_E'(L_s+L)D_1\int_0^{x_0}\dot{x}dx+k_{pa}R^{1/2}\eta_E'\int_0^{x_0}\dot{x}x^{1/2}dx$$
$$=\frac{\eta_E'(L_s+L)D_1\int_0^{x_0}\int dxdx+\eta_E'k_{pa}R^{1/2}\int_0^{x_0}\int dxx^{1/2}dx}{\int_0^{\tau_1}dt}=\frac{\eta_E'}{\tau_1}x_0^2\left(\frac{D_1}{2}(L_s+L)+\frac{2}{5}k_{pa}R^{1/2}x_0^{1/2}\right) \tag{3.389}$$

Now, by using the method of equivalent works, it follows as

$$A_{xc0}=C_x\int_0^{x_0}xdx=\frac{1}{2}C_xx_0^2=E'x_0^2\left(\frac{D_1}{2}(L_s+L)+\frac{2}{15}k_{pa}R^{1/2}x_0^{1/2}\right) \tag{3.390}$$

and

$$A_{xb0}=\int_0^{x_0}\dot{x}dx=B_x\frac{\int_0^{x_0}xdx}{\int_0^{\tau_1}dt}=B_x\frac{x_0^2}{2\tau_1}=\frac{\eta_E'}{\tau_1}x_0^2\left(\frac{D_1}{2}(L_s+L)+\frac{2}{5}k_{pa}R^{1/2}x_0^{1/2}\right) \tag{3.391}$$

Thus, in the case of a sliding contact between a circular cylindrical forward surface of a sled and a semi-space, we get the next expressions for the normal viscoelastic constants as follows

$$C_x=2E'\left(\frac{D_1}{2}(L_s+L)+\frac{2}{15}k_{pa}R^{1/2}x_0^{1/2}\right) \tag{3.392}$$

and

$$B_x=2\eta_E'\left(\frac{D_1}{2}(L_s+L)+\frac{2}{5}k_{pa}R^{1/2}x_0^{1/2}\right) \tag{3.393}$$

Then, according to Equations 2.546 and 2.547, see also Figure 2.13, the equations for the tangential viscoelastic forces can be written as

$$F_{c\tau y}=G'yP_{Lx} \tag{3.394}$$

$$F_{b\tau y}=\eta_G'\dot{y}P_{Lx} \tag{3.395}$$

where $P_{Lx}=L+D_1x$.

The works for the tangential forces can be found by integration as follows:

$$A_{yc0}=\int_0^{y_0}F_{cy}dy=G'\int_0^{P_{Lx0}}dP_{Lx}\int_0^{y_0}ydy=\frac{G'}{2}P_{Lx0}y_0^2 \tag{3.396}$$

$$A_{yb0}=\int_0^{y_0}F_{by}dy=\frac{G''}{\omega_y}\int_0^{P_{Lx0}}dP_{Lx}\int_0^{y_0}\dot{y}dy=\eta_G'P_{Lx0}\frac{\int_0^{y_0}\int dydy}{\int_0^{\tau_1}dt}=\frac{1}{2}\eta_G'P_{Lx0}\frac{y_0^2}{\tau_1} \tag{3.397}$$

where $P_{Lx0} = L + D_1 x_0$. Further, using the method of the equivalent work, we get

$$A_{yc0} = C_y \int_0^{y_0} y dy = \frac{1}{2} C_{yr} y_0^2 = \frac{1}{2} G' P_{Lx0} y_0^2 \tag{3.398}$$

$$A_{yb0} = B_y \int_0^{y_0} \dot{y} dy = B_y \frac{\int_0^{y_0} \int dy dy}{\int_0^{\tau_1} dt} = B_y \frac{y_0^2}{2\tau_1} = \frac{\eta_G'}{2\tau_1} P_{Lx0} y_0^2 \tag{3.399}$$

and

$$C_y = G' P_{Lx0} \tag{3.400}$$

$$B_y = \eta_G' P_{Lx0} \tag{3.401}$$

Since in this case $P_{Lx0} = L + D_1 x_0$, the sliding friction force, in Equation 3.271, can be expressed as

$$F_{\tau s} = V_y \eta_G' (L + D_1 x_0) \tag{3.402}$$

Also, using Equation 3.392, the equation for the coefficient of sliding friction in Equation 3.274 can be written as

$$f_s = V_y \frac{\eta_G' (L + D_1 x_0)}{2E' x_0 \left(\frac{D_1}{2}(L_s + L) + \frac{2}{15} k_{pa} R^{1/2} x_0^{1/2} \right)} \tag{3.403}$$

3.3.3.2 ARBITRARY CURVILINEAR CYLINDRICAL FORWARD SURFACE OF SLED

According to Equations 2.560 and 2.561, see also Figure 2.13, the equations for the normal viscoelastic forces can be written as

$$F_{cn} = F_{acn} + F_{bcn} = E'(L_s + L) D_1 x + E' \frac{C_r D_1^{1/n}}{n+1} x^{\frac{n+1}{n}} \tag{3.404}$$

$$F_{bn} = F_{abn} + F_{bbn} = \eta_E' (L_s + L) D_1 \dot{x} + \eta_E' C_r D_1^{1/n} \dot{x} x^{1/n} \tag{3.405}$$

The works for the normal forces for the equilibrium position $x = x_0$ can be found by integration as follows:

$$A_{xc0} = \int_0^{x_0} F_{cn} dx = D_1 E'(L_s + L) \int_0^{x_0} x dx + E' \frac{C_r D_1^{1/n}}{n+1} \int_0^{x_0} x^{\frac{n+1}{n}} dx = E' x_0^2 \left(\frac{D_1}{2}(L_s + L) + \frac{n C_r D_1^{1/n}}{(2n+1)(n+1)} x_0^{1/n} \right) \tag{3.406}$$

$$A_{xb0} = \int_0^{x_0} F_{bn} dx = \eta_E'(L_s + L)D_1 \int_0^{x_0} \dot{x} dx + \eta_E' C_r D_1^{1/n} \int_0^{x_0} \dot{x} x^{1/n} dx$$
$$= \frac{\eta_E'(L_s + L)D_1 \int_0^{x_0} \int dx dx + \eta_E' C_r D_1^{1/n} \int_0^{x_0} \int dx x^{1/n} dx}{\int_0^{\tau_1} dt} = \frac{\eta_E'}{\tau_1} x_0^2 \left(\frac{D_1}{2}(L_s + L) + \frac{nC_r D_1^{1/n}}{2n+1} x_0^{1/n} \right) \tag{3.407}$$

Now, by using the method of equivalent works, it follows as

$$A_{xc0} = C_x \int_0^{x_0} x dx = \frac{1}{2} C_x x_0^2 = E' x_0^2 \left(\frac{D_1}{2}(L_s + L) + \frac{nC_r D_1^{1/n}}{(2n+1)(n+1)} x_0^{1/n} \right) \tag{3.408}$$

and

$$A_{xb0} = \int_0^{x_0} \dot{x} dx = B_x \frac{\int_0^{x_0} x dx}{\int_0^{\tau_1} dt} = B_x \frac{x_0^2}{2\tau_1} = \frac{\eta_E'}{\tau_1} x_0^2 \left(\frac{D_1}{2}(L_s + L) + \frac{nC_r D_1^{1/n}}{2n+1} x_0^{1/n} \right) \tag{3.409}$$

Thus, in the case of a sliding contact between a circular cylindrical forward surface of a sled and a semi-space, we get the next expressions for the normal viscoelastic constants

$$C_x = 2E' \left(\frac{D_1}{2}(L_s + L) + \frac{nC_r D_1^{1/n}}{(2n+1)(n+1)} x_0^{1/n} \right) \tag{3.410}$$

and

$$B_x = 2\eta_E' \left(\frac{D_1}{2}(L_s + L) + \frac{nC_r D_1^{1/n}}{2n+1} x_0^{1/n} \right) \tag{3.411}$$

In this case, the equations for the effective tangential viscoelastic forces are same as in Equations 3.394 and 3.395, parameters of viscoelasticity C_y and B_y, respectively, are same as obtained in Equations 3.400 and 3.401. Thus, according to Equation 3.237, and since in this case $P_{Lx0} = L + D_1 x_0$, the sliding friction force in Equation 3.271 can be expressed as

$$F_{\tau s} = V_y \eta_G'(L + D_1 x_0) \tag{3.412}$$

Using Equation 3.410, the equation for the coefficient of sliding friction in Equation 3.274 can be written as

$$f_s = V_y \frac{\eta_G'(L + D_1 x_0)}{2E' x_0 \left(\frac{D_1}{2}(L_s + L) + \frac{nC_r D_1^{1/n}}{(2n+1)(n+1)} x_0^{1/n} \right)} \tag{3.413}$$

Let us remember that in a general case, the radius of generatrix of the curvilinear surface is expressed as function $r = C_r x^{1/n}$, see Equation 2.386, where $0 < n < \infty$ and the dimension of C_r is $[m^{\frac{n-1}{n}}]$.

In the case of the flat shape of the forward surface, when $n = 1$ and $C_r = tg\alpha_r$, we get

$$f_s = V_y \frac{\eta_G'(L + D_1 x_0)}{D_1 E' x_0 \left((L_s + L) + \frac{tg\alpha_r}{3} x_0 \right)} \tag{3.414}$$

3.3.4 VISCOELASTIC SLIDING BETWEEN A BODY OF VESSEL AND SEMI-SPACE

According to Equations 2.568 and 2.569, see also Figure 2.14, the equations for the normal viscoelastic forces can be written as

$$F_{cn} = E' L_s D_1 x + 4E' \frac{C_r D_1^{1/n}}{n+1} x^{\frac{n+1}{n}} \tag{3.415}$$

$$F_{bn} = \eta_E' L_s D_1 \dot{x} + 4\eta_E' C_r D_1^{1/n} \dot{x} x^{1/n} \tag{3.416}$$

The works for the normal forces for the equilibrium position $x = x_0$ can be found by integration as follows:

$$A_{xc0} = \int_0^{x_0} F_{cn} dx = D_1 E' L_s \int_0^{x_0} x dx + 4E' \frac{C_r D_1^{1/n}}{n+1} \int_0^{x_0} x^{\frac{n+1}{n}} dx = E' x_0^2 \left(\frac{D_1}{2} L_s + \frac{4nC_r D_1^{1/n}}{(2n+1)(n+1)} x_0^{1/n} \right) \tag{3.417}$$

$$A_{xb0} = \int_0^{x_0} F_{bn} dx = \eta_E' L_s D_1 \int_0^{x_0} \dot{x} dx + 4\eta_E' C_r D_1^{1/n} \int_0^{x_0} \dot{x} x^{1/n} dx$$
$$= \frac{\eta_E' L_s D_1 \int_0^{x_0} \int dx dx + 4\eta_E' C_r D_1^{1/n} \int_0^{x_0} \int dx x^{1/n} dx}{\int_0^{\tau_1} dt} = \frac{\eta_E'}{\tau_1} x_0^2 \left(\frac{D_1}{2} L_s + \frac{4nC_r D_1^{1/n}}{2n+1} x_0^{1/n} \right) \tag{3.418}$$

Now, by using the method of equivalent works, it follows as

$$A_{xc0} = C_x \int_0^{x_0} x dx = \frac{1}{2} C_x x_0^2 = E' x_0^2 \left(\frac{D_1}{2} L_s + \frac{4nC_r D_1^{1/n}}{(2n+1)(n+1)} x_0^{1/n} \right) \tag{3.419}$$

and

$$A_{xb0} = \int_0^{x_0} \dot{x} dx = B_x \frac{\int_0^{x_0} x dx}{\int_0^{\tau_1} dt} = B_x \frac{x_0^2}{2\tau_1} = \frac{\eta_E'}{\tau_1} x_0^2 \left(\frac{D_1}{2} L_s + \frac{4nC_r D_1^{1/n}}{2n+1} x_0^{1/n} \right) \quad (3.420)$$

Thus, in the case of a sliding contact between a circular cylindrical forward surface of a sled and a semi-space, we get the next expressions for the normal viscoelastic constants

$$C_x = 2E' \left(\frac{D_1}{2} L_s + \frac{4nC_r D_1^{1/n}}{(2n+1)(n+1)} x_0^{1/n} \right) \quad (3.421)$$

and

$$B_x = 2\eta_E' \left(\frac{D_1}{2} L_s + \frac{4nC_r D_1^{1/n}}{2n+1} x_0^{1/n} \right) \quad (3.422)$$

In this case, the equations for the tangential viscoelastic forces are same as in Equations 2.404 and 2.405, $F_{c\tau y} = G'P_{xn}y$ and $F_{b\tau y} = \eta_G' P_{xn}\dot{y}$, where $P_{xn} = 2C_r D_1^{1/n} x^{1/n} + D_1 x$, see Equation 2.406, parameters of viscoelasticity C_y and B_y are same as obtained in Equations 3.254 and 3.255. Also, since in this case $P_{n0} = 2C_r D_1^{1/n} x_0^{1/n} + D_1 x_0$, see Equation 3.251, the equation for sliding friction force according to Equation 3.271 can be written as

$$F_{\tau s} = V_y \frac{G''}{\omega_y} (2C_r D_1^{1/n} x_0^{1/n} + D_1 x_0) \quad (3.423)$$

Using Equation 3.421 for C_x, the equation for the coefficient of sliding friction in Equation 3.274 can be written as

$$f_s = V_y \frac{G''(D_1 x_0 + 2C_r D_1^{1/n} x_0^{1/n})}{2E' x_0 \omega_y \left(\frac{D_1}{2} L_s + \frac{4nC_r D_1^{1/n}}{(2n+1)(n+1)} x_0^{1/n} \right)} \quad (3.424)$$

In the case of the flat shape of the forward surface, when $n = 1$ and $C_r = tg\alpha_r$, we get

$$f_s = V_y \frac{G''(1 + 2tg\alpha_r)}{E' \omega_y \left(L_s + \frac{4tg\alpha_r}{3} x_0 \right)} \quad (3.425)$$

3.3.5 VISCOELASTIC ROLLING MOTION BETWEEN CYLINDRICAL WHEEL AND SEMI-SPACE

According to Equation 3.309, $B_{yr} = \frac{G''}{\omega_{yr}} P_{rw0}$, and Equation 3.305, $P_{rw0} = \frac{x_0}{R_w}(D_1 x_0 + L_w)$, also see Figure 2.12, the rolling friction force in Equation 3.361 can be expressed as

$$F_{\tau r} = V_y \frac{G''}{\omega_{yr}} \frac{x_0}{R_w}(D_1 x_0 + L_w) \tag{3.426}$$

Also, according to Equation 3.298 $C_x = 2E'\left(\frac{4}{15} k_p R_w^{1/2} x_0^{1/2} + \frac{1}{2} L_w\right)$, the coefficient of rolling friction in Equation 3.364 can be found as

$$f_r = \frac{V_y G''}{\omega_{yr} R_w E'} \times \frac{D_1 x_0 + L_w}{(8/15) k_p R_w^{1/2} x_0^{1/2} + L_w} \tag{3.427}$$

3.3.6 VISCOELASTIC ROLLING MOTION BETWEEN TOROIDAL SURFACE OF WHEEL AND SEMI-SPACE

According to Equation 3.331 it follows that $B_{yr} = \frac{G''}{\omega_{yr}} P_{tw0}$ and according to Equation 3.327 $P_{tw0} = \frac{x_0}{R_w}(D_1 x_0 + 2k_{pb} R_p^{1/2} x_0^{1/2})$, also see Figure 2.12, the rolling friction force in Equation 3.361 can be expressed as

$$F_{\tau r} = V_y \frac{G''}{\omega_{yr}} \frac{x_0}{R_w}(D_1 x_0 + 2k_{pb} R_b^{1/2} x_0^{1/2}) \tag{3.428}$$

Also, according to Equation 3.320 it follows that $C_x = \frac{8}{15} E' \psi_w x_0^{1/2}$, where $\psi_w = (k_{pa} R_w^{1/2} + k_{pb} R_p^{1/2})$, the coefficient of rolling friction in Equation 3.364 can be expressed as

$$f_r = \frac{V_y G''}{\omega_{yr} R_w E'} \times \frac{15(D_1 x_0^{1/2} + 2k_{pb} R_b^{1/2})}{8\psi_w} \tag{3.429}$$

3.3.7 VISCOELASTIC ROLLING MOTION BETWEEN AN ARBITRARY SHAPE OF DISK AND SEMI-SPACE

According to Equation 3.353, it follows that $B_{yr} = \frac{G''}{\omega_{yr}} P_{xd0}$ and according to Equation 3.349 $P_{xd0} = \frac{x_0}{R_a}(2C_r D_1^{1/n} x_0^{1/n} + D_1 x_0)$, also see Figure 2.15, the rolling friction force in Equation 3.361 can be expressed as

$$F_{\tau r} = V_y \frac{G''}{\omega_{yr}} \frac{x_0}{R_a}(2C_r D_1^{1/n} x_0^{1/n} + D_1 x_0) \tag{3.430}$$

According to Equation 3.342 $C_x = 4E'\left(\frac{2}{15} k_{pa} R_a^{1/2} x_0^{1/2} + \frac{nC_r D_1^{1/n}}{(n+1)(2n+1)} x_0^{\frac{1}{n}}\right)$, thus, the equation for the coefficient of rolling friction in Equation 3.364 can be written as

$$f_r = \frac{V_y G''(D_1 x_0 + 2C_r D_1^{1/n} x_0^{1/n})}{4E' R_a \omega_{yr}\left(\frac{2}{15} k_{pa} R_a^{1/2} x_0^{1/2} + \frac{nC_r D_1^{1/n}}{(2n+1)(n+1)} x_0^{1/n}\right)} \tag{3.431}$$

In the case of the flat shape of the forward surface, when $n = 1$ and $C_r = tg\alpha_r$, respectively we get

$$f_r = \frac{3V_y G'' D_1 x_0^{1/2}(1 + 2tg\alpha_r)}{E' R_a \omega_{yr}\left(\frac{8}{5} k_{pa} R_a^{1/2} + \frac{2}{3} D_1 x_0^{1/2} tg\alpha_r\right)} \tag{3.432}$$

3.4 EXAMPLES OF A VISCOELASTIC SLIDE AND ROLLING MOTION BETWEEN A BODY AND THE FLAT SEMI-SPACE UNDER ACTION OF DRIVING FORCE

3.4.1 VISCOELASTIC SLIDING AND ROLLING FRICTION BETWEEN A BODY AND THE FLAT SEMI-SPACE UNDER ACTION OF THE TANGENTIAL HARMONIC DRIVING FORCE

SLIDING FRICTION:

Let a solid body be in the steady equilibrium state in the flat semi-space, but if we apply the tangential harmonic driving force $F_y = F_d \sin(\omega_e t)$ to this body, where F_d is the amplitude of driving force and ω_e is the forced frequency, it is obvious that a body will start to oscillate. A body will be involved in the symmetrical cycle of the forced oscillation. Thus, we can write the next differential equation of the forced oscillation under action of this tangential force

$$m_i \frac{d^2 y}{dt^2} + B_y \dot{y} + C_y y = F_d \sin(\omega_e t) \tag{3.433}$$

This equation is the nonhomogeneous equation and its solution is the sum $y = y_1 + y_2$, where y_1 is the solution of the differential homogeneous equation with constant coefficients and y_2 is the private solution. The first member of y_1 decreases eventually, and it can be neglected through the relaxation period of time $\tau_{yr} = 1/\delta_y$ and; therefore, in the stationary conditions of motion we can take $y = y_2$. The private solution of Equation 3.433 is widely known as

$$y = A_y \sin(\omega_e t + \phi_y) \tag{3.434}$$

where A_y is the amplitude of displacement of oscillation, φ_y the phase angle of shift between y and a driving force. The expression for amplitude of displacement is widely known (Biryukov et al., 2011) as

$$A_y = \frac{F_d}{m_i \sqrt{\left(\omega_{0y}^2 - \omega_e^2\right)^2 + 4\delta_y^2 \omega_e^2}} \tag{3.435}$$

Also, the expressions for the phase angle is known as

$$\cos\phi_y = \frac{\omega_{0y}^2 - \omega_e^2}{\sqrt{(\omega_{0y}^2 - \omega_e^2)^2 + 4\delta_y^2 \omega_e^2}} \tag{3.436}$$

$$\sin\phi_y = -\frac{2\delta_y \omega_e}{\sqrt{(\omega_{0y}^2 - \omega_e^2)^2 + 4\delta_y^2 \omega_e^2}} \tag{3.437}$$

Now, since the dissipative force of internal friction in case of cyclic deformation between two contacting smooth surfaces is $F_{cs} = \dot{y}B_y$, and since as after differentiation, it follows that $\dot{y} = A_y \omega_e \cos(\omega_e t + \varphi_y)$, finally we get that

$$F_{cs} = B_y A_y \omega_e \cos(\omega_e t + \phi_y) \tag{3.438}$$

Then, by taking into account Equation 3.435 we find that

$$F_{cs} = \frac{B_y F_d \omega_e}{m_i \sqrt{\left(\omega_{0y}^2 - \omega_e^2\right)^2 + 4\delta_y^2 \omega_e^2}} \cos(\omega_e t + \phi_y) \tag{3.439}$$

and since $F_n = C_x x_0$, the coefficient of sliding friction can be expressed as

$$f_{cs}=\frac{F_{cs}}{F_n}=\frac{B_y F_d \omega_e}{m_i C_x x_0\sqrt{\left(\omega_{0y}^2-\omega_e^2\right)^2+4\delta_y^2\omega_e^2}}\cos(\omega_e t+\varphi_y) \tag{3.440}$$

As we can see that in this condition of sliding the coefficient of friction is not a constant value!

As we can see, there are possibly three boundaries variants of oscillations:

If $\omega_e >> \omega_{0y}$, it follows as

$$A_y=\frac{F_d}{m_i\omega_e\sqrt{\omega_e^2+4\delta_y^2}} \tag{3.441}$$

$$\cos\phi_y=\frac{-\omega_e}{\sqrt{\omega_e^2+4\delta_y^2}} \tag{3.442}$$

$$\sin\phi_y=-\frac{2\delta_y}{\sqrt{\omega_e^2+4\delta_y^2}} \tag{3.443}$$

If $\omega_e << \omega_{0y}$, it follows that

$$A_y=\frac{F_d}{m_i\omega_{0y}^2} \tag{3.444}$$

and it follows that $\cos\phi_y = 1$, $\sin\phi_y = 0$ and, thus, $\varphi_y = 0$.

But, in the case of resonance, if $\omega_e = \omega_{0y}$, we get

$$A_y=\frac{F_d}{2m_i\delta_y\omega_e} \tag{3.445}$$

and, it follows that $\cos\phi_y = 0$, $\sin\phi_y = -1$ and respectively $\phi_y = -\pi/2$. Also, since $\cos(\omega_e t - \pi/2) = \sin(\omega_e t)$ and $B_y = 2m_i\delta_y$, we get $F_{\tau s} = F_y = F_{dy}\sin(\omega_e t)$.

ROLLING FRICTION:

The differential equation of the forced oscillation under action of the tangential driving force can be expressed in the same manner as in the case of sliding friction as

$$m_i\frac{d^2y}{dt^2}+B_{yr}\dot{y}+C_{yr}y=F_d\sin(\omega_e t) \tag{3.446}$$

It is obvious that we should use here the parameters B_{yr} and C_{yr} for rolling motion. The private solution of Equation 3.446 can be written as

$$y=A_{yr}\sin(\omega_e t+\phi_{yr}) \tag{3.447}$$

where A_{yr} is the amplitude of displacement of oscillation at the rolling motion, φ_{yr} the phase angle of shift between y and a driving force at the rolling motion. The expression for amplitude of displacement can also be written as

$$A_{yr} = \frac{F_d}{m_i\sqrt{\left(\omega_{0yr}^2 - \omega_e^2\right)^2 + 4\delta_{yr}^2\omega_e^2}} \tag{3.448}$$

We should use here the parametres δ_{yr} and ω_{0yr} for rolling motion. Also, the expressions for the phase angle can be expressed as

$$\cos\phi_{yr} = \frac{\omega_{0yr}^2 - \omega_e^2}{\sqrt{(\omega_{0yr}^2 - \omega_e^2)^2 + 4\delta_{yr}^2\omega_e^2}} \tag{3.449}$$

$$\sin\phi_{yr} = -\frac{2\delta_{yr}\omega_e}{\sqrt{(\omega_{0yr}^2 - \omega_e^2)^2 + 4\delta_y^2\omega_e^2}} \tag{3.450}$$

Now, since in this case the rolling frictional force can be expressed as $F_{cr} = B_{yr}\dot{y}$ and after differentiation, it follows that $\dot{y} = A_{yr}\omega_e\cos(\omega_e t + \phi_{yr})$; therefore, finally we get that

$$F_{cr} = B_{yr}A_{yr}\omega_e\cos(\omega_e t + \phi_{yr}) \tag{3.451}$$

Then, taking into account Equation 3.448, we find that

$$F_{cr} = \frac{B_{yr}F_d\omega_e}{m_i\sqrt{\left(\omega_{0yr}^2 - \omega_e^2\right)^2 + 4\delta_{yr}^2\omega_e^2}}\cos(\omega_e t + \phi_{yr}) \tag{3.452}$$

and since $F_n = C_x x_0$, and the coefficient of rolling friction can be expressed as

$$f_{cr} = \frac{F_{cr}}{F_n} = \frac{B_{yr}F_d\omega_e}{m_i C_x x_0\sqrt{\left(\omega_{0yr}^2 - \omega_e^2\right)^2 + 4\delta_{yr}^2\omega_e^2}}\cos(\omega_e t + \phi_{yr}) \tag{3.453}$$

3.4.2 VISCOELASTIC SLIDING AND ROLLING BETWEEN A BODY AND THE FLAT SEMI-SPACE UNDER SUMMARY ACTIONS OF A STATIONARY MOTION WITH A CONSTANT VELOCITY AND THE TANGENTIAL HARMONIC DRIVING FORCE

3.4.2.1 SLIDING MOTION

Let a solid body slide with the constant velocity V_y at the flat semi-space. As we already know, this state is equivalent to the steady equilibrium state, and in this case, the constant driving force $F_{0y} = a_{0y}m_i$ equals to the frictional force $F_{\tau s} = B_y V_y$. But, if we apply the tangential harmonic driving force $F_y = F_d \sin(\omega_e t)$ to this body, where F_d is the amplitude of driving force and ω_e is the forced frequency; obviously, a body will start to oscillate. Thus, in the relative cyclic motion we get the same picture of loading as we already have in the case of loading by the tangential cyclic driving force. The action of the tangential cyclic driving force will be satisfied in Equation 3.433, and its solution is the sum $y = y_1 + y_2$; where y_1 is the solution of the differential homogeneous equation with constant coefficients and y_2 is the private solution. The first member of y_1 decreases eventually, and it can be neglected through the relaxation period of time $\tau_{yr} = 1/\delta_y$ and; therefore, in the stationary conditions of motion, we can take $y = y_2$. The private solution of Equation 3.433 is already known as $y = A_y \sin(\omega_e t + \varphi_y)$, see Equation 3.434. Thus, it is obvious that in this case, the general friction sliding force can be found at the sum

$$F_{\Sigma s} = F_{\tau s} + F_{cs} \tag{3.454}$$

By substituting expressions for $F_{\tau s}$ from Equation 3.271 and F_{cs} from Equation 3.439 into Equation 3.454 we get

$$F_{\Sigma s} = B_y(V_y + \frac{F_d\omega_e}{m_i\sqrt{(\omega_{0y}^2 - \omega_e^2)^2 + 4\delta_y^2\omega_e^2}}\cos(\omega_e t + \phi_y)) \tag{3.455}$$

The equation for a general coefficient of sliding friction can be written as

$$f_{\Sigma s} = \frac{F_{\Sigma s}}{F_n} = \frac{B_y}{C_x x_0}(V_y + \frac{F_d\omega_e}{m_i\sqrt{(\omega_{0y}^2 - \omega_e^2)^2 + 4\delta_y^2\omega_e^2}}\cos(\omega_e t + \phi_y)) \tag{3.456}$$

3.4.2.2 ROLLING MOTION

In the case of rolling motion, we get the similar result. The general friction rolling force can be found as the sum given below

$$F_{\Sigma r} = F_{\tau r} + F_{cr} \tag{3.457}$$

By substituting expressions for $F_{\tau r}$ from Equation 3.361 and F_{cr} from Equation 3.465 into Equation 3.459 we get

$$F_{\Sigma r} = B_{yr}(V_y + \frac{F_d\omega_e}{m_i\sqrt{(\omega_{0yr}^2 - \omega_e^2)^2 + 4\delta_{yr}^2\omega_e^2}}\cos(\omega_e t + \phi_{yr})) \tag{3.458}$$

The equation for a general coefficient of rolling friction can be written as

$$f_{\Sigma r} = \frac{F_{\Sigma r}}{F_n} = \frac{B_{yr}}{C_x x_0}(V_y + \frac{F_d \omega_e}{m_i \sqrt{(\omega_{0yr}^2 - \omega_e^2)^2 + 4\delta_{yr}^2 \omega_e^2}} \cos(\omega_e t + \phi_{yr})) \quad (3.459)$$

3.4.3 VISCOELASTIC SLIDING AND ROLLING MOTION BETWEEN A BODY AND FLAT SEMI-SPACE UNDER ACTION OF THE CENTRIFUGAL DRIVING FORCE

3.4.3.1 SLIDING MOTION

On a solid body, which is set on the flat semi-space, the rotating centrifugal driving force F_d will be applied to the centre of mass of this body in the vertical plane of coordinates *X, Y*. This driving force can be found as $F_d = m_c \omega_c^2 R_c$, where m_c denotes the rotating mass, R_c is the radius of rotation, ω_c is the angular frequency of a rotation. It is obvious that the projections of this force at the axis coordinates *X, Y* can be expressed as $F_y = F_d \sin(\omega_e t)$and $F_x = F_d \operatorname{con}(\omega_e t)$, where F_y is the tangential cyclic driving force and F_x is the normal cyclic driving force. The solution for the sliding friction was already found, see Equations 3.438 and 4.440, but as $F_d = m_c \omega_c^2 R_c$ we get that

$$F_{cs} = \frac{B_y m_c \omega_c^2 \omega_e R_c}{m_i \sqrt{\left(\omega_{0y}^2 - \omega_e^2\right)^2 + 4\delta_y^2 \omega_e^2}} \cos(\omega_e t + \phi_y) \quad (3.460)$$

since in this case $F_n = C_x x_0 + F_x$, where $F_x = F_d \cos(\omega_e t)$ and $F_d = m_c \omega_c^2 R_c$ it follows that

$$F_n = C_x x_0 + m_c \omega_c^2 R \cos(\omega_e t) \quad (3.461)$$

and respectively the coefficient of sliding friction can be expressed as

$$f_{cs} = \frac{F_{\tau s}}{F_n} = \frac{B_y m_c \omega_c^2 \omega_e R_c}{m_i (C_x x_0 + m_c \omega_c^2 R_c \cos(\omega_e t) \sqrt{\left(\omega_{0y}^2 - \omega_e^2\right)^2 + 4\delta_y^2 \omega_e^2}} \cos(\omega_e t + \phi_y) \quad (3.462)$$

3.4.3.2 ROLLING MOTION

The solution for rolling friction was already found, too, in Equations 3.452 and 4.453, thus, we get

$$F_{cr} = \frac{B_{yr} m_c \omega_c^2 \omega_e R_c}{m_i \sqrt{\left(\omega_{0yr}^2 - \omega_e^2\right)^2 + 4\delta_{yr}^2 \omega_e^2}} \cos(\omega_e t + \phi_{yr}) \tag{3.463}$$

and

$$f_{cr} = \frac{F_{cr}}{F_n} = \frac{B_{yr} m_c \omega_c^2 \omega_e R_c}{m_i (C_x x_0 + m_c \omega_c^2 R_c \cos(\omega_e t)) \sqrt{\left(\omega_{0yr}^2 - \omega_e^2\right)^2 + 4\delta_{yr}^2 \omega_e^2}} \cos(\omega_e t + \phi_{yr}) \tag{3.464}$$

3.4.4 VISCOELASTIC SLIDING AND ROLLING BETWEEN A BODY AND THE FLAT SEMI-SPACE UNDER SUMMARY ACTIONS OF STATIONARY MOTION WITH A CONSTANT VELOCITY AND THE CENTRIFUGAL DRIVING FORCE

Since in this case the solutions for sliding and rolling friction forces are the same as in Equations 3.455 and 3.458, also since $F_d = m_c \omega_c^2 R_c$ we can write that in case of sliding motion the friction force can be expressed as

$$F_{\Sigma s} = B_y \left(V_y + \frac{m_c \omega_c^2 \omega_e R_c}{m_i \sqrt{(\omega_{0y}^2 - \omega_e^2)^2 + 4\delta_y^2 \omega_e^2}} \cos(\omega_e t + \phi_y)\right) \tag{3.465}$$

In case of rolling motion the friction force can be expressed as

$$F_{\Sigma r} = B_{yr} \left(V_y + \frac{m_c \omega_c^2 \omega_e R_c}{m_i \sqrt{(\omega_{0yr}^2 - \omega_e^2)^2 + 4\delta_{yr}^2 \omega_e^2}} \cos(\omega_e t + \varphi_{yr})\right) \tag{3.466}$$

Also, since in this case, according to Equation 3.461, $F_n = C_x x_0 + m_c \omega_c^2 R \cos(\omega_e t)$, equations for the friction coefficients can be written, in case of sliding motion, as

$$f_{\Sigma s} = \frac{F_{\Sigma s}}{F_n} = \frac{B_y}{C_x x_0 + m_c \omega_c^2 R_c \cos(\omega_e t)} \left(V_y + \frac{m_c \omega_c^2 \omega_e R_c}{m_i \sqrt{(\omega_{0y}^2 - \omega_e^2)^2 + 4\delta_y^2 \omega_e^2}} \cos(\omega_e t + \phi_y)\right) \tag{3.467}$$

In case of rolling motion

$$f_{\Sigma r} = \frac{F_{\Sigma r}}{F_n} = \frac{B_{yr}}{C_x x_0 + m_c \omega_c^2 R_c \cos(\omega_e t)} \left(V_y + \frac{m_c \omega_c^2 \omega_e R_c}{m_i \sqrt{(\omega_{0yr}^2 - \omega_e^2)^2 + 4\delta_{yr}^2 \omega_e^2}} \cos(\omega_e t + \phi_{yr})\right) \tag{3.468}$$

3.5 SPECIFICITY OF HIGH-VISCOELASTIC CONTACT DURING SLIDING OR ROLLING MOTION

As we know, some viscoelastic materials named elastomers are the materials that are used in various products, machines, devices and equipment. They possess a specific complex of dynamic, mechanical and viscoelastic properties and have the ability to high-viscoelastic deformation. Their ability allows them to decrease wear and reduce vibration and noise; thus, increase durability and reliability compared to rigid constructions. The most typical examples of the elastomers are rubber obtained by vulcanization of different caoutchouc. One of the most specific properties of the elastomers is their high elastic deformation and, as a result, they have good damping capability. Also, we have to take into account the fact that in the process of a dynamic contact between a solid body and a high elastic surface of elastomers; deformations, loads and temperature in the contact zone change quickly. But as we know the behavior of the elastomers, which being in the high elastic or in the glassing conditions, is strongly dependent on the temperature, velocity and frequency of loading. It is necessary to state that when an elastomer, for example rubber, is in glassing condition it also possesses elasticity under a very small deformation (1 % maximum), but if the temperature exceeds the temperatures of mechanical glassing, elastic deformations may reach large amounts (50–300 %). Thus, if the deformation of compressing reaches big enough size, we have to account for the damped dissipative normal viscous force $\tilde{F}_{bn} = b_x V_x$, see Equation 1.54. Since the method of finding parameters of viscosity has already been found and since $b_x = B_x$, we can write that

$$F_{bn} = B_x V_x \tag{3.469}$$

And since $F_{cn} = C_x x_0$ the general viscoelastic force can be expressed as

$$F_n = F_{cn} + F_{bn} = C_x x_0 + B_x V_x \tag{3.470}$$

Since according to Equation 1.63 $V_x = \dfrac{x_0 V_y}{y_B}$, it follows that

$$F_{bn} = B_x x_0 \frac{V_y}{y_B} \tag{3.471}$$

therefore, Equation 3.470 can be rewritten as

$$F_n = x_0 \left(C_x + B_x \frac{V_y}{y_B} \right) \tag{3.472}$$

As we can see here, the problem is in the definition of the size y_B, which depends on the shape of the contact area. For example, in the cases of contact between a spherical or a cylindrical body and the flat semi-space $y_B = r = k_p \sqrt{x_0 R}$, it follows that

$$V_x = \frac{V_y}{k_p}\sqrt{\frac{x_0}{R}} \tag{3.473}$$

and

$$F_{bn} = B_x \frac{V_y}{k_p}\sqrt{\frac{x_0}{R}} \tag{3.474}$$

And also, it follows that

$$F_n = C_x x_0 + B_x \frac{V_y}{k_p}\sqrt{\frac{x_0}{R}} \tag{3.475}$$

Indeed, as we can see in the case of small velocity, normal deformation, small damping effect, when $C_x >> B_x$, $R_a >> x_0$, we can neglect the normal damping viscous force. But on the other hand, it is obvious that in case of contact with elastomers we have to take into account the effect of action of this force. Since $F_{\tau s} = B_{ys} V_y$, the coefficient of sliding friction in the cases of contact between a spherical or cylindrical body and the flat semi-space can be found as

$$f_s = \frac{B_y V_y}{x_0^{1/2}\left(C_x x_0^{1/2} + B_x \dfrac{V_y}{k_p R^{1/2}}\right)} \tag{3.476}$$

In the case of a rolling motion $V_x = \frac{y_B}{R} V_y$, see Equation 1.71, since $y_B = r = k_p\sqrt{x_0 R}$ it follows that

$$V_x = V_y k_p \sqrt{\frac{x_0}{R}} \tag{3.477}$$

and

$$F_{bn} = B_x V_y k_p \sqrt{\frac{x_0}{R}} \tag{3.478}$$

therefore, Equation 3.470 can be rewritten as

$$F_n = C_x x_0 + B_x V_y k_p \sqrt{\frac{x_0}{R}} \tag{3.479}$$

Since $F_{\tau s} = B_{yr} V_y$, the coefficient of rolling friction in the cases of contact between a spherical or cylindrical body and the flat semi-space can be found as

$$f_r = \frac{B_{ry} V_y}{x_0^{1/2}\left(C_x x_0^{1/2} + B_x k_p \dfrac{V_y}{R^{1/2}}\right)} \tag{3.480}$$

Also, it is very important that the viscoelasticity parameters C_x and B_x must be led to the conditions of contact with the velocity V_y by using, for example the method of the temperature-velocity or temperature-time superposition, see Chapter 5. Also, as we know all the parameters of viscoelasticity are defined by using the maximum of indentation $x = x_m$ or $x = x_0$. And since in case of high-viscoelastic deformation, we have to use definition maximum of indentation to take into account the normal viscous damping force as well, as it was made in the case of impact between spherical bodies. For example, in case of sliding or rolling motion on the horizontal surface of the flat semi-space with constant velocity V_y, since $x = x_0$ $\dot{x} = V_x$ and $F_n = F_{cn} + F_{bn} = mg$ and according to Equations 2.355 and 2.356, it follows $F_{cn} = \frac{4}{3}E'R^{1/2}x_0^{3/2}$, $F_{bn} = 4k_p\eta_E' R^{1/2}\dot{x}x_0^{1/2}V_x$, and since $V_x = \dfrac{x_0 V_y}{y_B}$ and $y_B = r = k_p\sqrt{x_0 R}$, the normal viscoelastic force can be expressed as

$$F_n = \mathrm{E}_x x_0^{3/2} + \mathrm{H}_x x_0 = mg \tag{3.481}$$

where the constants $\mathrm{E}_x = \frac{4}{3}k_p E'R^{12}$ and $\mathrm{H}_x = \dfrac{4V_y}{k_p}\eta_E'$. Thus, we get the next algebraic equation

$$\mathrm{E}_x^2 x_0^3 - \mathrm{H}_x^2 x_0^2 + 2\mathrm{H}_x x_0 - (mg)^2 = 0 \tag{3.482}$$

The solution of this equation can be obtained in different ways. After deformations $x = x_0$ and y_B are found, we can find the parameters of viscoelasticity C_x, B_x and C_y, B_y for this concrete case by using already obtained formulas for them in this chapter. If, for example, we have to find viscoelasticity parameters for any other case, we will have to derivate out again the similar equation and to solve it relatively $x = x_0$.

3.6 VISCOELASTIC LUBRICATION

3.6.1 INTRODUCTION

As we know, the substance of lubrication in the static state can be a liquid, solid body or gas. But, the behavior of liquid lubricants or plastic lubricants strongly depends on the dynamic conditions of loading, such as temperature, frequency and velocity of loading. Indeed, if the velocity and frequency are high enough, the behavior of lubrication is similar to the behavior of elastomers—high-viscoelastic materials. This process of lubrication is named usually as the elasto-hydrodynamic process. It is obvious that lubricants under the influence of rather high speed in a zone of contact pass into a viscoelastic state. Lubricants can be considered as a third body being in the contact zone. Thus, we can find all specific dynamic viscosities and specific viscosity and elasticity module (the specific loss module and storage module, and specific loss viscosity and specific storage viscosity) by taking into account the viscoelastic properties of three components such as the lubricant and materials of contacting surfaces, which are subjected to a process of deformation of compression and shear in the contact area. The schematic illustration of the contact between two lubricated curvilinear surfaces of solids is depicted in Figure 3.5.

First of all, we can find the specific module and specific viscosities between lubrication and solid surfaces as it was found in Chapter 2. It is obvious that for the determination of the effective dynamic viscosity and the effective dynamic elasticity modules, we can summarize the elastic and viscous compliances as shown below:

$$\frac{1}{E'} = \frac{1}{E'_1} + \frac{1}{E'_3} + \frac{1}{E'_2} \tag{3.483}$$

$$\frac{1}{\eta'_E} = \frac{1}{\eta'_{1E}} + \frac{1}{\eta'_{3E}} + \frac{1}{\eta'_{2E}} \tag{3.484}$$

$$\frac{1}{G'} = \frac{1}{G'_1} + \frac{1}{G'_3} + \frac{1}{G'_2} \tag{3.485}$$

$$\frac{1}{\eta'_G} = \frac{1}{\eta'_{1G}} + \frac{1}{\eta'_{3G}} + \frac{1}{\eta'_{2G}} \tag{3.486}$$

where E'_3 is the dynamic elasticity modulus at the compression of lubrication, η'_{3E} i is the dynamic viscosity at the compression of lubrication, G'_3 is the dynamic elasticity modulus at the shear of lubrication, η'_{3G} is the dynamic viscosity at the shear of lubrication. As we remember, the subscript $i=1$ is used for a softer surface, $i=2$ is used for a harder surface, and therefore, the subscript $i=3$ is used here for a lubricant.

Hence, the expressions for a calculation of the effective dynamic viscosities and the effective dynamic module can be written as:

$$E' = \frac{E'_1 E'_3 E'_2}{E'_1 E'_2 + E'_1 E'_3 + E'_2 E'_3} \tag{3.487}$$

$$\eta'_E = \frac{\eta'_{1E}\eta'_{3E}\eta'_{2E}}{\eta'_{1E}\eta'_{2E} + \eta'_{1E}\eta'_{3E} + \eta'_{2E}\eta'_{3E}} \tag{3.488}$$

$$G'_{1.3} = \frac{G'_1 G'_3 G'_2}{G'_1 G'_2 + G'_1 G'_3 + G'_2 G'_3} \tag{3.489}$$

$$\eta'_{1.3G} = \frac{\eta'_{1G}\eta'_{3G}\eta'_{3G}}{\eta'_{1G}\eta'_{2G} + \eta'_{1G}\eta'_{3G} + \eta'_{2G}\eta'_{3G}} \tag{3.490}$$

It is obvious that $x = x_1 + x_3 + x_2$, where $x_3 = h_1 + h_2$ is the compression of lubrication in the contact zone, see Figure 3.6; and h_1 is the thickness of film layer of lubricant on a softer surface; h_2 is the thickness of film layer of lubricant on a harder surface. Also, since $F_{xc} = E'x = E'_1 x_1 = E'_3 x_3 = E'_2 x_2$ we can write that

$$x_1 = D_1 x \tag{3.491}$$

$$x_2 = D_2 x \tag{3.492}$$

$$x_3 = D_3 x \tag{3.493}$$

where $D_1 = \frac{E'}{E'_1}$, $D_2 = \frac{E'}{E'_2}$, and $D_3 = \frac{E'}{E'_3}$ are the coefficients of deformations.

You can tell that liquids including lubricants are not squeezed because Poisson's coefficient of liquid is equal to 0.5. Yes, for motionless liquid it is equal to zero, but in a mobile condition of liquid, in any liquid compresses, it is less than 0.5. The increases of speed or frequency of loading of liquid lead to decrease the coefficient of Poisson. The modules of viscosity and elasticity of any liquid can be found by means of, for example the rotational viscometer or other ways.

Thus, as we can see that the contact zone in the conditions of lubricant increases. The radius of a zone of contact can be determined by a formula already known as $r = k_p\sqrt{Rx}$. But it is necessary to consider, see Figure 3.5, that $x = x_1 + x_3 + x_2$, $R = \frac{R_1 R_2}{R_1 \pm R_2} = \frac{(R_{01} + h_1)(R_{02} + h_2)}{(R_{01} + h_1) + (R_{02} + h_2)}$ and taking into account Equations 3.491–3.493, we get

$$r = k_p\sqrt{x(D_1 + D_2 + D_3)\frac{(R_{01} + h_1)(R_{02} + h_2)}{(R_{01} + h_1) + (R_{02} + h_2)}} \tag{3.494}$$

The problem here is to find thicknesses of film layers of lubricants h_1, h_2, which depends on their adhesive properties and also temperature; the initial velocity of compression V_x and of the circumferential velocity $V_\tau = \omega_1 R_1 = \omega_2 R_2$.

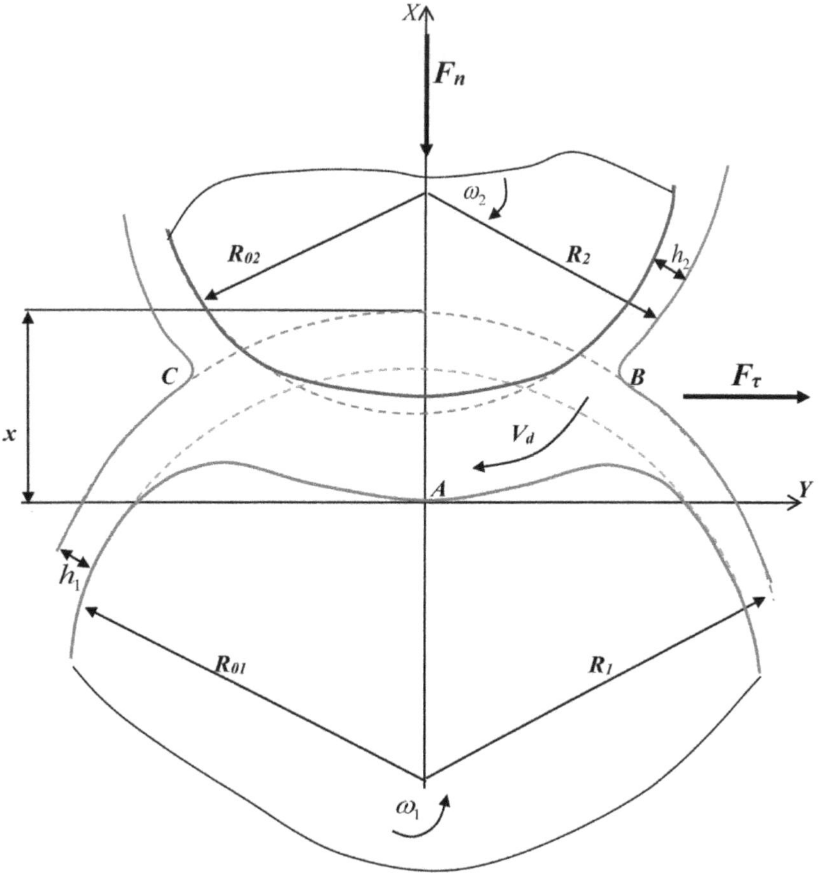

FIGURE 3.5 (See color insert.) Schematic illustration of the contact between two lubricated curvilinear surfaces of solids.

3.6.2 ROLLING CONTACT BETWEEN TWO CYLINDERS

Also, for example, in case of a rolling contact between two cylinders, for finding tangential force we can rewrite Equation 2.380 by taking into account that $\dot{y} = V_y$ and $x = x_0$

$$F_{b\tau} = F_{\tau r} = \eta'_G V_y P_{xLr} \tag{3.495}$$

where

$$P_{xLr} = \frac{x_0}{R}(k_h x_0 + L\ ) \tag{3.496}$$

On the other hand, using Equations 2.381 and 3.477, taking into account that $\dot{x} = V_x$ and $x = x_0$, (as we remember, x_0 is the constant equilibrium size of maximum compression between contacting surfaces), we get the following equation for the normal viscoelastic force

$$F_n = 2\eta'_E V_y k_p \frac{x_0^{1/2}}{R^{1/2}}(k_p R^{1/2} x_0^{1/2} + L) + E' x_0 \left(\frac{2}{3} k_p R^{1/2} x_0^{1/2} + L \right) \tag{3.497}$$

Finally, the expression for calculation of the coefficient of rolling friction can be written as

$$f_r = \frac{\eta'_G V_y P_{xLr}}{2\eta'_E V_y k_p \frac{x_0^{1/2}}{R^{1/2}}(k_p R^{1/2} x_0^{1/2} + L) + E' x_0 (\frac{2}{3} k_p R^{1/2} x_0^{1/2} + L)} \tag{3.498}$$

where these expressions $V_y = V_{y2} - V_{y1} = V_\tau(\cos\alpha_2 - \cos\alpha_1)$, and sin $\alpha_1 = r/R_1$ and sin $\alpha_2 = r/R_2$, see Figure 3.5, and also as we know that $r = a/2$, see Figure 2.7.

3.7 DYNAMICS OF AN ARBITRARY VISCOELASTIC COLLISION BETWEEN A SLED, WHICH HAS AN ARBITRARY CURVILINEAR CYLINDRICAL FORWARD SURFACE AND SEMI-SPACE

3.7.1 NORMAL DISPLACEMENT

3.7.1.1 DIFFERENTIAL EQUATIONS OF DISPLACEMENT AT THE COMPRESSION

For consideration of an arbitrary viscoelastic collision between a sled that has an arbitrary curvilinear cylindrical forward surface and the flat semi-space, we can use the illustration in Figure 2.13. At the initial instance of the time of an arbitrary viscoelastic collision between a sled and a semi-space, a sled has the initial velocities $\dot{x}\,(t = 0) = V_x$ and $\dot{y}\,(t = 0) = V_y$, and the mass m_2 of a sled is equal to m. But according to Equation 3.4 it follows that $\frac{1}{m} = \frac{1}{m_2} + \frac{1}{m_1} = 0$ and as in the case of the contact between of a body and a semi-space $m_1 = \infty$, and hence, we can take that the mass of a sled $m_2 = m$.

The equations for the normal viscous force and the elastic force already known as $F_{cn} = E'\{(L_s + L)D_1 x + \frac{C_r D_1^{1/n}}{n+1} x^{\frac{n+1}{n}}\}$, see Equation 3.404, and

$F_{bn} = \eta'_E\{(L_s + L)D_1\dot{x} + C_r D_1^{1/n}\dot{x}x^{1/n}\}$, see Equation 3.405. And since $\frac{E''}{\omega_x} = \eta'_E$, it follows

$$F_{bn} = \frac{E''}{\omega_x}\{(L_s + L)D_1\dot{x} + C_r D_1^{1/n}\dot{x}x^{1/n}\} \tag{3.499}$$

Thus, the general normal viscoelastic force can be written as

$$\begin{aligned} F_n = F_{bn} + F_{cn} = b_x\dot{x} + c_x x = \frac{E''}{\omega_x}\{(L_s + L)D_1 + C_r D_1^{1/n}x^{1/n}\} \\ \dot{x} + E'\{(L_s + L)D_1 + \frac{C_r D_1^{1/n}}{n+1}x^{\frac{1}{n}}\}x \end{aligned} \tag{3.500}$$

where the expressions for the variable viscoelasticity parameters can be written as

$$b_x = \frac{E''}{\omega_x}\{(L_s + L)D_1 + C_r D_1^{1/n}x^{1/n}\} \tag{3.501}$$

$$c_x = E'\{(L_s + L)D_1 + \frac{C_r D_1^{1/n}}{n+1}x^{1/n}\} \tag{3.502}$$

According to Newton's Second Law, see Equation 3.1 and 3.500, the differential equation of the movement (displacement) of the centre of mass of a body of the sled by axis X can be expressed as

$$m\ddot{x} = \frac{E''}{\omega_x}\{(L_s + L)D_1 + C_r D_1^{1/n}x^{1/n}\}\dot{x} + E'\{(L_s + L)D_1 + \frac{C_r D_1^{1/n}}{n+1}x^{\frac{1}{n}}\}x = 0 \tag{3.503}$$

or it can be also written in the canonical form as $m\dot{x} + b_x\dot{x} + c_x x = 0$, see Equation 3.28.

3.7.1.2 WORK AND ENERGY IN THE PHASES OF COMPRESSION AND RESTITUTION

As we know, the normal initial kinetic energy $W_x = \frac{mV_x^2}{2}$ and the kinetic energy at the instance of rebound $W_{tx} = \frac{mV_{tx}^2}{2}$, where $V_{\tau x} = \dot{x}(\tau_x)$ is the normal relative velocity of the centres of mass of sled at the instance of the rebound time $t = \tau_x$; where τ_x is the time period of the contact. The graphical illustration of the functional dependences between the normal viscoelastic forces and the displacement of the centre of mass of a body (in this case, it is a sled) is depicted in Figure 3.2a. Also, the 'rheological model of Kelvin–Voigt', which usually is used for the viscoelastic contact is represented in Figure 3.2b.

The works A_{xcm} and A_{xbm} at the maximum compression $x = x_m$ can be found by integration similar as was done in Equations 3.406 and 3.407, respectively

$$A_{xcm} = \int_0^{x_m} F_{cn} dx = D_1 E'(L_s + L)\int_0^{x_m} x dx + E' \frac{C_r D_1^{1/n}}{n+1}\int_0^{x_m} x^{\frac{n+1}{n}} dx$$
$$= E' x_m^2 \left(\frac{D_1}{2}(L_s + L) + \frac{nC_r D_1^{1/n}}{(2n+1)(n+1)} x_m^{1/n} \right) \tag{3.504}$$

and

$$A_{xbm} = \int_0^{x_m} F_{bn} dx = \frac{E''}{\omega_x}(L_s + L)D_1 \int_0^{x_m} \dot{x} dx + \frac{E''}{\omega_x} C_r D_1^{1/n} \int_0^{x_m} \dot{x} x^{1/n} dx$$
$$= \frac{E''}{\omega_x} \frac{(L_s + L)D_1 \int_0^{x_m} \int dx dx + C_r D_1^{1/n} \int_0^{x_m} \int dx x^{1/n} dx}{\int_0^{\tau_1} dt} = \frac{E''}{\omega_x \tau_1} x_m^2 \left(\frac{D_1}{2}(L_s + L) + \frac{nC_r D_1^{1/n}}{2n+1} x_m^{1/n} \right) \tag{3.505}$$

Analogically, the works A_{xct} and A_{xbt} in the restitution phase can be found as follows

$$A_{xct} = -\int_{x_m}^{0} F_{cn} dx = -D_1 E'(L_s + L)\int_{x_m}^{0} x dx - E' \frac{C_r D_1^{1/n}}{n+1}\int_{x_m}^{0} x^{\frac{n+1}{n}} dx$$
$$= E' x_m^2 \left(\frac{D_1}{2}(L_s + L) + \frac{nC_r D_1^{1/n}}{(2n+1)(n+1)} x_m^{1/n} \right) \tag{3.506}$$

and

$$A_{xbt} = -\int_{x_m}^{0} F_{bn} dx = \frac{E''}{\omega_x}(L_s + L)D_1 \int_{x_m}^{0} \dot{x} dx - \frac{E''}{\omega_x} C_r D_1^{1/n} \int_{x_m}^{0} \dot{x} x^{1/n} dx$$
$$= -\frac{E''}{\omega_x} \frac{(L_s + L)D_1 \int_{x_m}^{0} \int dx dx + C_r D_1^{1/n} \int_{x_m}^{0} \int dx x^{1/n} dx}{\int_{\tau_1}^{\tau_x} dt} = \frac{E''}{\omega_x \tau_2} x_m^2 \left(\frac{D_1}{2}(L_s + L) + \frac{nC_r D_1^{1/n}}{2n+1} x_m^{1/n} \right) \tag{3.507}$$

Denoting $\frac{D_1}{2}(L_s + L) = \kappa_1$, $\frac{nC_r D_1^{1/n}}{(2n+1)(n+1)} = \kappa_2$ and $\frac{nC_r D_1^{1/n}}{2n+1} = \kappa_3$ equations for viscoelastic works can be written respectively as:

$$A_{xcm} = E' x_m^2 (\kappa_1 + \kappa_2 x_m^{1/n}) \tag{3.508}$$

$$A_{xbm} = \frac{E''}{\omega_x \tau_1} x_m^2 (\kappa_1 + \kappa_3 x_m^{1/n}) \tag{3.509}$$

and

$$A_{xct} = E'x_m^2(\kappa_1 + \kappa_2 x_m^{1/n}) \tag{3.510}$$

$$A_{xbt} = \frac{E''}{\omega_x \tau_2} x_m^2(\kappa_1 + \kappa_3 x_m^{1/n}) \tag{3.511}$$

Since $A_{xm} = A_{xcm} + A_{xbm} = \frac{mV_x^2}{2}$ and $\frac{E''}{E'} = tg\beta_E$, we can find the work in the maximum of indentation (compression) of a sled into a semi-space as

$$A_{xm} = E'x_m^2\left((\kappa_1 + \kappa_2 x_m^{1/n}) + \frac{tg\beta_E}{\omega_x \tau_1}(\kappa_1 + \kappa_3 x_m^{1/n})\right) = \frac{mV_x^2}{2} \tag{3.512}$$

Also, since $A_{xt} = A_{xct} - A_{xbt} = \frac{mV_{tx}^2}{2}$, we can find the work in the restitution phase as

$$A_{xt} = E'x_m^2\left((\kappa_1 + \kappa_2 x_m^{1/n}) - \frac{tg\beta_E}{\omega_x \tau_2}(\kappa_1 + \kappa_3 x_m^{1/n})\right) = \frac{mV_{tx}^2}{2} \tag{3.513}$$

But also, on the other hand, usually $\kappa_1 >> \kappa_2 x_m^{1/n}$ and $\kappa_1 >> \kappa_3 x_m^{1/n}$, and, therefore, Equations 3.512 and 3.513 can be simplified as

$$A_{xm} = E'x_m^2 \kappa_1\left(1 + \frac{tg\beta_E}{\omega_x \tau_1}\right) = \frac{mV_x^2}{2} \tag{3.514}$$

$$A_{xt} = E'x_m^2 \kappa_1\left(1 - \frac{tg\beta_E}{\omega_x \tau_2}\right) = \frac{mV_{tx}^2}{2} \tag{3.515}$$

Using Equation 3.514, we get the formula for the approximate size of maximum indentation x_m as

$$x_m = \left[\frac{mV_x^2 \omega_x \tau_1}{2\kappa_1 E'(tg\beta_E + \omega_x \tau_1)}\right]^{1/2} \tag{3.516}$$

The energetic coefficient of restitution can be found as the ratio between A_{tx} and A_{xm}:

$$e_x = k_x^2 = \frac{V_{tx}^2}{V_x^2} = \left(\frac{\omega_x \tau_2 - tg\beta}{\omega_x \tau_1 + tg\beta}\right)\frac{\tau_1}{\tau_2} \tag{3.517}$$

Since $k_x = \frac{\tau_1}{\tau_2}$, see Equation 3.43, we get that

$$tg\beta_E = \omega_x \tau_1 \times \frac{1 - k_x}{k_x} \tag{3.518}$$

Thus, we have got the equation that binds the coefficient of restitution and the tangent of the angle of mechanical losses. So, if $k_x = 1$, $tg\beta \rightarrow 0$ we get the totally elastic impact, but if $k_x = 0$, $tg\beta \rightarrow \infty$ then we get the totally viscous impact. Using Equation 3.518 we can write the formula for the restitution coefficient as

$$k_x = \left[\frac{\omega_x \tau_1}{(tg\beta_E + \omega_x \tau_1)} \right] \tag{3.519}$$

By comparing Equations 3.516 and 3.519 we can finally get the expression for the maximum magnitude of the compression between a sled and semi-space respectively as

$$x_m = \left[\frac{mV_x^2}{2\kappa_1 E'} k_x \right]^{1/2} \tag{3.520}$$

Since $\kappa_1 = \frac{D_1}{2}(L_s + L)$ we get

$$x_m = \left[\frac{mV_x^2}{D_1(L_s + L)E'} k_x \right]^{1/2} \tag{3.520*}$$

On the other hand, since $\tau_x = \tau_1 + \tau_2$, $k_x = \frac{\tau_1}{\tau_2}$, see Equation 3.43, and $\tau_x = \frac{\pi}{\omega_x}$, see Equation 3.57, it follows that

$$\tau_1 = \tau_x \frac{k_x}{1 + k_x} \tag{3.521}$$

Substitution of τ_1 from Equation 3.521 in Equation 3.518 gives

$$tg\beta_E = \pi \times \frac{1 - k_x}{1 + k_x} \tag{3.522}$$

Hence, now we can write the equation for the restitution coefficient as follows

$$k_x = \frac{(\pi - tg\beta_E)}{(\pi + tg\beta_E)} \tag{3.523}$$

Finally, according to Equations 3.520* and 3.523, we can write the following

$$x_m = \left[\frac{mV_x^2}{D_1(L_s + L)E'} \times \frac{(\pi - tg\beta_E)}{(\pi + tg\beta_E)} \right]^{1/2} \tag{3.524}$$

3.7.1.3 APPROXIMATE SOLUTION TO THE DIFFERENTIAL EQUATIONS OF NORMAL DISPLACEMENT BY THE METHOD OF EQUIVALENT WORKS

According to the boundary conditions $t = \tau_1$, $x = x_m$ and using the known expressions for works A_{xcm} and A_{xbm} from Equations 3.508 and 3.509, we can write that

$$A_{xcm} = C_x \int_0^{x_m} x dx = \frac{1}{2} C_x x_m^2 = E' x_m^2 (\kappa_1 + \kappa_2 x_m^{1/n}) \quad (3.525)$$

and

$$A_{xbm} = B_x \int_0^{x_m} \dot{x} dx = B_x \frac{\int_0^{x_m} x dx}{\int_0^{\tau_1} dt} = B_x \frac{x_m^2}{2\tau_1} = \frac{E''}{\omega_x \tau_1} x_m^2 (\kappa_1 + \kappa_3 x_m^{1/n}) \quad (3.526)$$

Hence, we can write the expressions for the equivalent constant viscoelasticity parameters as

$$C_x = 2E'(\kappa_1 + \kappa_2 x_m^{1/n}) \quad (3.527)$$

$$B_x = \frac{2E''}{\omega_x}(\kappa_1 + \kappa_3 x_m^{1/n}) \quad (3.528)$$

Thus, the differential Equation 3.503 can be written as the equation with constant parameters as $m\dot{x} + B_x \dot{x} + C_x x = 0$ or in the full form as

$$m\ddot{x} + \frac{2E}{\omega_x}(\kappa_1 + \kappa_3 x_m^{1/n})\dot{x} + 2E'(\kappa_1 + \kappa_2 x_m^{1/n})x = 0 \quad (3.529)$$

The solution to this equation is already known as

$$x = \frac{V_x}{\omega_x} e^{-\delta_x t} \sin(\omega_x t) \quad (3.530)$$

where $\omega_x = \sqrt{\omega_{0x}^2 - \delta_x^2}$, $\delta_x = \frac{B_x}{2m}$ is the normal damping factor, $\omega_{0x} = \sqrt{\frac{C_x}{m}}$ is the angular frequency of free harmonic oscillations by axis X.

Then, after integration we get

$$\dot{x} = \frac{V_x}{\omega_x} e^{-\delta_x t}[\omega_x \cos(\omega_x t) - \delta_x \sin(\omega_x t)] \quad (3.531)$$

Using Equation 3.531 for the velocity, the duration of the time of the impact equals to the period of the time of the contact and it can be found now from the conditions $\dot{x} = V_{tx}$ and $t = \tau_x$ as $\tau_x = -\frac{\ln k_x}{\delta_x}$, and since $\delta_x = \frac{B_x}{2m}$, $\omega_x = \frac{\pi}{\tau_x}$ and $k_x = \frac{(\pi - tg\beta_E)}{(\pi + tg\beta_E)}$, see Equation 3.523, we get

$$\tau_x^2 = -\ln\left(\frac{\pi - tg\beta_E}{\pi + tg\beta_E}\right) \times \frac{\pi m}{E'(\kappa_1 + \kappa_3 x_m^{1/n})} \tag{3.532}$$

3.7.2 TANGENTIAL DISPLACEMENT

The equations for the tangential viscoelastic forces already known in this case as $F_{c\tau y} = G'yP_{Lx}$, see Equation 3.394, and $F_{b\tau y} = \eta'_G \dot{y} P_{Lx}$, see Equation 3.395, where $P_{Lx} = L + D_1 x$. Thus, according to the boundary conditions $t = \tau_1$, $x = x_m$ and $y = y_m$, the works for the tangential forces can be found by integration as follows:

$$A_{ycm} = \int_0^{y_m} F_{cy} dy = G' \int_0^{P_{Lxm}} dP_{Lx} \int_0^{y_m} y dy = \frac{G'}{2} P_{Lxm} y_m^2 \tag{3.533}$$

$$A_{ybm} = \int_0^{y_m} F_{by} dy = \frac{G''}{\omega_y} \int_0^{P_{Lxm}} dP_{Lx} \int_0^{y_m} \dot{y} dy = \eta'_G P_{Lx0} \frac{\int_0^{y_m} \int dy dy}{\int_0^{\tau_1} dt} = \frac{1}{2} \eta'_G P_{Lxm} \frac{y_m^2}{\tau_1} \tag{3.534}$$

where $P_{Lxm} = L + D_1 x_m$. Further, using the method of the equivalent work, we get

$$A_{yc0} = C_y \int_0^{y_0} y dy = \frac{1}{2} C_{yr} y_m^2 = \frac{1}{2} G' P_{Lxm} y_m^2 \tag{3.535}$$

$$A_{yb0} = B_y \int_0^{y_m} \dot{y} dy = B_y \frac{\int_0^{y_m} \int dy dy}{\int_0^{\tau_1} dt} = B_y \frac{y_0^2}{2\tau_1} = \frac{\eta'_G}{2\tau_1} P_{Lxm} y_m^2 \tag{3.536}$$

and

$$C_y = G' P_{Lxm} = G'(L + D_1 x_m) \tag{3.537}$$

$$B_y = \eta'_G (L + D_1 x_m) \tag{3.538}$$

Thus, the differential equation for the tangential displacement between the centres mass of a sled, can be written as the equation with constant parameters as follows $m\ddot{y} + B_y \dot{y} + C_y y = 0$, and as well in the common expression as

$$m\ddot{y} + \eta'_G (L + D_1 x_m)\dot{y} + G'(L + D_1 x_m) y = 0 \tag{3.539}$$

Equation 3.539 is the equation of the damped oscillations and according to the initial condition $y = 0$ and $\dot{y} = V_y$, the solutions to this equation is known as

$$y = \frac{V_y}{\omega_y} e^{-\delta_y t} \sin(\omega_y t) \tag{3.540}$$

where $\omega_{yr} = \sqrt{\omega_{0y}^2 - \delta_y^2}$, $\delta_y = \frac{B_y}{2m}$ is the tangential damping factor, $\omega_{0y} = \sqrt{\frac{C_y}{m}}$ is the angular frequency of the harmonic oscillations by axis Y.

As well the equation for the tangential velocity of the centre of mass of a sled can be received by differentiation of Equation 3.540 as

$$\dot{y} = \frac{V_y}{\omega_y} e^{-\delta_y t} [\omega_y \cos(\omega_y t) - \delta_y \sin(\omega_y t)] \tag{3.541}$$

Now, since $V_{ty} = y_c (t = \tau_x)$, the energetic coefficient of restitution for tangential displacement can be found as

$$e_y = k_y^2 = \frac{V_{ty}^2}{V_y^2} = \exp\left(-\frac{B_y}{m}\tau_x\right) \tag{3.542}$$

Or on the other hand, since $\tau_x = \pi / \omega_x = \pi / \sqrt{\omega_{0x}^2 - \delta_x^2}$, we get

$$e_y = k_y^2 = \frac{V_{ty}^2}{V_y^2} = \exp\left(-\frac{\pi B_y}{m\sqrt{\omega_{0x}^2 - \delta_x^2}}\right) \tag{4.543}$$

3.8 DYNAMICS OF AN ARBITRARY VISCOELASTIC COLLISION BETWEEN VESSEL AND SEMI-SPACE

3.8.1 NORMAL DISPLACEMENT

3.8.1.1 DIFFERENTIAL EQUATIONS OF DISPLACEMENT AT THE COMPRESSION

For consideration of an arbitrary viscoelastic collision between a vessel and the flat semi-space, we can use the illustration in Figure 2.14. At the initial instance of the time of an arbitrary viscoelastic collision between a vessel and a semi-space, a vessel has the initial velocities $\dot{x}(t = 0) = V_x$ and $\dot{y}(t = 0) = V_y$ and the mass m.

The equations for the normal viscous force and the elastic force already known as $F_{cn} = E'L_s D_1 x + 4E'\frac{C_r D_1^{1/n}}{n+1} x^{\frac{n+1}{n}}$, see Equation 3.415, and $F_{bn} = \eta_E' L_s D_1 \dot{x} + 4\eta_E' C_r D_1^{1/n} \dot{x} x^{1/n}$, see Equation 3.405; and since $\frac{E''}{\omega_x} = \eta_E'$, it follows as

$$F_{bn} = \frac{E''}{\omega_x}\{(L_s D_1 \dot{x} + 4C_r D_1^{1/n} \dot{x} x^{1/n}\} \tag{3.544}$$

Thus, the general normal viscoelastic force can be written as sum given below

$$\begin{aligned} F_n = F_{bn} + F_{cn} = b_x \dot{x} + c_x x = \frac{E''}{\omega_x}\{(L_s D_1 + 4C_r D_1^{1/n} x^{1/n}\} \\ \dot{x} + E'\{(L_s D_1 + 4\frac{C_r D_1^{1/n}}{n+1} x^{\frac{1}{n}}\}x \end{aligned} \tag{3.545}$$

where the expressions for the variable viscoelasticity parameters can be written as

$$b_x = \frac{E''}{\omega_x}\{(L_s D_1 + 4C_r D_1^{1/n} x^{1/n}\} \tag{3.546}$$

$$c_x = E'\{(L_s D_1 + 4\frac{C_r D_1^{1/n}}{n+1} x^{1/n}\} \tag{3.547}$$

According to Newton's Second Law, see Equations 3.1 and 3.545, the differential equation of the movement (displacement) of the centre of mass of a vessel by axis X can be expressed as

$$m\ddot{x} = \frac{E''}{\omega_x}\{(L_s D_1 + 4C_r D_1^{1/n} x^{1/n}\}\dot{x} + E'\{(L_s D_1 + 4\frac{C_r D_1^{1/n}}{n+1} x^{\frac{1}{n}}\}x = 0 \tag{3.548}$$

or it can be also written in the canonical form as $m\ddot{x} + b_x\dot{x} + c_x x = 0$, see Equation 3.28, and where m denotes the mass of a vessel.

3.8.1.2 WORK AND ENERGY IN THE PHASES OF COMPRESSION AND RESTITUTION

The works A_{xcm} and A_{xbm} at the maximum compression $x = x_m$ can be found by integration similarly as was done in Equations 3.417 and 3.418, respectively as

$$
\begin{aligned}
A_{xcm} &= \int_0^{x_m} F_{cn}dx = D_1 E' L_s \int_0^{x_m} xdx + 4E' \frac{C_r D_1^{1/n}}{n+1} \int_0^{x_m} x^{\frac{n+1}{n}} dx \\
&= E' x_m^2 \left(\frac{D_1}{2} L_s + \frac{4nC_r D_1^{1/n}}{(2n+1)(n+1)} x_m^{1/n} \right)
\end{aligned}
\tag{3.549}
$$

$$
\begin{aligned}
A_{xbm} &= \int_0^{x_m} F_{bn}dx = \frac{E''}{\omega_x} L_s D_1 \int_0^{x_m} \dot{x}dx + 4\frac{E''}{\omega_x} C_r D_1^{1/n} \int_0^{x_m} \dot{x}x^{1/n}dx \\
&= \frac{E'' L_s D_1 \int_0^{x_m} \int dxdx + 4E'' C_r D_1^{1/n} \int_0^{x_m} \int dxx^{1/n}dx}{\omega_x \int_0^{\tau_1} dt} = \frac{E''}{\omega_x \tau_1} x_m^2 \left(\frac{D_1}{2} L_s + \frac{4nC_r D_1^{1/n}}{2n+1} x_m^{1/n} \right)
\end{aligned}
\tag{3.550}
$$

Analogically, the works A_{xct} and A_{xbt} in the restitution phase can be found as follows

$$
\begin{aligned}
A_{xct} &= -\int_{x_m}^{0} F_{cn}dx = -D_1 E' L_s \int_{x_m}^{0} xdx - 4E' \frac{C_r D_1^{1/n}}{n+1} \int_{x_m}^{x0} x^{\frac{n+1}{n}} dx \\
&= E' x_m^2 \left(\frac{D_1}{2} L_s + \frac{4nC_r D_1^{1/n}}{(2n+1)(n+1)} x_m^{1/n} \right)
\end{aligned}
\tag{3.551}
$$

and

$$
\begin{aligned}
A_{xbt} &= -\int_{x_m}^{0} F_{bn}dx = -\eta_E' L_s D_1 \int_0^{x_m} \dot{x}dx + 4\eta_E' C_r D_1^{1/n} \int_0^{x_m} \dot{x}x^{1/n}dx \\
&= \frac{\eta_E' L_s D_1 \int_0^{x_m} \int dxdx + 4\eta_E' C_r D_1^{1/n} \int_0^{x_m} \int dxx^{1/n}dx}{\int_{\tau_1}^{\tau_x} dt} = \frac{\eta_E'}{\tau_2} x_m^2 \left(\frac{D_1}{2} L_s + \frac{4nC_r D_1^{1/n}}{2n+1} x_m^{1/n} \right)
\end{aligned}
\tag{3.552}
$$

Denoting $\frac{D_1}{2} L_s = q_1$, $\frac{4nC_r D_1^{1/n}}{(2n+1)(n+1)} = q_2$ and $\frac{4nC_r D_1^{1/n}}{2n+1} = q_3$, we get

$$A_{xcm} = E' x_m^2 (q_1 + q_2 x_m^{1/n}) \tag{3.553}$$

$$A_{xbm} = \frac{E''}{\omega_x \tau_1} x_m^2 (q_1 + q_3 x_m^{1/n}) \tag{3.554}$$

and

$$A_{xct} = E' x_m^2 (q_1 + q_2 x_m^{1/n}) \tag{3.555}$$

$$A_{xbt} = \frac{E''}{\omega_x \tau_2} x_m^2 (q_1 + q_3 x_m^{1/n}) \tag{3.556}$$

Since $A_{xm} = A_{xcm} + A_{xbm} = \frac{mV_x^2}{2}$ and $\frac{E''}{E'} = tg\beta_E$, we get for the full work in the maximum of indentation a vessel into a semi-space as

$$A_{xm} = E'x_m^2\left((q_1 + q_2x_m^{1/n}) + \frac{tg\beta_E}{\omega_x\tau_1}(q_1 + q_3x_m^{1/n})\right) = \frac{mV_x^2}{2} \tag{3.557}$$

Also, since $A_{xt} = A_{xct} - A_{xbt} = \frac{mV_{tx}^2}{2}$, we can find the full work in the restitution phase, respectively as

$$A_{xt} = E'x_m^2\left((q_1 + q_2x_m^{1/n}) - \frac{tg\beta_E}{\omega_x\tau_2}(q_1 + q_3x_m^{1/n})\right) = \frac{mV_{tx}^2}{2} \tag{3.558}$$

But also, on the other hand, usually $q_1 >> q_2x_m^{1/n}$ and $q_1 >> q_3x_m^{1/n}$, and; therefore, Equations 3.557 and 3.558 can be simplified as

$$A_{xm} = E'x_m^2q_1\left(1 + \frac{tg\beta_E}{\omega_x\tau_1}\right) = \frac{mV_x^2}{2} \tag{3.559}$$

$$A_{xt} = E'x_m^2q_1\left(1 - \frac{tg\beta_E}{\omega_x\tau_2}\right) = \frac{mV_{tx}^2}{2} \tag{3.560}$$

Using Equation 3.559, we get the formula for the approximate size of maximum indentation x_m

$$x_m = \left[\frac{mV_x^2\omega_x\tau_1}{2q_1E'(tg\beta_E + \omega_x\tau_1)}\right]^{1/2} \tag{3.561}$$

As we can see, the approximate solution gives the similar result as we have got at the impact of a sled.

The equation for energetic coefficient of restitution can be found such as the ratio between A_{tx} and A_{xm} and; therefore, we get the same result as in Equation 3.517. Also, since $k_x = \frac{\tau_1}{\tau_2}$, see Equation 3.43, we get the same result as in Equations 3.518 and 3.519, respectively, $tg\beta_E = \omega_x\tau_1 \times \frac{1-k_x}{k_x}$ and $k_x = \left[\frac{\omega_x\tau_1}{(tg\beta_E + \omega_x\tau_1)}\right]$. When we compare Equations 3.561 and 3.519, we can finally get the expression for the maximum magnitude of the compression between a vessel and semi-space respectively as

$$x_m = \left[\frac{mV_x^2}{2q_1E'}k_x\right]^{1/2} \quad (3.562)$$

Since $q_1 = \frac{D_1}{2}(L_s + L)$ we get the following equation

$$x_m = \left[\frac{mV_x^2}{D_1L_sE'}k_x\right]^{1/2} \quad (3.563)$$

Also, since in this case, we can use Equation 3.523, where $k_x = \frac{(\pi - tg\beta_E)}{(\pi + tg\beta_E)}$, we can write that

$$x_m = \left[\frac{mV_x^2}{D_1L_sE'} \times \frac{(\pi - tg\beta_E)}{(\pi + tg\beta_E)}\right]^{1/2} \quad (3.564)$$

3.8.1.3 APPROXIMATE SOLUTION TO THE DIFFERENTIAL EQUATIONS OF THE NORMAL DISPLACEMENT

According to the boundary conditions $t = \tau_1$ and $x = x_{m,}$ and using the known expressions for works A_{xcm} and A_{xbm} from Equations 3.553 and 3.554, we can write that

$$A_{xcm} = C_x \int_0^{x_m} x dx = \frac{1}{2}C_x x_m^2 = E' x_m^2 (q_1 + q_2 x_m^{1/n}) \quad (3.565)$$

and

$$A_{xbm} = B_x \int_0^{x_m} \dot{x} dx = B_x \frac{\int_0^{x_m} x dx}{\int_0^{\tau_1} dt} = B_x \frac{x_m^2}{2\tau_1} = \frac{E''}{\omega_x \tau_1} x_m^2 (q_1 + q_3 x_m^{1/n}) \quad (3.566)$$

Hence, we can write the expressions for the equivalent constant viscoelasticity parameters, respectively as

$$C_x = 2E'(q_1 + q_3 x_m^{1/n}) \quad (3.567)$$

$$B_x = \frac{2E''}{\omega_x}(q_1 + q_3 x_m^{1/n}) \quad (3.568)$$

Thus, the differential equation with constant parameters can be expressed as $m\ddot{x} + B_x\dot{x} + C_x x = 0$, and it follows

$$m\ddot{x} + \frac{2E}{\omega_x}(q_1 + q_3 x_m^{1/n})\dot{x} + 2E'(q_1 + q_2 x_m^{1/n})x = 0 \quad (3.569)$$

The solution to this equation is already known as

$$x = \frac{V_x}{\omega_x} e^{-\delta_x t} \sin(\omega_x t) \tag{3.570}$$

where $\omega_x = \sqrt{\omega_{0x}^2 - \delta_x^2}$, $\delta_x = \frac{B_x}{2m}$ is the normal damping factor, $\omega_{0x} = \sqrt{\frac{C_x}{m}}$ is the angular frequency of free harmonic oscillations by axis X. Then, after integration of Equation 3.570, we get the following

$$\dot{x} = \frac{V_x}{\omega_x} e^{-\delta_x t}[\omega_x \cos(\omega_x t) - \delta_x \sin(\omega_x t)] \tag{3.571}$$

Using Equation 3.571 for the velocity, the duration of the impact equals to the time period of the contact, which can be found now from the conditions $\dot{x} = V_{tx}$ and $t = \tau_x$ as $\tau_x = -\frac{\ln k_x}{\delta_x}$, and since $\delta_x = \frac{B_x}{2m}$, $\omega_x = \frac{\pi}{\tau_x}$ and $k_x = \frac{(\pi - tg\beta_E)}{(\pi + tg\beta_E)}$ is known from Equation 3.523, we get

$$\tau_x^2 = -\ln\left(\frac{\pi - tg\beta_E}{\pi + tg\beta_E}\right) \times \frac{\pi m}{E'(q_1 + q_3 x_m^{1/n})} \tag{3.572}$$

3.8.2 TANGENTIAL DISPLACEMENT

In the case of indentation at the impact of the curvilinear surface of a vessel into the flat surface of a semi-space, the equations for the tangential viscoelastic forces are same as in Equations 2.404 and 2.405, namely $F_{c\tau y} = G'P_{xn}y$ and $F_{b\tau y} = \eta'_G P_{xn}\dot{y}$, where $P_{xn} = 2C_r D_1^{1/n} x^{1/n} + D_1 x$, see Equation 2.406. Also, since $\eta'_G = \frac{G''}{\omega_y}$, it follows $F_{b\tau y} = \eta'_G P_{xn}\dot{y}$. Thus, according to the boundary conditions $t = \tau_1$, $x = x_m$ and $y = y_m$, the works for the tangential forces can be found by integration as follows:

$$A_{ycm} = \int_0^{y_m} F_{c\tau y} dy = G' \int_0^{P_{nm}} dP_{xn} \int_0^{y_m} y dy = \frac{G'}{2} P_{nm} y_m^2 \tag{3.573}$$

$$A_{ybm} = \int_0^{y_m} F_{b\tau y} dy = \frac{G''}{\omega_y} \int_0^{P_{nm}} dP_{xn} \int_0^{y_m} \dot{y} dy = \frac{G''}{\omega_y} P_{nm} \frac{\int_0^{y_m} \int dy dy}{\int_0^{\tau_1} dt} = \frac{G''}{2\omega_y} P_{nm} \frac{y_m^2}{\tau_1} \tag{3.574}$$

where

$$P_{nm} = 2C_r D_1^{1/n} x_m^{1/n} + D_1 x_m \tag{3.575}$$

Also, by using the already known method of the equivalent work in the phase of the shear, we get

$$A_{ycm} = C_y \int_0^{y_m} y dy = \frac{1}{2} C_y y_m^2 = \frac{1}{2} G' P_{nm} y_m^2 \tag{3.576}$$

$$A_{ybm} = B_y \int_0^{y_m} \dot{y} dy = B_y \frac{\int_0^{y_m} \int dydy}{\int_0^{\tau_1} dt} = B_y \frac{y_m^2}{2\tau_1} = \frac{G''}{2\omega_y \tau_1} P_{nm} y_m^2 \tag{3.577}$$

and we get

$$C_y = G' P_{nm} = G'(2C_r D_1^{1/n} x_m^{1/n} + D_1 x_m) \tag{3.578}$$

$$B_y = \frac{G''}{\omega_y} P_{nm} = \frac{G''}{\omega_y}(2C_r D_1^{1/n} x_m^{1/n} + D_1 x_m) \tag{3.579}$$

Thus, the differential equation for the tangential displacement of the centres mass of a vessel can be written as the equation with constant parameters as $m\ddot{y} + B_y \dot{y} + C_y y = 0$, and in the common expression as

$$m\ddot{y} + \frac{G''}{\omega_y}(2C_r D_1^{1/n} x_m^{1/n} + D_1 x_m)\dot{y} + G'(2C_r D_1^{1/n} x_m^{1/n} + D_1 x_m) y = 0 \tag{3.580}$$

Equation 3.580 is the equation of the damped oscillations and according to the initial condition $y = 0$ and $\dot{y} = V_y$, the solutions to this equation is known as

$$y = \frac{V_y}{\omega_y} e^{-\delta_y t} \sin(\omega_y t) \tag{3.581}$$

where $\omega_{yr} = \sqrt{\omega_{0y}^2 - \delta_y^2}$, $\delta_y = \frac{B_y}{2m}$ is the tangential damping factor, $\omega_{0y} = \sqrt{\frac{C_y}{m}}$ is the angular frequency of the harmonic oscillations by axis Y.

The equation for the tangential velocity of the centre of mass of a vessel can be received by differentiation of Equation 3.581 as

$$\dot{y} = \frac{V_y}{\omega_y} e^{-\delta_y t} [\omega_y \cos(\omega_y t) - \delta_y \sin(\omega_y t)] \tag{3.582}$$

Now, since $e_y = k_y^2$, by using Equation 3.582 and taking into account Equation 3.579, the energetic coefficient of restitution for tangential displacement can be found as

$$e_y = k_y^2 = \frac{V_{ty}^2}{V_y} = \exp\left(-\frac{G''(2C_r D_1^{1/n} x_m^{1/n} + D_1 x_m)}{\omega_y m}\tau_x\right) \tag{3.583}$$

Or on the other hand, since $\tau_x = \pi/\omega_x$, we get

$$e_y = k_y^2 = \frac{V_{ty}^2}{V_y} = \exp\left(-\frac{\pi G''(2C_r D_1^{1/n} x_m^{1/n} + D_1 x_m)}{\omega_y \omega_x m}\right) \tag{3.584}$$

3.9 DYNAMICS OF AN ARBITRARY VISCOELASTIC COLLISION BETWEEN CYLINDRICAL WHEEL AND SEMI-SPACE

3.9.1 NORMAL DISPLACEMENT

3.9.1.1 DIFFERENTIAL EQUATIONS OF DISPLACEMENT AT THE COMPRESSION

For consideration of an arbitrary viscoelastic collision between cylindrical wheel that has the mass m and the initial velocities $\dot{x}(t = 0) = V_x$ and $\dot{y}(t=0) = V_y$ and the flat semi-space, we can use the illustration in Figure 2.12.

The equations for the normal viscoelastic forces for this case of contact, already known as $F_{cn} = E'\left(\frac{2}{3}k_p R_w^{1/2} x^{312} + L_w\right)x$, see Equation 3.290, and $F_{bn} = 2\frac{E''}{\omega_x}(k_p R_w^{1/2} x^{1/2} + L_w)\dot{x}$, see Equation 3.291

Thus, the general normal viscoelastic force can be written as the following sum

$$F_n = F_{bn} + F_{cn} = b_x \dot{x} + c_x x = 2\frac{E''}{\omega_x}(k_p R_w^{1/2} x^{1/2} + L_w)\dot{x} + E'\left(\frac{2}{3}k_p R_w^{1/2} x^{1/2} + L_w\right)x \tag{3.585}$$

where the expressions for the variable viscoelasticity parameters can be written as

$$b_x = 2\frac{E''}{\omega_x}(k_p R_w^{1/2} x^{1/2} + L_w) \tag{3.586}$$

$$c_x = E'\left(\frac{2}{3}k_p R_w^{1/2} x^{1/2} + L_w\right)x \tag{3.587}$$

According to Newton's Second Law, see Equation 3.1, and also according to Equation 3.585, the differential equation of the movement (displacement) of the centre of mass of a wheel by axis X can be expressed as

$$m\ddot{x} = 2\frac{E''}{\omega_x}(k_p R_w^{1/2} x^{1/2} + L_w)\dot{x} + E'\left(\frac{2}{3}k_p R_w^{1/2} x^{1/2} + L_w\right)x \tag{3.588}$$

or it can be also written in the canonical form as $m\ddot{x} + b_x\dot{x} + c_x x = 0$, see Equation 3.28, where m denotes the mass of a wheel.

3.9.1.2 WORK AND ENERGY IN THE PHASES OF COMPRESSION AND RESTITUTION

The works A_{xcm} and A_{xbm} at the maximum compression $x = x_m$ can be found by integration similar to that done in Equations 3.294 and 3.295, respectively

$$A_{xcm} = \int_0^{x_m} F_{cn}dx = E'\int_0^{x_m}\left(\frac{2}{3}k_p R_w^{1/2} x^{3/2} + L_w x\right)dx = E'x_m^2\frac{1}{2}\left(\frac{8}{15}k_p R_w^{1/2} x_m^{1/2} + L_w\right) \tag{3.589}$$

and

$$A_{xbm} = \int_0^{x_m} F_{bn}dx = \frac{2E''}{\omega_x}\int_0^{x_m}\left(k_p R_w^{1/2} x^{1/2} + L_w\right)\dot{x}dx = \frac{2E''}{\omega_x}\frac{\int_0^{x_m}\int dx\left(k_p R_w^{1/2} x^{1/2} + L_w\right)dx}{\int_0^{\tau_1} dt}$$
$$= \frac{E''}{\omega_x \tau_1}x_m^2\left(\frac{4}{5}k_p R_w^{1/2} x_m^{1/2} + L_w\right) \tag{3.590}$$

Analogically, the works A_{xct} and A_{xbt} in the restitution phase can be found as follows

$$A_{xct} = -\int_{x_m}^{0} F_{cn}dx = -E'\int_{x_m}^{0}\left(\frac{2}{3}k_p R_w^{1/2} x^{3/2} + L_w x\right)dx = E'x_m^2\frac{1}{2}\left(\frac{8}{15}k_p R_w^{1/2} x_m^{1/2} + L_w\right) \tag{3.591}$$

$$A_{xbt} = -\int_{x_m}^{0} F_{bn}dx = -\frac{2E''}{\omega_x}\int_{x_m}^{0}\left(k_p R_w^{1/2} x^{1/2} + L_w\right)\dot{x}dx = \frac{2E''}{\omega_x}\frac{\int_0^{x_m}\int dx\left(k_p R_w^{1/2} x^{1/2} + L_w\right)dx}{\int_{\tau_1}^{\tau_x} dt}$$
$$= \frac{E''}{\omega_x \tau_2}x_m^2\left(\frac{4}{5}k_p R_w^{1/2} x_m^{1/2} + L_w\right) \tag{3.592}$$

It is obvious that usually $L \gg \frac{4}{5}k_p R_w^{1/2} x_m^{1/2}$ and $L \gg \frac{8}{15}k_p R_w^{1/2} x_m^{1/2}$, thus approximately we can take that

$$A_{xcm} = \frac{1}{2}E'x_m^2 L_w \tag{3.593}$$

$$A_{xbm} = \frac{E''}{\omega_x \tau_1} x_m^2 L_w \tag{3.594}$$

and

$$A_{xct} = \frac{1}{2} E' x_m^2 L_w \tag{3.595}$$

$$A_{xbt} = \frac{E''}{\omega_x \tau_2} x_m^2 L_w \tag{3.596}$$

Since $A_{xm} = A_{xcm} + A_{xbm} = \frac{mV_x^2}{2}$ and $\frac{E''}{E'} = tg\beta_E$, we get for the full work in the maximum of indentation cylindrical wheel into a semi-space

$$A_{xm} = E' x_m^2 L_w \left(\frac{1}{2} + \frac{tg\beta_E}{\omega_x \tau_1} \right) = \frac{mV_x^2}{2} \tag{3.597}$$

Also, since $A_{xt} = A_{xct} - A_{xbt} = \frac{mV_{tx}^2}{2}$, we can find the full work in the restitution phase as

$$A_{xm} = E' x_m^2 L_w \left(\frac{1}{2} - \frac{tg\beta_E}{\omega_x \tau_2} \right) = \frac{mV_{tx}^2}{2} \tag{3.598}$$

Using Equation 3.597, we get the formula for x_m as

$$x_m = \left[\frac{mV_x^2 \omega_x \tau_1}{L_w E'(2tg\beta_E + \omega_x \tau_1)} \right]^{1/2} \tag{3.599}$$

The energetic coefficient of restitution can be found like the ratio between A_{tx} and A_{xm}:

$$e_x = k_x^2 = \frac{V_{tx}^2}{V_x^2} = \left(\frac{\omega_x \tau_2 - 2tg\beta}{\omega_x \tau_1 + 2tg\beta} \right) \frac{\tau_1}{\tau_2} \tag{3.600}$$

Since $k_x = \frac{\tau_1}{\tau_2}$, see Equation 3.43, we get in this case that

$$tg\beta_E = \frac{\omega_x \tau_1}{2} \times \frac{1 - k_x}{k_x} \tag{3.601}$$

Using Equation 3.601, we can write the formula for the restitution coefficient as

$$k_x = \left[\frac{\omega_x \tau_1}{(2tg\beta_E + \omega_x \tau_1)} \right] \tag{3.602}$$

When we compare Equations 3.602 and 3.599, we can finally get the expression for the maximum magnitude of the compression between a cylindrical wheel and a semi-space respectively as

$$x_m = \left[\frac{mV_x^2}{L_w E'} k_x \right]^{1/2} \tag{3.603}$$

Also, since as we know $\tau_x = \frac{\pi}{\omega_x}$ and $\tau_x = \tau_1 + \tau_2$ we get

$$tg\beta_E = \frac{\pi}{2} \times \frac{(1-k_x)}{(1+k_x)} \tag{3.604}$$

For the equation of the restitution coefficient we can write now as follows

$$k_x = \frac{(\pi - 2tg\beta_E)}{(\pi + 2tg\beta_E)} \tag{3.605}$$

Finally, according to Equations 3.603 and 3.605, it follows that

$$x_m = \left[\frac{mV_x^2}{E'L_w} \times \frac{(\pi - 2tg\beta_E)}{(\pi + 2tg\beta_E)} \right]^{1/2} \tag{3.606}$$

3.9.1.3 APPROXIMATE SOLUTION TO THE DIFFERENTIAL EQUATIONS OF THE NORMAL DISPLACEMENT

According to the boundary conditions $t = \tau_1$, $x = x_m$, and using the known expressions for works A_{xcm} and A_{xbm} from Equations 3.589 and 3.590, we can write that

$$A_{xcm} = C_x \int_0^{x_m} x dx = \frac{1}{2} C_x x_m^2 = E' x_m^2 \frac{1}{2} \left(\frac{8}{15} k_p R_w^{1/2} x_m^{1/2} + L_w \right) \tag{3.607}$$

and

$$A_{xbm} = B_x \int_0^{x_m} \dot{x} dx = B_x \frac{\int_0^{x_m} x dx}{\int_0^{\tau_1} dt} = B_x \frac{x_m^2}{2\tau_1} = \frac{E''}{\omega_x \tau_1} x_m^2 \left(\frac{4}{5} k_p R_w^{1/2} x_m^{1/2} + L_w \right) \tag{3.608}$$

Hence, we can write the expressions for the equivalent constant viscoelasticity parameters as

$$C_x = E' \left(\frac{8}{15} k_p R_w^{1/2} x_m^{1/2} + L_w \right) \tag{3.609}$$

$$B_x = \frac{2E''}{\omega_x}\left(\frac{4}{5}k_p R_w^{1/2} x_m^{1/2} + L_w\right) \tag{3.610}$$

Thus, the differential equation with constant parameters can be expressed as $m\ddot{x} + B_x\dot{x} + C_x x = 0$, and it follows that

$$m\ddot{x} + \frac{2E''}{\omega_x}\left(\frac{4}{5}k_p R_w^{1/2} x_m^{1/2} + L_w\right)\dot{x} + E'\left(\frac{8}{15}k_p R_w^{1/2} x_m^{1/2} + L_w\right)x = 0 \tag{3.611}$$

The solution to this equation is already known as

$$x = \frac{V_x}{\omega_x} e^{-\delta_x t}\sin(\omega_x t) \tag{3.612}$$

where $\omega_x = \sqrt{\omega_{0x}^2 - \delta_x^2}$, $\delta_x = \frac{B_x}{2m}$ is the normal damping factor, $\omega_{0x} = \sqrt{\frac{C_x}{m}}$ is the angular frequency of free harmonic oscillations by axis X. Then, after integration of Equation 3.612, we get

$$\dot{x} = \frac{V_x}{\omega_x} e^{-\delta_x t}[\omega_x \cos(\omega_x t) - \delta_x \sin(\omega_x t)] \tag{3.613}$$

Using Equation 3.613 for the velocity, the duration of the time of the impact is equal to the period of the time of the contact, which can be found now from the conditions $\dot{x} = V_{tx}$ and $t = \tau_x$ as $\tau_x = -\frac{\ln k_x}{\delta_x}$, and since $\delta_x = \frac{B_x}{2m}$, $\omega_x = \frac{\pi}{\tau_x}$, and since in this case $k_x = \frac{(\pi - 2tg\beta_E)}{(\pi + 2tg\beta_E)}$ is known from Equation 3.523, we get

$$\tau_x^2 = \pi m \frac{-\ln\left(\frac{\pi - 2tg\beta_E}{\pi + 2tg\beta_E}\right)}{E'\left(\frac{4}{5}k_p R_w^{1/2} x_m^{1/2} + L_w\right)} \tag{3.614}$$

3.9.2 TANGENTIAL DISPLACEMENT

In the case of a rolling motion between a cylindrical wheel and the flat semi-space, see Figure 2.12, the equations for viscoelastic tangential forces already known as $F_{cty} = G'P_{rw}y$, see Equation 3.300, and $F_{b\tau y} = \frac{G''}{\omega_{yr}}P_{rw}\dot{y}$, see Equation 3.301, and where according to Equation 3.302 $P_{rw} = \frac{x}{R_w}(D_1 x + L_w)$.

Thus, the works for the tangential forces can be found by integration as follows:

$$A_{ycm} = \int_0^{y_m} F_{c\tau y} dy = G' \int_0^{P_{rwm}} dP_{rw} \int_0^{y_m} y dy = \frac{G'}{2} P_{rwm} y_m^2 \tag{3.615}$$

$$A_{ybm} = \int_0^{y_m} F_{b\tau y} dy = \frac{G''}{\omega_y} \int_0^{P_{rwm}} dP_{rw} \int_0^{y_m} \dot{y} dy = \frac{G''}{\omega_{yr}} P_{rwm} \frac{\int_0^{y_m} \int dy dy}{\int_0^{\tau_1} dt} = \frac{G''}{2\omega_{yr}} P_{rwm} \frac{y_m^2}{\tau_1} \tag{3.616}$$

where

$$P_{rwm} = \frac{x_m}{R_w}(D_1 x_m + L_w) \tag{3.617}$$

Using the method of the equivalent work in the phase of the shear we get

$$A_{ycm} = C_y \int_0^{y_m} y dy = \frac{1}{2} C_{yr} y_m^2 = \frac{1}{2} G' P_{rwm} y_m^2 \tag{3.618}$$

$$A_{ybm} = B_y \int_0^{y_m} \dot{y} dy = B_y \frac{\int_0^{y_m} \int dy dy}{\int_0^{\tau_1} dt} = B_{yr} \frac{y_m^2}{2\tau_1} = \frac{G''}{2\omega_y \tau_1} P_{rwm} y_m^2 \tag{3.619}$$

and, thus, we get

$$C_{yr} = G' P_{rwm} = G' \frac{x_m}{R_w}(D_1 x_m + L_w) \tag{3.620}$$

and

$$B_{yr} = \frac{G''}{\omega_{yr}} P_{rwm} = \frac{G''}{\omega_{yr}} \frac{x_m}{R_w}(D_1 x_m + L_w) \tag{3.621}$$

Thus, the differential equation for the tangential displacement of the centre masses of a cylindrical wheel can be written as the equation with constant parameters $m\ddot{x} + B_{yr}\dot{y} + C_{yr}y = 0$, and in the common expression as

$$m\ddot{y} + \frac{G''}{\omega_{yr}} \frac{x_m}{R_w}(D_1 x_m + L_w)\dot{y} + G' \frac{x_m}{R_w}(D_1 x_m + L_w)y = 0 \tag{3.622}$$

Equation 3.623 is the equation of the damped oscillations and according to the initial condition $y = 0$ and $\dot{y} = V_y$, the solutions to this equation is known as

$$y = \frac{V_y}{\omega_{yr}} e^{-\delta_{yr} t} \sin(\omega_{yr} t) \tag{3.623}$$

where $\omega_{yr} = \sqrt{\omega_{0y}^2 - \delta_{yr}^2}$, $\delta_{yr} = \dfrac{B_{yr}}{2m}$ is the tangential damping factor, $\omega_{0yr} = \sqrt{\dfrac{C_{yr}}{m}}$ is the angular frequency of the harmonic oscillations by axis Y.

The equation for the tangential velocity of the centres of mass of a wheel can be received by differentiation of Equation 3.624 as

$$\dot{y} = \frac{V_y}{\omega_{yr}} e^{-\delta_{yr} t}[\omega_{yr}\cos(\omega_y t) - \delta_{yr}\sin(\omega_{yr} t)] \tag{3.624}$$

Now, since $e_y = k_y^2$ and $\delta_{yr} = \dfrac{B_{yr}}{2m}$, using Equation 3.624 and taking into account Equation 3.621, the energetic coefficient of restitution for tangential displacement can be found as

$$e_y = k_y^2 = \frac{V_{ty}^2}{V_y^2} = \exp\left(-\frac{G''x_m(D_1 x_m + L_w)}{m\omega_{yr} R_w}\tau_x\right) \tag{3.625}$$

Or on the other hand, since $\tau_x = \pi/\omega_x$, we get

$$e_y = k_y^2 = \frac{V_{ty}^2}{V_y^2} = \exp\left(-\frac{G''x_m(D_1 x_m + L_w)}{2m\omega_{yr} R_w}\tau_x\right) \tag{3.626}$$

Also, since the work A_{yc} is transformed into the kinetic energy of the relative rotation of a wheel and also according to Equations 3.97 and 3.99 we can write that

$$A_{yc} = C_{yr}\int_0^{y_t} y\,dy = \frac{1}{2}C_{yr} y_t^2 = \frac{J_z \omega_t^2}{2} \tag{3.627}$$

Respectively, the equation for the relative angular velocity at the instance of rebound can be expressed as follows

$$\omega_t = \left(\frac{C_{yr}}{J_z}\right)^{\frac{1}{2}} y_t \tag{3.628}$$

Taking into account Equation 3.620 and that $y_t = \dfrac{V_y}{\omega_{yr}} e^{-\delta_y \tau_x}\sin(\omega_{yr}\tau_x)$, we get

$$\omega_t = \left(\frac{G'x_m}{J_z R_w}(D_1 x_m + L_w)\right)^{\frac{1}{2}} \frac{V_y}{\omega_{yr}} e^{-\delta_{yr}\tau_x}\sin(\omega_{yr}\tau_x) \tag{3.629}$$

3.10 DYNAMICS OF AN ARBITRARY VISCOELASTIC COLLISION BETWEEN TOROIDAL SURFACE OF WHEEL AND SEMI-SPACE

3.10.1 NORMAL DISPLACEMENT

3.10.1.1 DIFFERENTIAL EQUATIONS OF DISPLACEMENT AT THE COMPRESSION

For consideration of an arbitrary viscoelastic collision between a toroidal surface of the wheel and the flat semi-space, we can use the illustration in Figure 2.12. At the initial instance of time of an arbitrary viscoelastic collision between a toroidal wheel and the flat semi-space, a wheel has the initial velocities $\dot{x}(t=0)\ V_x$ and $\dot{y}(t=0)\ V_y$, and the mass m.

The equations for the normal viscoelastic forces for this case of contact are already known as $F_{cn} = \frac{2}{3}E'\psi_w x^{3/2}$, see Equation 3.312, and $F_{bn} = 2\frac{E''}{\omega_x}\psi_w \dot{x}x^{1/2}$, see Equation 3.313. Thus, the general normal viscoelastic force can be written as sum given below

$$F_n = F_{bn} + F_{cn} = b_x\dot{x} + c_x x = 2\frac{E''}{\omega_x}\psi_w \dot{x}x^{1/2} + \frac{2}{3}E'\psi_w x^{3/2} \tag{3.630}$$

where the expressions for the variable viscoelasticity parameters can be written as

$$b_x = 2\frac{E''}{\omega_x}\psi_w \dot{x} \tag{3.631}$$

$$c_x = \frac{2}{3}E'\psi_w x^{1/2} \tag{3.632}$$

According to Newton's Second Law, see Equation 3.1 and 3.630, the differential equation of the movement (displacement) of the centre of mass of a toroidal wheel by axis X can be expressed as

$$m\ddot{x} + 2\frac{E''}{\omega_x}\psi_w \dot{x}x^{1/2} + \frac{2}{3}E'\psi x^{3/2} = 0 \tag{3.633}$$

where $\psi_w = (k_{pa}R_w^{1/2} + k_{pb}R_p^{1/2})$ and m denotes the mass of a wheel.

Equation 3.633 can be also written in the canonical form as $m\ddot{x} + b_x\dot{x} + c_x x = 0$ see also Equation 3.28.

3.10.1.2 WORK AND ENERGY IN THE PHASES OF COMPRESSION AND RESTITUTION

The works A_{xcm} and A_{xbm} at the maximum compression $x = x_m$ can be found by integration as follows

$$A_{xcm} = \int_0^{x_m} F_{cn} dx = \frac{2}{3} E' k_p \psi_w \int_0^{x_m} x^{3/2} dx = \frac{4}{15} E' \psi_w x_m^{5/2} \tag{3.634}$$

$$A_{xbm} = \int_0^{x_m} F_{bn} dx = \frac{2E''}{\omega_X} \psi_w \int_0^{x_m} \dot{x} x^{1/2} dx = \frac{2E''}{\omega_x} \psi_w \frac{\int_0^{x_m} \int dx x^{1/2} dx}{\int_0^{\tau_1} dt} = \frac{4E''}{5\omega_x \tau_1} \psi_w x_m^{5/2} \tag{3.635}$$

Analogically, the works A_{xct} and A_{xbt} in the restitution phase can be found as follows

$$A_{xct} = -\int_{x_m}^{0} F_{cn} dx = -\frac{2}{3} E' k_p \psi_w \int_{x_m}^{0} x^{3/2} dx = \frac{4}{15} E' \psi_w x_m^{5/2} \tag{3.636}$$

$$A_{xbt} = -\int_{x_m}^{0} F_{bn} dx = -\frac{2E''}{\omega_X} \psi_w \int_{x_m}^{0} \dot{x} x^{1/2} dx = \frac{2E''}{\omega_x} \psi_w \frac{\int_{x_m}^{0} \int dx x^{1/2} dx}{\int_{\tau_1}^{\tau_x} dt} = \frac{4E''}{5\omega_x \tau_2} \psi_w x_m^{5/2} \tag{3.637}$$

Since $A_{xm} = A_{xcm} + A_{xbm} = \frac{mV_x^2}{2}$ and $\frac{E''}{E'} = tg\beta_E$, the equation for full work in the maximum of indentation of a toroidal wheel into a semi-space can be expressed as

$$A_{xm} = E' x_m^{5/2} \psi_w \frac{4}{15} \left(1 + \frac{3tg\beta_E}{\omega_x \tau_1} \right) = \frac{mV_x^2}{2} \tag{3.638}$$

Also, since $A_{xt} = A_{xct} - A_{xbt} = \frac{mV_{tx}^2}{2}$, we can find the full work in the restitution phase as

$$A_{xt} = E' x_m^{5/2} \psi_w \frac{4}{15} \left(1 - \frac{3tg\beta_E}{\omega_x \tau_2} \right) = \frac{mV_{tx}^2}{2} \tag{3.639}$$

Using Equation 3.638, we get the formula for x_m as follows

$$x_m = \left[\frac{15 m V_x^2 \omega_x \tau_1}{8E' \psi_w (3tg\beta_E + \omega_x \tau_1)} \right]^{5/2} \tag{3.640}$$

The energetic coefficient of restitution can be found like the ratio between A_{tx} and A_{xm}:

$$e_x = k_x^2 = \frac{V_{tx}^2}{V_x^2} = \left(\frac{\omega_x\tau_2 - 3tg\beta}{\omega_x\tau_1 + 3tg\beta}\right)\frac{\tau_1}{\tau_2} \tag{3.641}$$

Since $k_x = \frac{\tau_1}{\tau_2}$, see Equation 3.43, we get in this case that

$$tg\beta_E = \frac{\omega_x\tau_1}{3} \times \frac{1-k_x}{k_x} \tag{3.642}$$

Using Equation 3.601, we can write the formula for the restitution coefficient as

$$k_x = \left[\frac{\omega_x\tau_1}{(3tg\beta_E + \omega_x\tau_1)}\right] \tag{3.643}$$

If we compare Equations 3.640 and 3.643, we can finally get the expression for the maximum magnitude of the compression between a toroidal wheel and a semi-space as

$$x_m = \left[\frac{15mV_x^2}{8E'\psi_w}k_x\right]^{5/2} \tag{3.644}$$

Also, since as we know $\tau_x = \frac{\pi}{\omega_x}$ and $\tau_x = \tau_1 + \tau_2$ we get

$$tg\beta_E = \frac{\pi}{3} \times \frac{(1-k_x)}{(1+k_x)} \tag{3.645}$$

The equation for the restitution coefficient we can write now as follows

$$k_x = \frac{(\pi - 3tg\beta_E)}{(\pi + 3tg\beta_E)} \tag{3.646}$$

Finally, according to Equations 3.603 and 3.605, it follows that

$$x_m = \left[\frac{15mV_x^2}{8E'\psi_w} \times \frac{(\pi - 3tg\beta_E)}{(\pi + 3tg\beta_E)}\right]^{5/2} \tag{3.647}$$

3.10.1.3 APPROXIMATE SOLUTION TO THE DIFFERENTIAL EQUATIONS OF THE NORMAL DISPLACEMENT

According to the boundary conditions $t = \tau_1$, $x = x_m$ and using the known expressions for works A_{xcm} and A_{xbm} from Equations 3.634 and 3.635, we can write that

$$A_{xcm} = C_x \int_0^{x_m} x dx = \frac{1}{2} C_x x_m^2 = \frac{4}{15} E' \psi_w x_m^{5/2} \tag{3.648}$$

and

$$A_{xbm} = B_x \int_0^{x_m} \dot{x} dx = B_x \frac{\int_0^{x_m} x dx}{\int_0^{\tau_1} dt} = B_x \frac{x_m^2}{2\tau_1} = \frac{4E''}{5\omega_x \tau_1} \psi_w x_m^{5/2} \tag{3.649}$$

Hence, we can write the expressions for the equivalent constant viscoelasticity parameters as

$$C_x = \frac{8}{15} E' \psi_w x_m^{1/2} \tag{3.650}$$

$$B_x = \frac{8E''}{5\omega_x} \psi_w x_m^{1/2} \tag{3.651}$$

Thus, the differential equation with constant parameters can be expressed as $m\ddot{x} + B_x \dot{x} + C_x x = 0$, and it follows that

$$m\ddot{x} + \frac{8E''}{5\omega_x} \psi_w x_m^{1/2} \dot{x} + \frac{8}{15} E' \psi_w x_m^{1/2} x = 0 \tag{3.652}$$

The solution to this equation is already known as

$$x = \frac{V_x}{\omega_x} e^{-\delta_x t} \sin(\omega_x t) \tag{3.653}$$

where $\omega_x = \sqrt{\omega_{0x}^2 - \delta_x^2}$, $\delta_x = \frac{B_x}{2m}$ is the normal damping factor, $\omega_{0x} = \sqrt{\frac{C_x}{m}}$ is the angular frequency of free harmonic oscillations by axis X. Then, after integration of Equation 3.653 we get

$$\dot{x} = \frac{V_x}{\omega_x} e^{-\delta_x t} [\omega_x \cos(\omega_x t) - \delta_x \sin(\omega_x t)] \tag{3.654}$$

Further, using Equation 3.654 for the velocity, the duration of the impact equals to the time period of the contact, which can be found now from the conditions $\dot{x} = V_{tx}$ and $t = \tau_x$ as $\tau_x = -\frac{\ln k_x}{\delta_x}$, and since $\delta_x = \frac{B_x}{2m}$, $\omega_x = \frac{\pi}{\tau_x}$ and since in this case $k_x = \frac{(\pi - 3tg\beta_E)}{(\pi + 3tg\beta_E)}$, it is known from Equation 3.523, we get

$$\tau_x^2 = -\ln\left(\frac{\pi - 3tg\beta_E}{\pi + 3tg\beta_E}\right) \times \frac{5\pi m}{4E'\psi_w x_m^{1/2}} \tag{3.655}$$

3.10.2 TANGENTIAL DISPLACEMENT

In the case of a rolling motion between a toroidal wheel and the flat semi-space, see Figure 2.12, the equations for viscoelastic tangential forces are already known as $F_{c\tau y} = G'P_{twy}$, see Equation 3.322, and $F_{b\tau y} = \frac{G''}{\omega_{yr}} P_{tw}\dot{y}$, see Equation 3.323, and according to Equation 3.324 $P_{tw} = \frac{x}{R_w}(D_1 x + 2k_{pb}R_p^{1/2}x^{1/2})$.
Thus, the works for the tangential forces can be found by integration as follows:

$$A_{ycm} = \int_0^{y_m} F_{c\tau y}dy = G'\int_0^{P_{twm}} dP_{rw}\int_0^{y_m} ydy = \frac{G'}{2}P_{twm}y_m^2 \tag{3.656}$$

$$A_{ybm} = \int_0^{y_m} F_{b\tau y}dy = \frac{G''}{\omega_y}\int_0^{P_{twm}} dP_{tw}\int_0^{y_m} \dot{y}dy = \frac{G''}{\omega_{yr}}P_{twm}\frac{\int_0^{y_m}\int dydy}{\int_0^{\tau_1} dt} = \frac{G''}{2\omega_{yr}}P_{twm}\frac{y_m^2}{\tau_1} \tag{3.657}$$

where

$$P_{twm} = \frac{x_m}{R_w}(D_1 x_m + 2k_{pb}R_p^{1/2}x_m^{1/2}) \tag{3.658}$$

By using the method of equivalent work in the phase of shear, we get

$$A_{ycm} = C_y\int_0^{y_m} ydy = \frac{1}{2}C_{yr}y_m^2 = \frac{1}{2}G'P_{twm}y_m^2 \tag{3.659}$$

$$A_{ybm} = B_y\int_0^{y_m} \dot{y}dy = B_y\frac{\int_0^{y_m}\int dydy}{\int_0^{\tau_1} dt} = B_{yr}\frac{y_m^2}{2\tau_1} = \frac{G''}{2\omega_y\tau_1}P_{twm}y_m^2 \tag{3.660}$$

and, therefore, we get

$$C_{yr} = G'P_{twm} = G'\frac{x_m}{R_w}(D_1 x_m + 2k_{pb}R_p^{1/2}x_m^{1/2}) \tag{3.661}$$

and

$$B_{yr} = \frac{G''}{\omega_{yr}}P_{rwm} = \frac{G''}{\omega_{yr}}\frac{x_m}{R_w}(D_1 x_m + 2k_{pb}R_p^{1/2}x_m^{1/2}) \tag{3.662}$$

Thus, the differential equation for the tangential displacement of the centre masses of a toroidal wheel can be written as the equation with constant parameters $m\ddot{y} + B_{yr}\dot{y} + C_{yr}y = 0$, and in common expression as

$$m\ddot{y}+\frac{G''}{\omega_{yr}}\frac{x_m}{R_w}(D_1x_m+2k_{pb}R_p^{1/2}x_m^{1/2})\dot{y}+G'\frac{x_m}{R_w}(D_1x_m+2k_{pb}R_p^{1/2}x_m^{1/2})y=0 \quad (3.663)$$

Equation 3.6633 is the equation of the damped oscillations and according to the initial condition $y = 0$ and $\dot{y} = V_y$, the solution to this equation is known as

$$y=\frac{V_y}{\omega_{yr}}e^{-\delta_{yr}t}\sin(\omega_{yr}t) \quad (3.664)$$

where $\omega_{yr}=\sqrt{\omega_{0y}^2-\delta_{yr}^2}$, $\delta_{yr}=\frac{B_{yr}}{2m}$ the tangential damping factor, $\omega_{0yr}=\sqrt{\frac{C_{yr}}{m}}$ is the angular frequency of the harmonic oscillations by axis Y.

The equation for the tangential velocity of the centres of mass of a toroidal wheel can be received by differentiation of Equation 3.664 as given below

$$\dot{y}=\frac{V_y}{\omega_{yr}}e^{-\delta_{yr}t}[\omega_{yr}\cos(\omega_y t)-\delta_{yr}\sin(\omega_{yr}t)] \quad (3.665)$$

Now, since $e_y=k_y^2$ and $\delta_{yr}=\frac{B_{yr}}{2m}$, by using Equation 3.665 and taking into account Equation 3.662, the energetic coefficient of restitution for tangential displacement can be found respectively as

$$e_y=k_y^2=\frac{V_{ty}^2}{V_y^2}=\exp\left(-\frac{G''x_m(D_1x_m+2k_pR_p^{1/2}x_m^{1/2})}{m\omega_{yr}R_w}\tau_x\right) \quad (3.666)$$

Or on the other hand, since $\tau_x = \pi/\omega_x$, we get

$$e_y=k_y^2=\frac{V_{ty}^2}{V_y^2}=\exp\left(-\frac{\pi G''x_m(D_1x_m+2k_pR_p^{1/2}x_m^{1/2})}{m\omega_x\omega_{yr}R_w}\right) \quad (3.667)$$

Also, as we already know that the work A_{yc} is transformed into the kinetic energy of the relative rotation of a wheel, and we can use this in case of Equations 3.627 and 3.629 as well, and by taking into account Equation 3.661 and that $y_t=\frac{V_y}{\omega_{yr}}e^{-\delta_y\tau_x}\sin(\omega_{yr}\tau_x)$, we get

$$\omega_t=\left(\frac{G'x_m}{J_zR_w}(D_1x_m+2k_{pb}R_p^{1/2}x_m^{1/2})\right)^{\frac{1}{2}}\frac{V_y}{\omega_{yr}}e^{-\delta_{yr}}\sin(\omega_{yr}\tau_x) \quad (3.668)$$

KEYWORDS

- **harmonic oscillations**
- **viscoelastic forces**
- **Kelvin–Voigt model**
- **Poisson's coefficient**
- **tangential elastic stress**

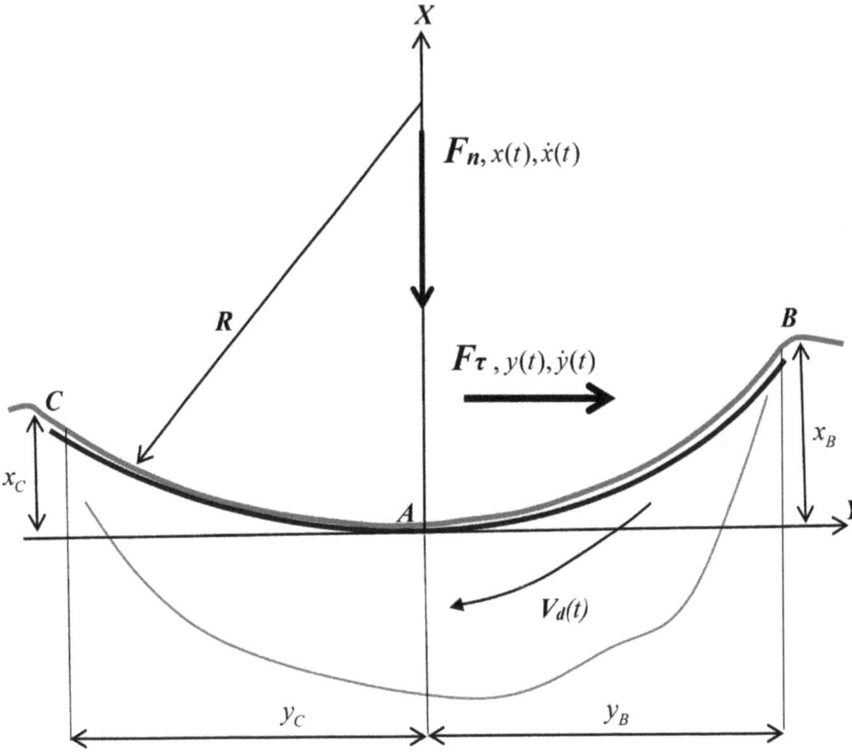

FIGURE 1.2 Illustration of a sliding or rolling contact between the surface of the rigid body and a semi-space.

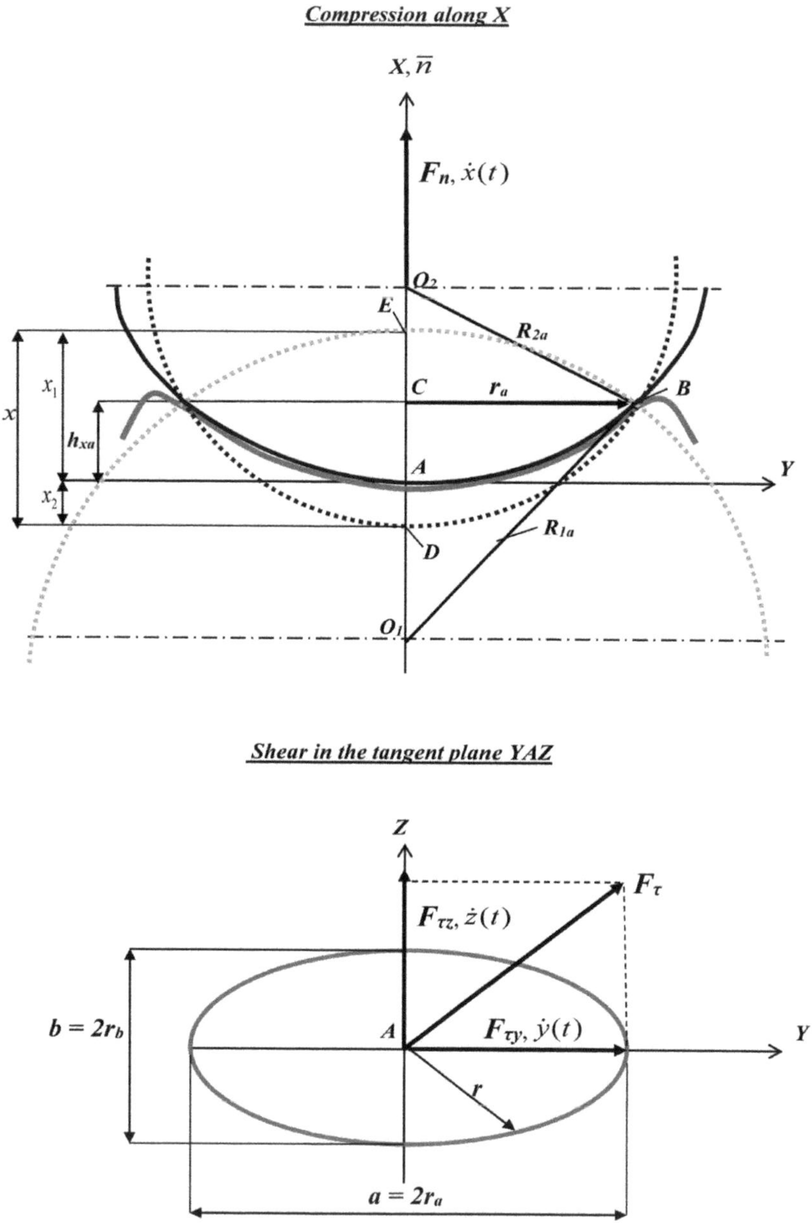

FIGURE 2.1 Schematic illustration of the mutual approach between two curvilinear surfaces of two solid bodies along the axis X relative to the initial point of contact A.

Source: Source: Reprinted from Goloshchapov, 2015b, http://dx.doi.org/10.5539/mer.v5n2p59 https://creativecommons.org/licenses/by/3.0/.

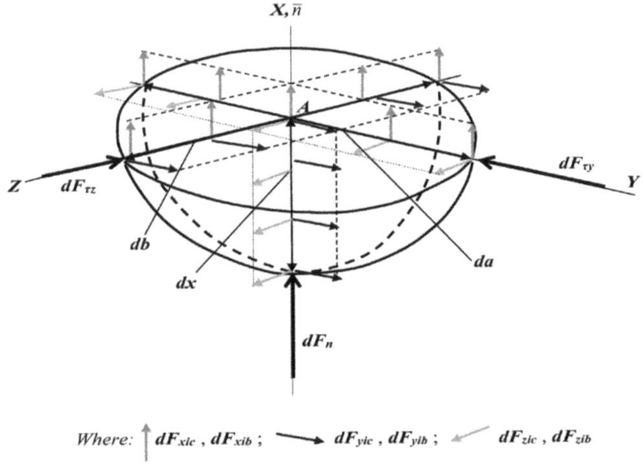

FIGURE 2.2 Illustration of the action of the infinitesimal differential specific elastic and viscous forces inside the infinitesimal volume dV in the vicinity of point A.

Source: Adapted from Goloshchapov, 2015, with permission from SAGE Publications. http://journals.sagepub.com/doi/abs/10.1177/1056789514560912).

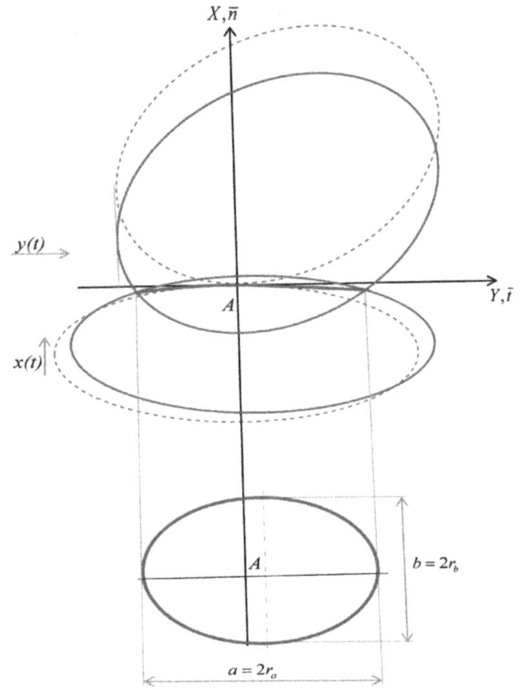

FIGURE 2.6 Illustration of an arbitrary contact between two ellipsoids.

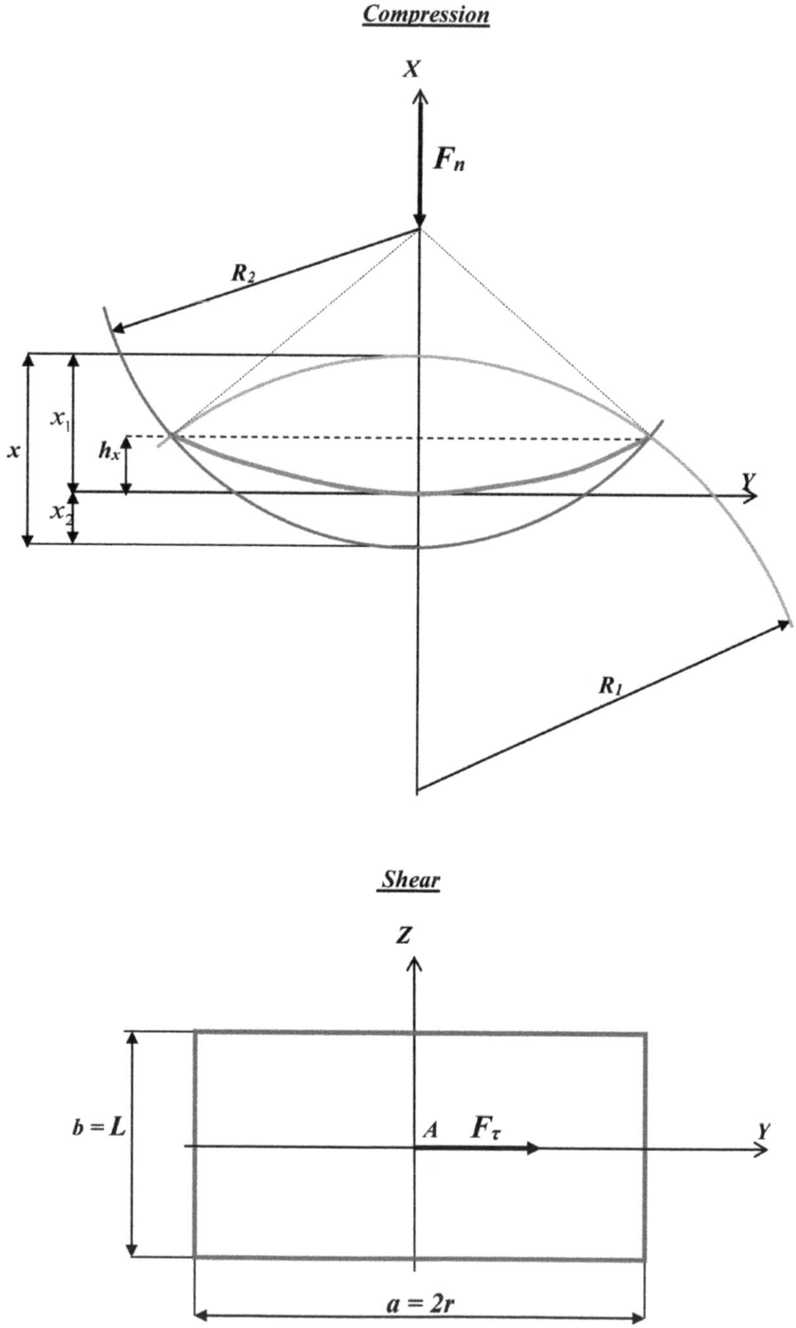

FIGURE 2.7 Illustration of the contact between two cylinders with parallel axes.

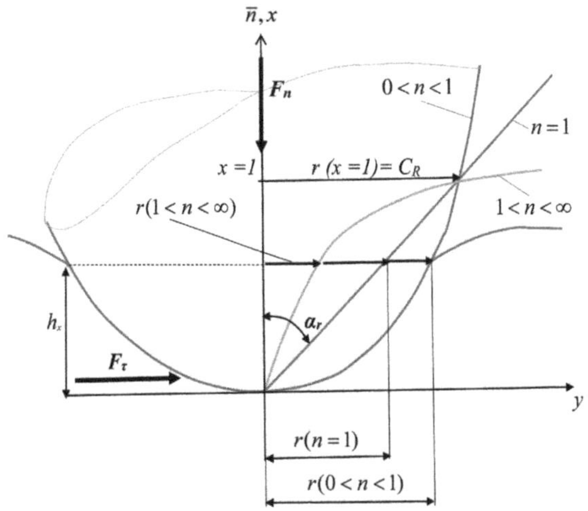

FIGURE 2.8 Illustration of the contact between an axis-symmetrical curvilinear surface of indenter or asperity and the flat surface of a semi-space.

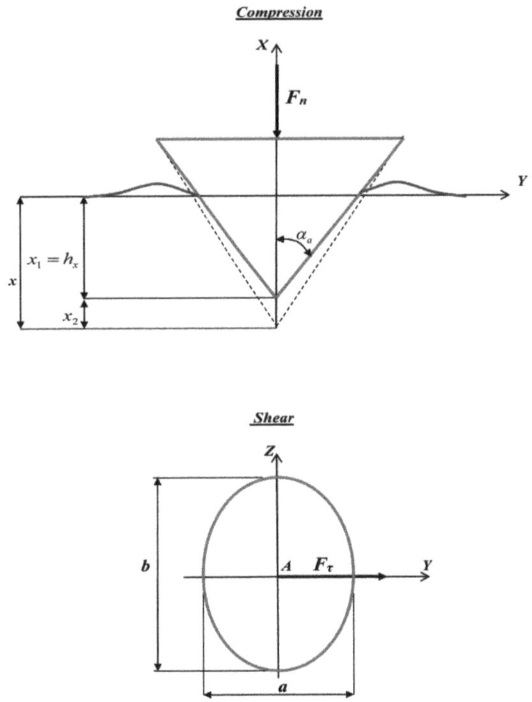

FIGURE 2.9 Illustration of the contact between an elliptical cone and the flat surface of a semi-space.

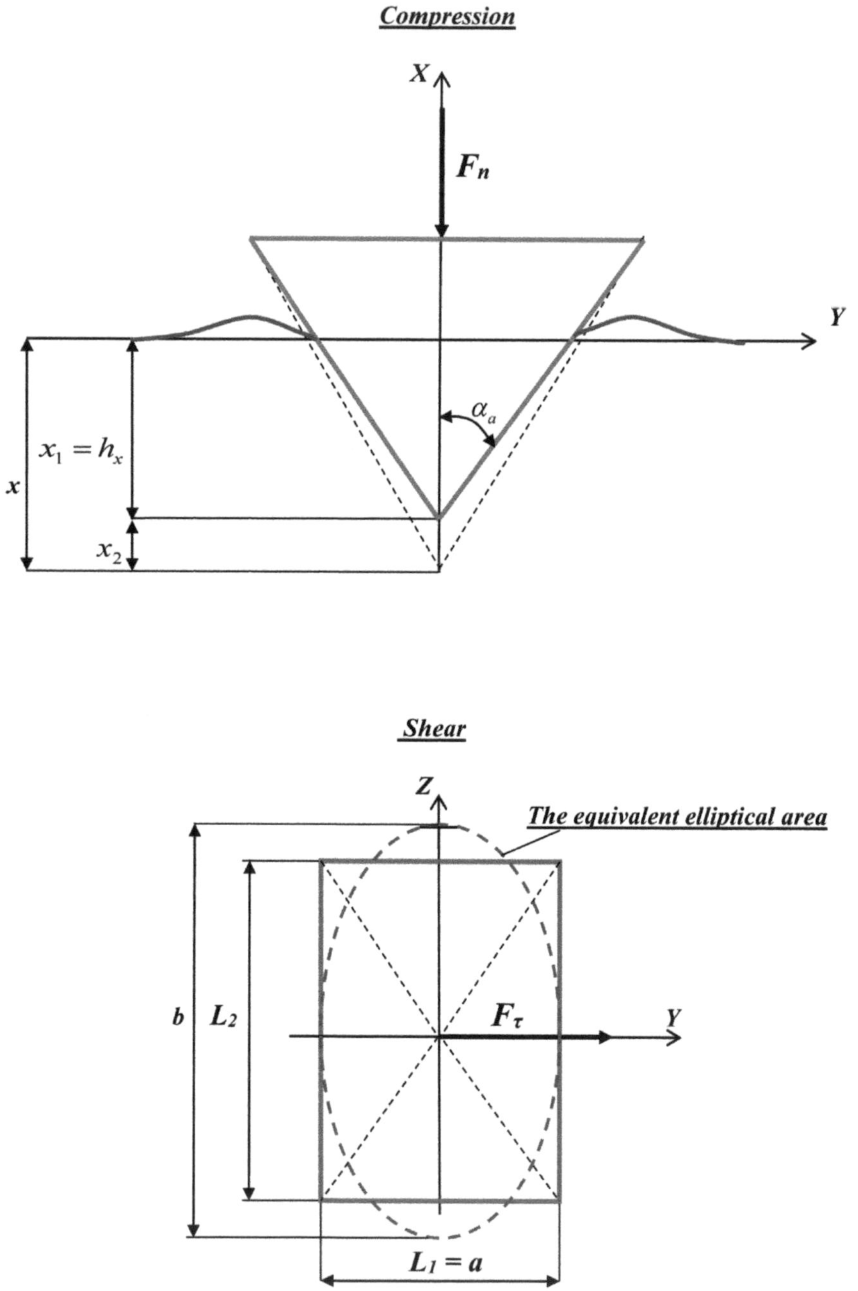

FIGURE 2.10 Illustrations of the contact between a pyramid and the flat surface of a semi-space.

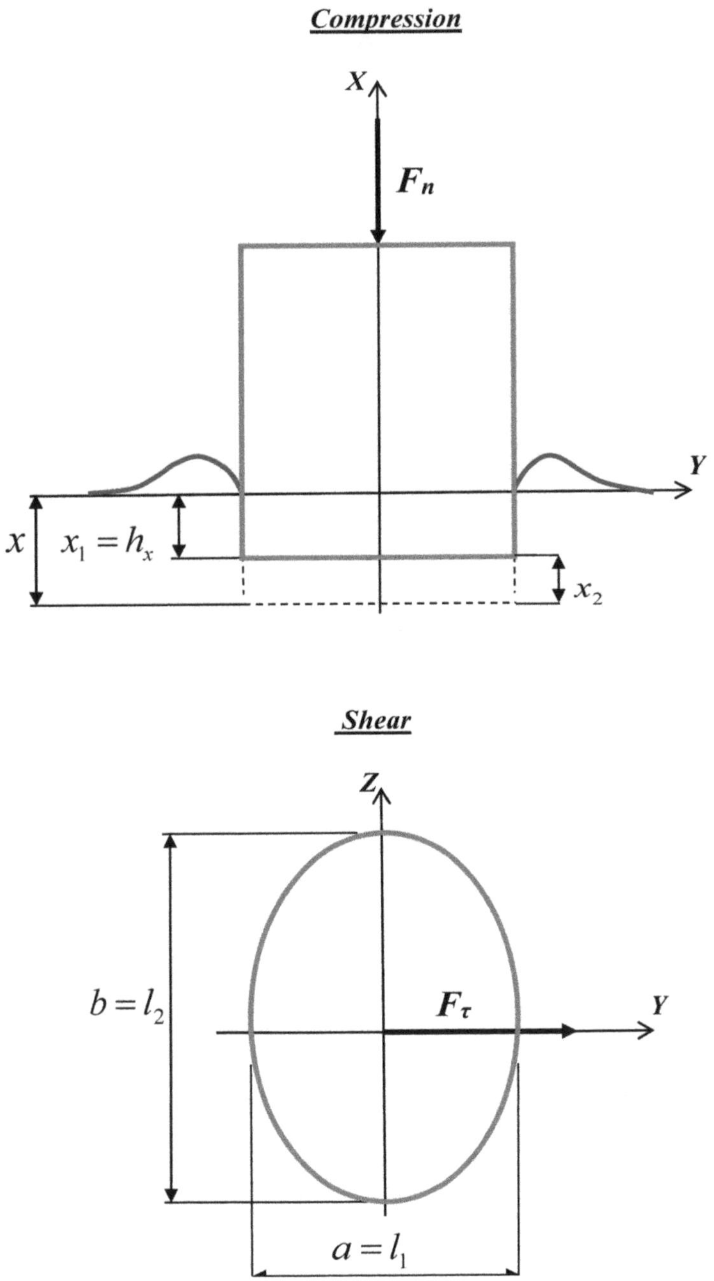

FIGURE 2.11 Illustrations of the contact between an elliptical cylinder and the flat surface of a semi-space.

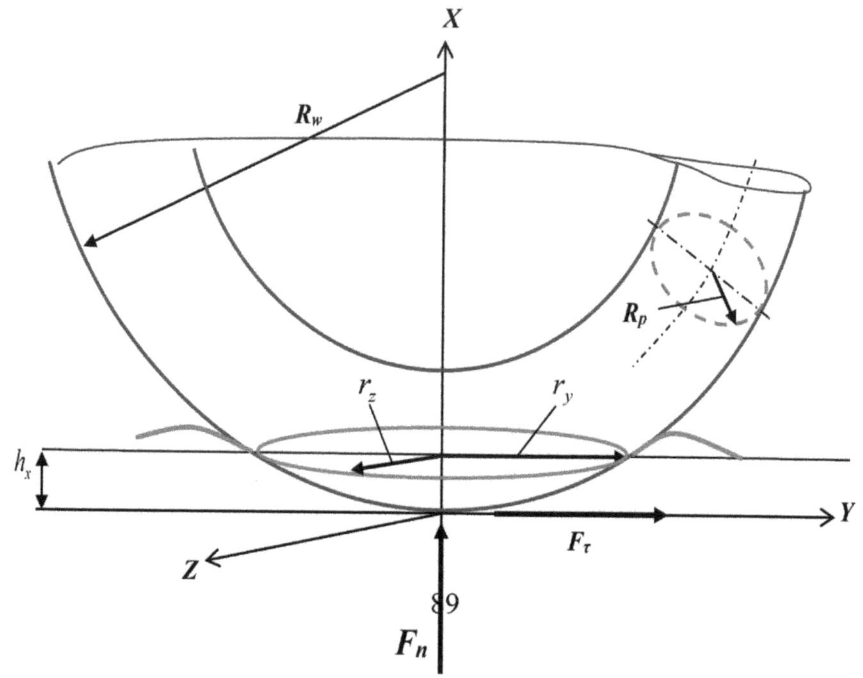

FIGURE 2.12 Illustration of the contact a toroidal surface of a wheel and the flat surface of a semi-space.

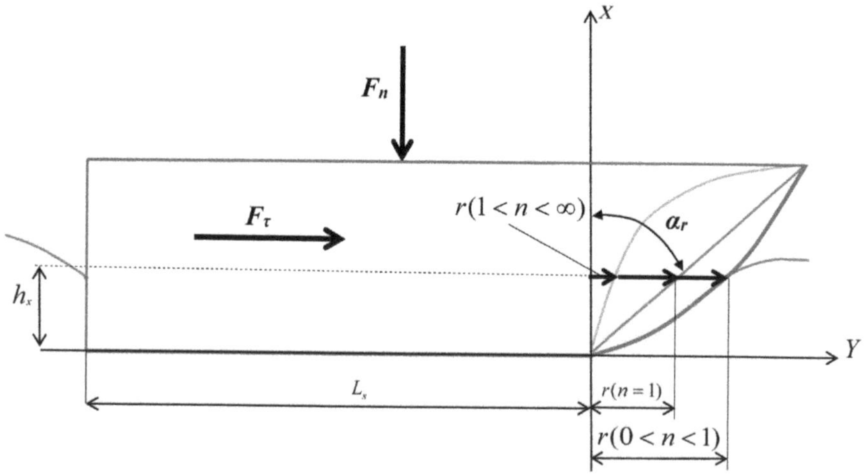

FIGURE 2.13 Illustration of the contact between the surface of sled and the flat surface of a semi-space.

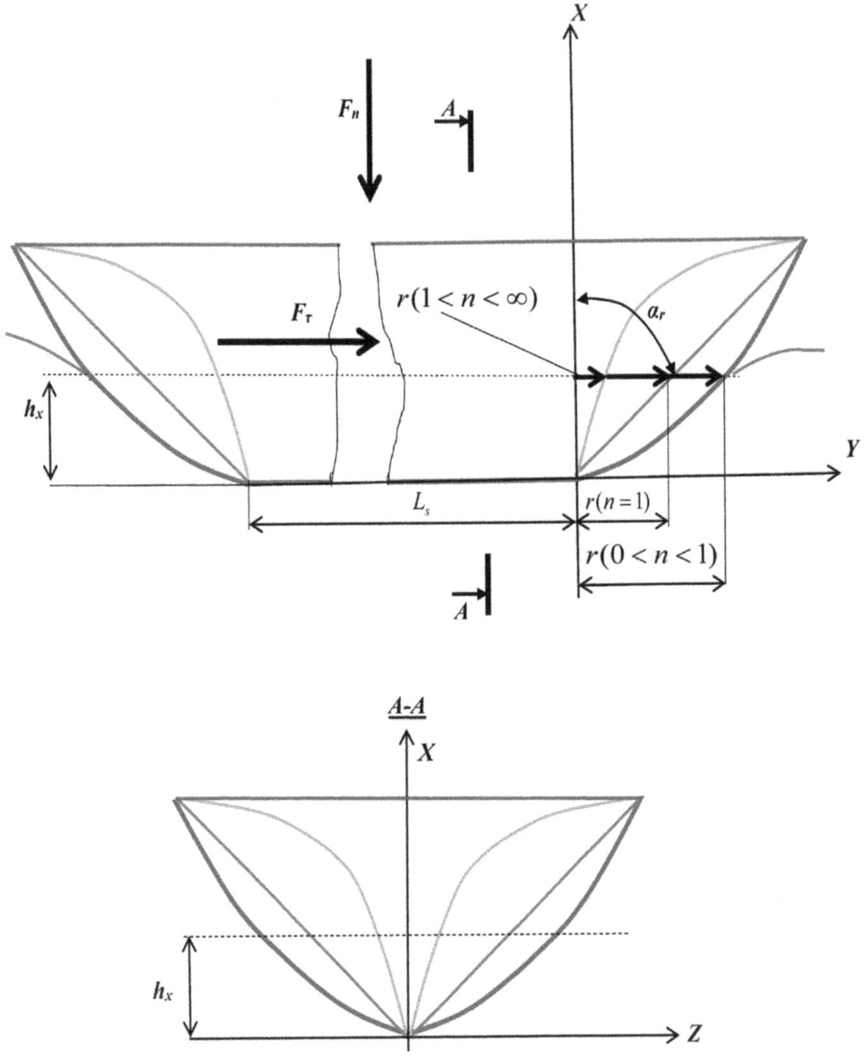

FIGURE 2.14 Illustrations of the contact between a curvilinear surface of a body of vessel and the flat surface of a semi-space.

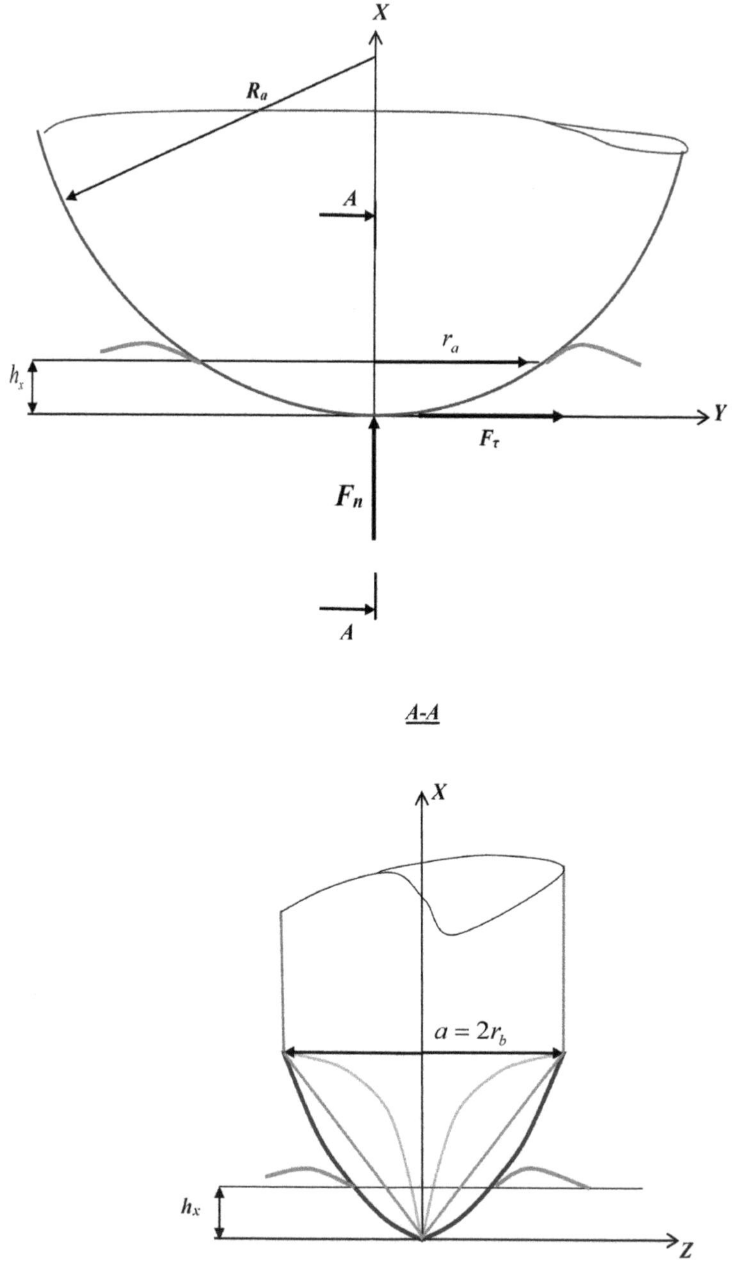

FIGURE 2.15 Illustrations of the contact between an arbitrary shape of disk and the flat surface of a semi-space.

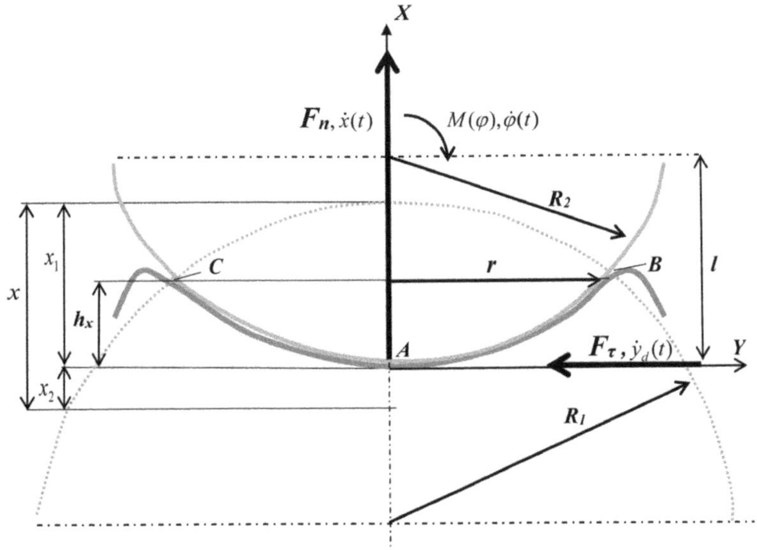

FIGURE 3.1 Schematic illustration of the contact at impact between two spherical solid bodies.

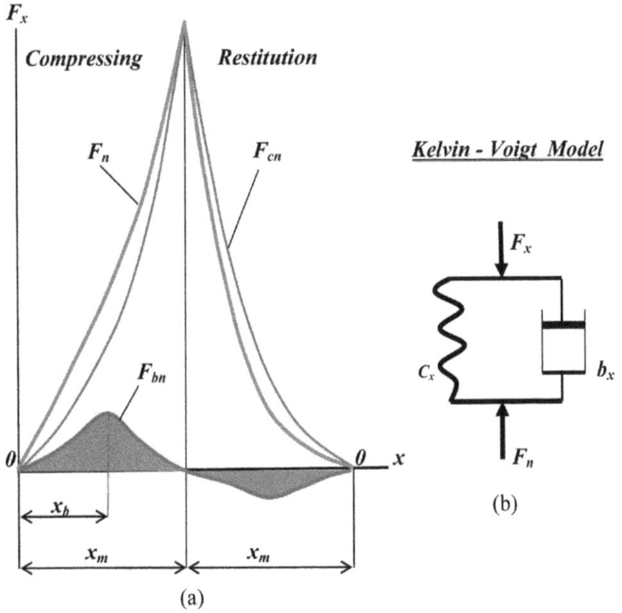

FIGURE 3.2 (a) The graphical illustration of the functional dependences between the normal viscoelastic forces and the displacement *x(t)* of the centre of mass of a body, (b) The 'nonlinear rheological model of Kelvin–Voigt', where c_x and b_x magnitudes are not constant.

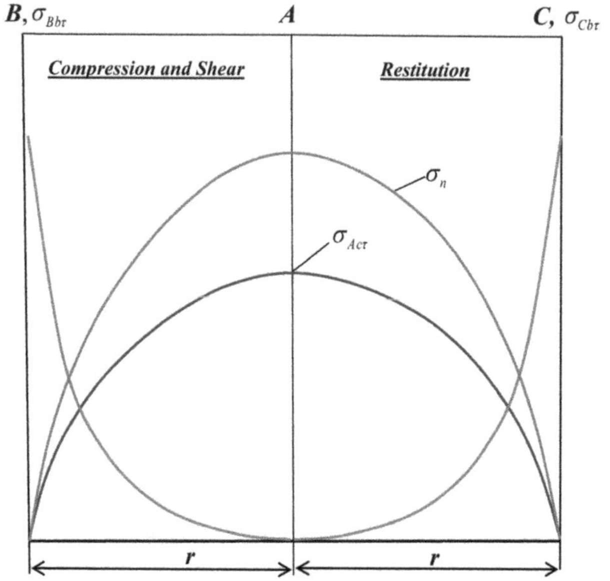

FIGURE 3.3 Distribution of the normal and the tangential stresses in the area of the contact.

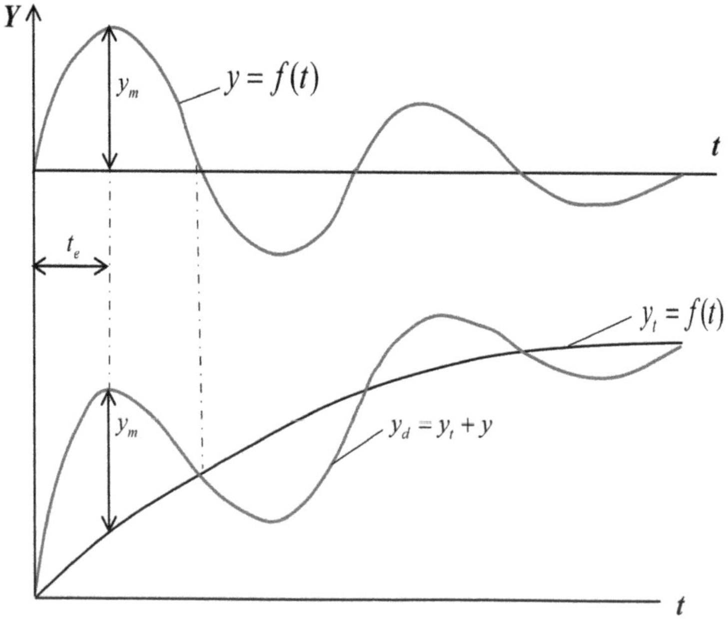

FIGURE 3.4 The graphical illustrations of a free damped oscillation, graphs $y=f(t)$, $y_t=f(t)$ and $y_d=f(t)$, at a sliding or rolling impact under initial conditions $t=0$, $\dot{y}=V_y$, $y=0$.

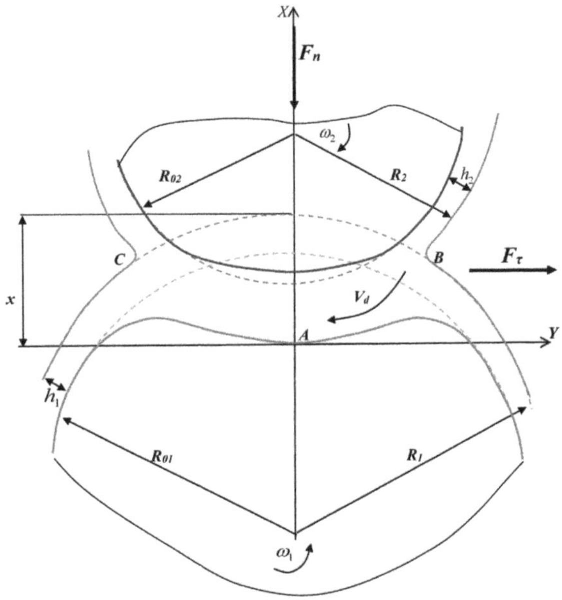

FIGURE 3.5 Schematic illustration of the contact between two lubricated curvilinear surfaces of solids.

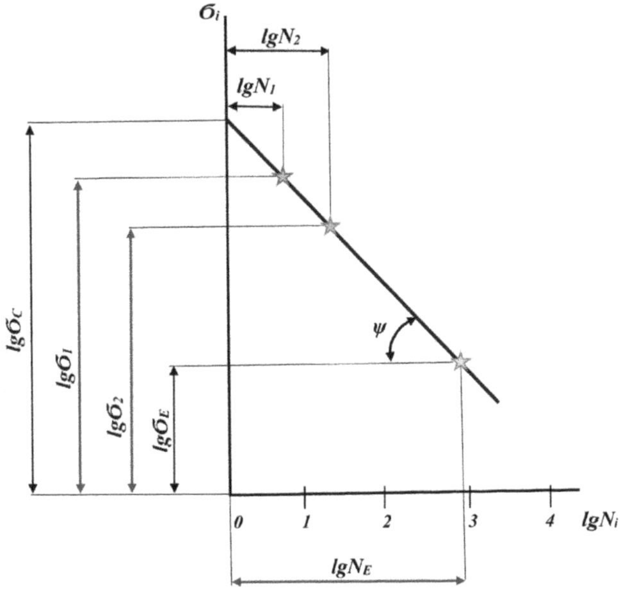

FIGURE 5.1 Illustration of the stress-number of cycles (S-N) Voeller diagram of the high acceleration fatigue life test for the definition of the exponent λ of the curve of a fatigue life.

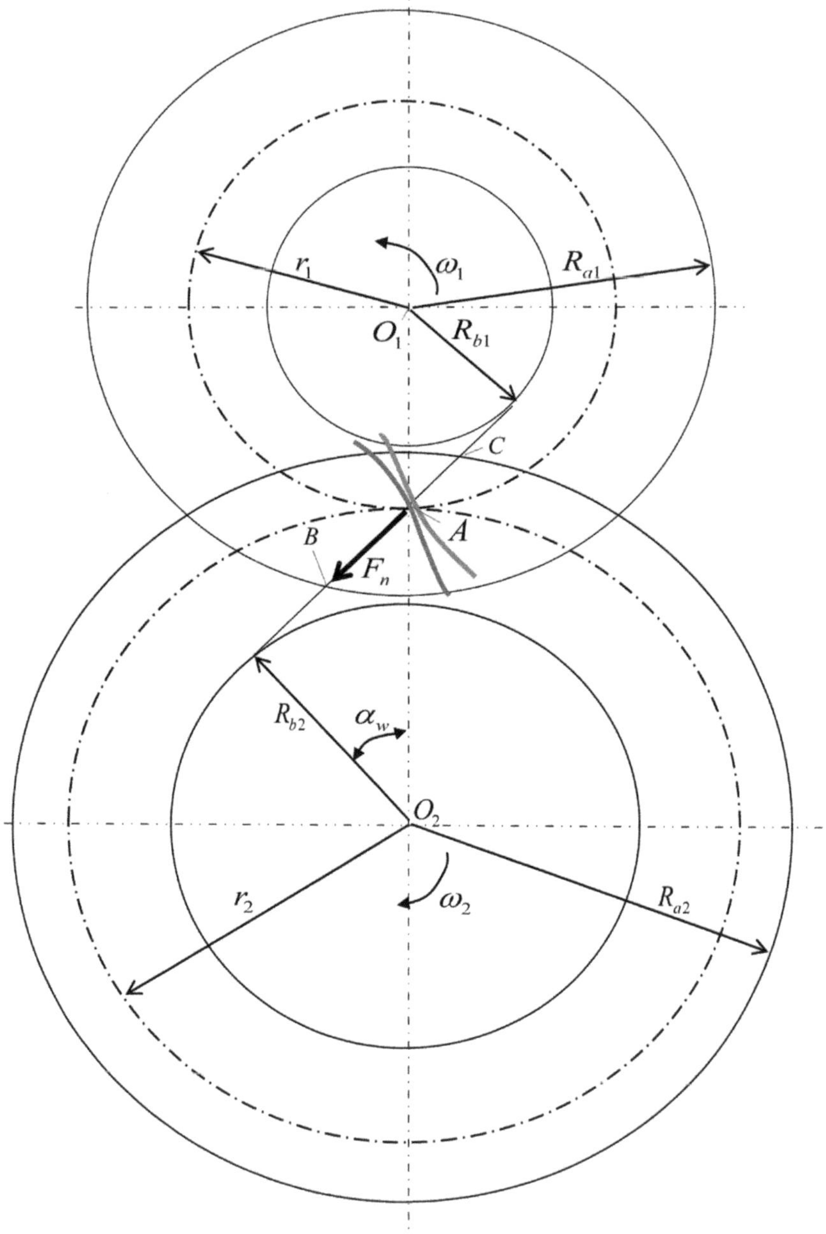

FIGURE 5.2 Schematic illustration of the contact at impact between involute surfaces of two teeth of two gears.

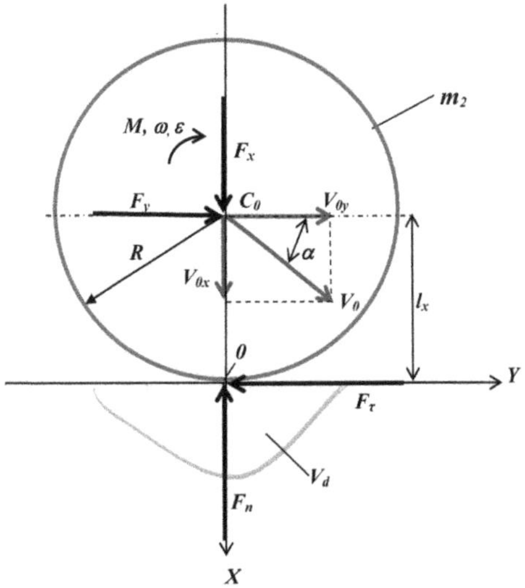

FIGURE 5.3 Schematic illustration of the mechanics of a viscoelastic contact between a spherical solid body and a semi-space at impact.

Source: Reprinted from Goloshchapov, 2015, with permission from SAGE Publications. http://journals.sagepub.com/doi/abs/10.1177/1056789514560912).

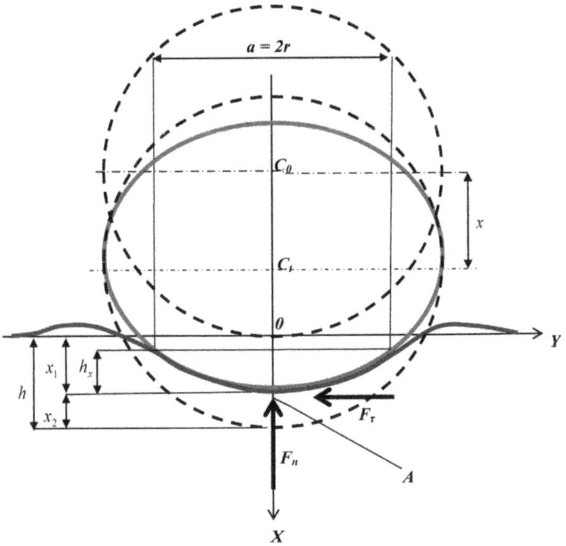

FIGURE 5.4 The schematic illustration of the indentation of a spherical body into a semi-space.

Source: Reprinted from Goloshchapov, 2015a, with permission from SAGE Publications. http://journals.sagepub.com/doi/abs/10.1177/1056789514560912).

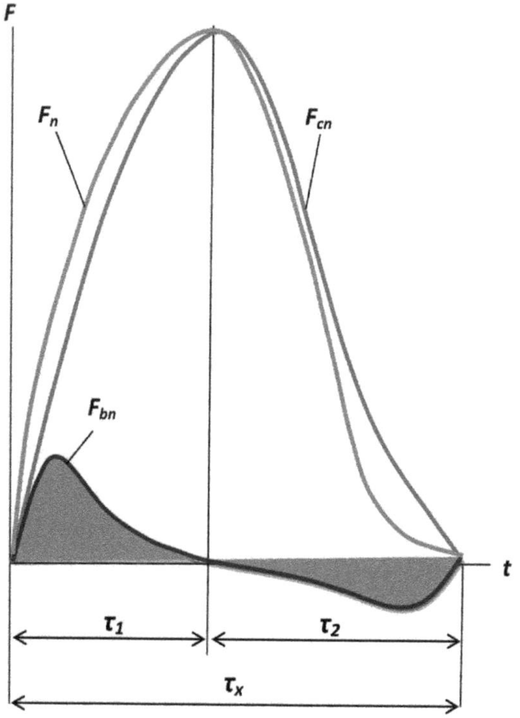

FIGURE 5.6 Graphical illustration of the results and the forms of curves obtained theoretically for viscoelastic forces.

CHAPTER 4

METHOD OF THE DIFFERENTIAL SPECIFIC FORCES IN DYNAMICS OF ELASTOPLASTIC CONTACTS

ABSTRACT

The derivation of the basic principles of the 'method of the differential specific forces' in dynamics of elastoplastic contacts between smooth curvilinear surfaces of two solid bodies have been proposed in this chapter, and then the method of definition of the elastic and the viscous forces have been developed. The equations for work and energy for general viscous and elastic forces in the phases of compression and restitution, at an elastoplastic sliding and at the rolling shear have been obtained. The examples of solutions of elastoplastic contact problems between two smooth surfaces of solids have been examined here. For different cases of dynamics of elastoplastic collisions such as in cases of collision between two spherical bodies, between a sled and a semi-space, between a vessel and a semi-space, between a cylindrical and a toroidal wheel and a semi-space, the differential equations of displacement at the compression, work and energy in the phases of compression and restitution, and the approximate solution to the differential equations for the normal and tangential displacements by using the method of the equivalent works have been discussed here. Also, techniques of finding the normal and tangential viscoelastic constants in different cases of elastoplastic indentations have been given in this chapter.

4.1 THE METHOD OF THE DIFFERENTIAL SPECIFIC FORCES FOR AN ELASTOPLASTIC CONTACT

4.1.1 SPECIFIC FORCES, DYNAMIC ELASTICITY MODULI AND DYNAMIC VISCOSITIES

The illustration of the mutual approach during elastoplastic contact between two smooth curvilinear surfaces of two solid bodies along axes X, Y and Z

related to the initial point of contact A is similar to already been considered for a viscoelastic contact, as it is depicted in Figure 2.1.

As we already know that in the infinitesimal period of the time dt, when the mutual approach between surfaces of two bodies is the infinitesimal magnitude dx, inside the elementary infinitesimal volume dV, which is surfacing around the vicinity of the initial point of the contact A (see Figs. 2.1 and 2.2), the differential specific elastic forces dF_{xic}, dF_{yic} and dF_{zic} and the differential specific viscous forces dF_{yib}, and dF_{zib} are beginning to act independently, and they are distributed on the infinitesimal sizes da, db and dx parallel to axes X, Y and Z.

But, in the case of *the elastoplastic contact* according to the *Maxwell model*, see Figure 4.1, when the specific elastic forces and viscous forces act along one line and they are equal to each other and we can write the following:

$$F_x = F_{xb} = F_{xc} = F_{x1} = F_{x1b} = F_{x1c} = F_{x2} = F_{x2b} = F_{x2c} \tag{4.1}$$

$$F_y = F_{yb} = F_{yc} = F_{y1} = F_{y1b} = F_{y1c} = F_{y2} = F_{y2b} = F_{y2c} \tag{4.2}$$

$$F_z = F_{zb} = F_{zc} = F_{z1} = F_{z1b} = F_{z1c} = F_{z2} = F_{z2b} = F_{z2c} \tag{4.3}$$

These specific forces can be found as well by integration of the series of Equations 2.106–2.117, by x_c, y_c, z_c, and x_b, $\dot{x}_b$, y_b, $\dot{y}_b$, z_b, $\dot{z}_b$, where x_c is an elastic deformation of the compression between contacting surfaces, y_c, z_c, are elastic deformations at the shear between contacting surfaces, x_b, is a viscous deformation of the compression between contacting surfaces, y_b, z_b, are viscous deformations at the shear between contacting surfaces, $\dot{x}_b$, is the velocity of a viscous deformation of the compression between contacting surfaces, $\dot{y}_b$, $\dot{z}_b$ are the velocity of viscous deformations at the shear between contacting surfaces. Or on the other hand, at the same time, x_c, y_c, z_c are the elastic displacements between the centres of mass of the contacting bodies, x_b, y_b, z_b, are the plastic displacements between the centres of mass of the contacting bodies, $\dot{y}_b$, $\dot{x}_b$, $\dot{z}_b$ are the relative velocities of the plastic displacements between the centres of mass of the bodies.

Since for a characterization of an elastoplastic contact between solid bodies we use the Maxwell model, we have to take in mind that according to this model, the dynamic elasticity moduli are equal to the Young's elasticity moduli E, E_1, E_2 and G, G_1, G_2. And also, on the other hand, we have to understand that in the case of using of the Maxwell Model, the dynamic Young's elasticity moduli E, E_1, E_2 and G, G_1, G_2 are not equal to the storage moduli E', E'_1, E'_2 and G', G'_1, G'_2 (see Section 4.2).

Let the elementary deformation between two bodies develop analogically like the deformation of the discrete element, which is depicted in Figure 4.1a. It is a simple case of the linear model of deformations of discrete elements, and instead of this model with four elements we can use its analogy—the model with two effective elements depicted in Figure 4.1b. Also, the 'elementary discrete elements model' for the normal forces can be used for the tangential forces in the same manner.

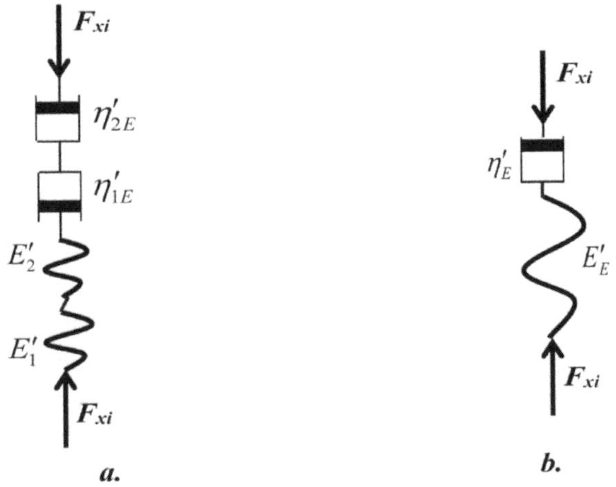

FIGURE 4.1 Illustration of the 'elementary discrete elements Maxwell model (EDEM)': (a) the elementary discrete element of the normal contact between two bodies; (b) the effective elementary discrete element of the normal contact.

In the case of a linear Maxwell model of elastoplastic deformation, for the determination of the effective viscosities and the effective elasticity moduli, we can summarize the elastic and viscous compliances as shown below:

$$\frac{1}{E} = \frac{1}{E_1} + \frac{1}{E_2} \tag{4.4}$$

$$\frac{1}{\eta'_E} = \frac{1}{\eta'_{1E}} + \frac{1}{\eta'_{2E}} \tag{4.5}$$

$$\frac{1}{G} = \frac{1}{G_1} + \frac{1}{G_2} \tag{4.6}$$

$$\frac{1}{\eta'_G} = \frac{1}{\eta'_{1G}} + \frac{1}{\eta'_{2G}} \tag{4.7}$$

Finally, the expressions for a calculation of the effective viscosities and the effective modulus of elasticity can be written as:

$$E = \frac{E_1 E_2}{E_1 + E_2} \tag{4.8}$$

$$\eta'_E = \frac{\eta'_{1E}\eta'_{2E}}{\eta'_{1E} + \eta'_{2E}} \tag{4.9}$$

$$E = \frac{E_1 E_2}{E_1 + E_2} \tag{4.10}$$

$$\eta'_G = \frac{\eta'_{1G}\eta'_{2G}}{\eta'_{1G} + \eta'_{2G}} \tag{4.11}$$

where respectively in the case of elastoplastic contact (Maxwell model): E_1 is the dynamic Young's elasticity modulus at the compression of the surface of a softer body; η'_{1E} is the dynamic viscosity at the compression of the surface of a softer body; G_1 is the dynamic Young's elasticity modulus at the shear of the surface of a softer body; η'_{1G} is the viscosity at the shear of the surface of a softer body; E_2 is the elasticity Young's modulus at the compression of the surface of a harder body; η'_{2E} is the viscosity at the compression of the surface of a harder body; G_2 is the elasticity Young's modulus at the shear of the surface of a harder body; η'_{2E} is the viscosity at the shear of the surface of a harder body; E is the effective dynamic Young's elasticity modulus at the compression; η'_E is the effective dynamic viscosity at the compression; G is the effective dynamic Young's elasticity modulus at the shear; η'_G is the effective dynamic viscosity at the shear.

Here and further in this book for characterization of a elastoplastic contact between solids we use the Maxwell model, and; therefore, we have to understand that according to this model the dynamic elasticity moduli are equal to the Young's moduli of elasticity E, E_1, E_2 and G, G_1, G_2. But also, we have to understand that in this case, using the Maxwell model, the dynamic Young's elasticity moduli E, E_1, E_2 and G, G_1, G_2 are not equal to the storage moduli E', E'_1, E'_2 and G', G'_1, G'_2

Also it is obvious for the Maxwell model that

$$F_{xc} = F_{x1c} = F_{x2c} \tag{4.12}$$

$$F_{xb} = F_{x1b} = F_{x2b} \tag{4.13}$$

$$F_{yc} = F_{y1c} = F_{y2c} \tag{4.14}$$

$$F_{yb} = F_{y1b} = F_{y2b} \tag{4.15}$$

$$F_{zc} = F_{z1c} = F_{z2c} \tag{4.16}$$

$$F_{zb} = F_{z1b} = F_{z2b} \tag{4.17}$$

Thus, consequently we can write:

$$F_{xc} = Ex_c = F_{x1c} = E_1x_{1c} = F_{x2c} = E_2x_{2c} \tag{4.18}$$

$$F_{xb} = \eta'_E\dot{x}_b = F_{x1b} = \eta'_{1E}\dot{x}_{1b} = F_{x2b} = \eta'_{2E}\dot{x}_{2b} \tag{4.19}$$

$$F_{yc} = Gy_c = F_{y1c} = G_1y_{1c} = F_{y2c} = G_2y_{2c} \tag{4.20}$$

$$F_{yb} = \eta'_G\dot{y}_b = F_{y1b} = \eta'_{1G}\dot{y}_{1b} = F_{y2b} = \eta'_{2G}\dot{y}_{2b} \tag{4.21}$$

$$F_{zc} = Gz_c = F_{z1c} = G_1z_{1c} = F_{z2c} = G_2z_{2c} \tag{4.22}$$

$$F_{zb} = \eta'_G\dot{z}_b = F_{z1b} = \eta'_{1G}\dot{z}_{1b} = F_{y2b} = \eta'_{2G}\dot{z}_{2b} \tag{4.23}$$

where x_{1c} is an elastic deformation of the compression of a softer surface; y_{1c} z_{1c} are elastic deformations at the shear of a softer surface; x_{1b} is a viscous deformation of the compression of a softer surface; y_{1b}, z_{1b} are viscous deformations at the shear of a softer surface; $\dot{x}_{1b}$ is the velocity of a viscous deformation of the compression of a softer surface; $\dot{y}_{1b}$, $\dot{z}_{1b}$ are the velocity of viscous deformations at the shear of a softer surface, and where x_{2c} is an elastic deformation of the compression of a harder surface; y_{2c}, z_{2c} are elastic deformations at the shear of a harder surface; x_{2b} is a viscous deformation of the compression of a harder surface; y_{2b}, z_{2b} are viscous deformations at the shear of a harder surface, $\dot{x}_{1b}$ is the velocity of a viscous deformation of the compression of a harder surface, $\dot{y}_{2b}$, $\dot{z}_{2b}$ are the velocity of viscous deformations at the shear of a harder surface.

As $x_c = x_{1c} + x_{2c}$, $y_c = y_{1c} + y_{2c}$, $z_c = z_{1c} + z_{2c}$ according to Equations 4.8, 4.10, 4.18, 4.20 and 4.22, we can write the next series of the equations:

For normal deformations,

$$x_{1c} = D_1x_c \tag{4.24}$$

and

$$x_{2c} = D_2x_c \tag{4.25}$$

where D_1 is the coefficient of the normal deformation of the softer surface, D_2 is the coefficient of the normal deformation of the harder surface, which can be calculated respectively as

$$D_1 = \frac{E_2}{E_1 + E'_2} \tag{4.26}$$

$$D_2 = \frac{E_2}{E_1 + E_2'} \tag{4.27}$$

For tangential deformations the following equations are given,

$$y_{1c} = D_{1\tau} y_c \tag{4.28}$$

$$z_{1c} = D_{1\tau} z_c \tag{4.29}$$

$$y_{2c} = D_{2\tau} . y_c \tag{4.30}$$

$$z_{2c} = D_{2\tau} z_c \tag{4.31}$$

where $D_{1\tau}$ is the coefficient of the tangential elastic deformation of the softer surface, $D_{2\tau}$ is the coefficient of the tangential elastic deformation of the harder surface, which can be calculated, respectively, as

$$D_{1\tau} = \frac{G_2}{G_1 + G_2} \tag{4.32}$$

$$D_{2\tau} = \frac{G_1}{G_1 + G_2} \tag{4.33}$$

4.1.2 DEFINITION OF THE VISCOELASTIC FORCES FOR AN ELASTOPLASTIC CONTACT

As we know, in the case of elastoplastic contact, viscous and elastic deformations are not equal to each other, but the total deformation is equal to their sums $x_c + x_b = x$, $y_c + y_b = y$, $z_c + z_b = z$, and; therefore, an integration of all differential equations in series 2.158–2.169 have to be taken separately for the elastic and viscous forces by x_c, x_b, y_c, y_b, z_c, z_b, $\dot{x}_c$, $\dot{x}_b$, $\dot{y}_c$, $\dot{y}_b$, $\dot{z}_c$, $\dot{z}_b$. Also, in case of Maxwell model, instead of the storage moduli E', E_1', E_2' and G', G_1', G_2', we have to apply the dynamic Young's elasticity moduli E, E_1, E_2 and G, G_1, G_2. Thus, first using differential Equations 2.158 and 2.160 for dF_{acn} and dF_{bcn}, we get for the normal elastic forces as below:

$$F_{acn} = k_{pa} E R_a^{1/2} \int^{x_c} \int dx \frac{dx}{x^{1/2}} = \frac{2}{3} k_{pa} E R_a^{1/2} x_c^{3/2} \tag{4.34}$$

$$F_{bcn} = k_{pb} E R_b^{1/2} \int^{x_c} \int dx \frac{dx}{x^{1/2}} = \frac{2}{3} k_{pb} E R_b^{1/2} x_c^{3/2} \tag{4.35}$$

Since $dF_{cn} = dF_{acn} + F_{bcn}$ follows that

$$F_{cn} = \frac{2}{3} E x_c^{3/2} (k_{pa} R_a^{1/2} + k_{pb} R_b^{1/2}) = \frac{2}{3} E \psi x_c^{3/2} \tag{4.36}$$

The equations for the normal viscous forces can be found in the same way by an integration of Equations 2.159 and 2.161 by x_b and $\dot{x}_b$, respectively, as:

$$F_{abn} = k_{pa} \eta_E' \int^{\dot{x}_b} d\dot{x} \int^{x_b} \frac{R_a^{1/2}}{x^{1/2}} dx = 2 k_{pa} \eta_E' R_a^{1/2} \dot{x}_b x_b^{1/2} \tag{4.37}$$

$$F_{bbn} = k_{pb} \eta_E' \int^{\dot{x}_b} d\dot{x} \int^{x_b} \frac{R_b^{1/2}}{x^{1/2}} dx = 2 k_{pb} \eta_E' R_b^{1/2} \dot{x}_b x_b^{1/2} \tag{4.38}$$

And since $dF_{bn} = dF_{abn} + F_{bbn}$ follows that

$$F_{bn} = 2 \eta_E' \dot{x}_b x_b^{1/2} (k_{pa} R_a^{1/2} + k_{pb} R_b^{1/2}) = 2 \eta_E' \psi \dot{x}_b x_b^{1/2} \tag{4.39}$$

where $\psi = (k_{pa} R_a^{1/2} + k_{pb} R_b^{1/2})$, ($m^{1/2}$) is the parameter of curvature.

Since x and y are linearly independent and $h_{ax} = k_{ha} x$, $h_{xb} = k_{hb} x$, after an integration of the equations for the effective tangential forces from series of Equations 2.162–2.169 their solutions can be written as follows:

$$F_{bcy} = k_{pb} R_b^{1/2} G \int^{y_c} dy \int \frac{dx}{x^{1/2}} = 2 k_{pb} G R_b^{1/2} x^{1/2} y_c \tag{4.40}$$

$$F_{bby} = k_{pb} R_b^{1/2} \eta_G' \int^{\dot{y}_b} d\dot{y} \int \frac{dx}{x^{1/2}} = 2 k_{pb} \eta_G' R_b^{1/2} x^{1/2} \dot{y}_b \tag{4.41}$$

$$F_{xcy} = G \int^{y_c} dy \int_0^{h_{xa}} dx = k_{ha} G x y_c \tag{4.42}$$

$$F_{xby} = \eta_G' \int^{\dot{y}_b} d\dot{y} \int^{h_{xa}} dx = k_{ha} \eta_G' x \dot{y}_b \tag{4.43}$$

$$F_{acz} = k_{pa} R_a^{1/2} G \int^{z_c} dz \int \frac{dx}{x^{1/2}} = 2 k_{pa} G R_a^{1/2} x^{1/2} z_c \tag{4.44}$$

$$F_{abz} = k_{pa} R_a^{1/2} \eta_G' \int^{\dot{z}_b} d\dot{z} \int \frac{dx}{x^{1/2}} = 2 k_{pa} \eta_G' R_a^{1/2} x^{1/2} \dot{z}_b \tag{4.45}$$

$$F_{xcz} = G \int^{z} dz \int_0^{h_{xb}} dx = k_{hb} G x z_c \tag{4.46}$$

$$F_{xbz} = \eta'_G \int^{\dot{z}_b} d\dot{z} \int_0^{h_{xb}} dx = \eta'_G x \dot{z}_b \tag{4.47}$$

The equations for the tangential elastic and viscous forces can be written as the sum of the elastic and viscous tangential forces from Equations 4.40–4.47 as:

$$F_{c\tau y} = F_{xcy} + F_{bcy} = GP_{bx} y_c \tag{4.48}$$

$$F_{b\tau y} = F_{xby} + F_{bby} = \eta'_G P_{bx} \dot{y}_b \tag{4.49}$$

$$F_{c\tau z} = F_{xcz} + F_{acz} = GP_{ax} z_c \tag{4.50}$$

$$F_{b\tau z} = F_{xbz} + F_{abz} = \eta'_G P_{ax} \dot{z}_b \tag{4.51}$$

where

$$P_{bx} = k_{ha} x + 2k_{pb} R_b^{1/2} x^{1/2} \tag{4.52}$$

$$P_{ax} = k_{hb} x + 2k_{pa} R_a^{1/2} x^{1/2} \tag{4.53}$$

and where

$$k_{ha} = \left(\frac{R_a - D_2 R_{2a}}{R_{2a}} \right) \tag{4.54}$$

$$k_{hb} = \left(\frac{R_b - D_2 R_{2b}}{R_{2b}} \right) \tag{4.55}$$

Thus, finally, next relations between general viscous and elastic forces can be written as:

$$F_n = F_{bn} = F_{cn} = 2\eta'_E \psi \dot{x}_b x_b^{1/2} = \frac{2}{3} E \psi x_c^{3/2} \tag{4.56}$$

$$F_{\tau y} = F_{b\tau y} = F_{c\tau y} = \eta'_G P_{bx} \dot{y}_b = GP_{bx} y_c \tag{4.57}$$

$$F_{\tau z} = F_{b\tau z} = F_{c\tau z} = \eta'_G P_{ax} \dot{z}_b = GP_{ax} z_c \tag{4.58}$$

4.2 MAIN CONSTITUTIVE RELATIONS IN THE MAXWELL MODEL

Now, it is the time to find some constitutive relations in the Maxwell model. It is known that for this model $tg\beta_E = \frac{E''}{E'}$ and $tg\beta_G = \frac{G''}{G'}$ (Lee, 1962; Flügge, 1975; McCrum, et al. 1997; Lakes, 1998; Menard, 1999; Hosford, 2005; Roylance, 2001). Also, it is known that in the Maxwell model

$$\tau_E = \frac{\eta'_E}{E} \tag{4.59}$$

$$\tau_G = \frac{\eta'_G}{G} \tag{4.60}$$

where τ_E and τ_G are the times of relaxation. But, on the other hand, since in the Maxwell model $E(t) = E\exp(-t/\tau_E)$, $G(t) = G\exp(-t/\tau_G)$, and also since

$$E'(\omega_x) = E\frac{\omega_x^2\tau_E^2}{1+\omega_x^2\tau_E^2} \tag{4.61}$$

$$G'(\omega_y) = G\frac{\omega_y^2\tau_G^2}{1+\omega_y^2\tau_G^2} \tag{4.62}$$

$$E''(\omega_x) = E\frac{\omega_x\tau_E}{1+\omega_x^2\tau_E^2} \tag{4.63}$$

$$G''(\omega_y) = G\frac{\omega_y\tau_G}{1+\omega_y^2\tau_G^2} \tag{4.64}$$

It follows that

$$tg\beta_E = \frac{1}{\omega_x\tau_E} \tag{4.65}$$

$$tg\beta_G = \frac{1}{\omega_y\tau_G} \tag{4.66}$$

and

$$\frac{tg\beta_E}{tg\beta_G} = \frac{\omega_y\tau_G}{\omega_x\tau_E} \tag{4.67}$$

Let us remember that ω_y is the tangential angular frequency of oscillations (frequency of loading) by axes Y and Z.

And also, since in the Maxwell model $\tau_E = \frac{\eta'_E}{E}$ and $\tau_G = \frac{\eta'_G}{G}$, taking into account Equations 4.65 and 4.66 we can rewrite Equations 4.61–4.64 as follows:

$$E'(\omega_x) = \eta'_E\frac{\omega_x tg\beta_E}{1+tg^2\beta_E} \tag{4.68}$$

$$G'(\omega_y) = \eta'_G\frac{\omega_y tg\beta_G}{1+tg^2\beta_G} \tag{4.69}$$

$$E''(\omega_x) = \eta'_E \frac{\omega_x tg^2 \beta_E}{1 + tg^2 \beta_E} \tag{4.70}$$

$$G''(\omega_y) = \eta'_G \frac{\omega_y tg^2 \beta_G}{1 + tg^2 \beta_G} \tag{4.71}$$

And also, taking into account Equations 4.59, 4.60, 4.65 and 4.66, we get the following

$$\eta'_E = \frac{E}{\omega_x tg \beta_E} \tag{4.72}$$

$$\eta'_G = \frac{G}{\omega_x tg \beta_G} \tag{4.73}$$

Also, since $\frac{tg^2 \beta}{1 + tg^2 \beta} = \sin^2 \beta$, we can write the next important attitudes between the dynamic viscosities and the dynamic viscosity modules in the Maxwell model, respectively as:

$$E''(\omega_x) = \eta'_E \omega_x \sin^2 \beta_E \tag{4.74}$$

$$G''(\omega_y) = \eta'_G \omega_y \sin^2 \beta_G \tag{4.75}$$

As we can see, in the case of an elastoplastic contact (Maxwell model), the dependencies between the dynamic Young's elasticity moduli E, E_1, E_2 and G, G_1, G_2 and the expressions for storage moduli E', E'_1, E'_2 and G', G'_1, G'_2 are more complicated than in the case of a viscoelastic contact (Kelvin–Voigt model).

4.3 WORK OF VISCOUS AND ELASTIC FORCES IN AN ELASTOPLASTIC SLIDING CONTACT BETWEEN CURVILINEAR SURFACES OF TWO SOLID BODIES

4.3.1 WORK OF THE NORMAL FORCES

Since the limits of integration in this case are x_c and x_b, the work A_{xcm} of elastic force F_{cn}, see Equation 4.36, in the compression phase can be found as follows

$$A_{xcm} = \int_0^{x_{cm}} F_{cn} dx_c = \frac{2}{3} E'\psi \int_0^{x_{cm}} x_c^{3/2} dx_c = \frac{4}{15} E\psi x_{cm}^{5/2} \tag{4.76}$$

where x_{cm} is the maximum of the elastic displacement between contacting surfaces.

The work A_{xbm} of viscous force F_{bn} at the compression can be found as

$$A_{xbm} = \int_0^{x_{bm}} F_{bn} dx_b = \int_0^{x_{bm}} 2\eta'_E \psi \dot{x}_b x_b^{1/2} dx_b = 2\eta'_E \psi \frac{\int_0^{x_{bm}} \int dx_b x_b^{1/2} dx_b}{\int_0^{\tau_1} dt} = \frac{4\eta'_E \psi x_{bm}^{5/2}}{5\tau_1} \quad (4.77)$$

where x_{bm} is the plastic displacement between contacting surfaces in the moment of the time $t = \tau_1$, when $x_c = x_{cm}$.

Analogically, the works A_{xct} and A_{xbt} in the restitution phase, when $x_c = x_{ct} = 0$, can be found as follows

$$A_{xct} = -\int_{x_{cm}}^{0} F_{cn} dx_c = -\int_{x_{cm}}^{0} \frac{2}{3} E\psi x_c^{3/2} dx_c = \frac{4}{15} E\psi x_{cm}^{5/2} \quad (4.78)$$

and

$$A_{xbt} = \int_{x_{bm}}^{x_{bt}} F_{bn} dx = \int_{x_{bm}}^{x_{bt}} 2\eta'_E \psi \dot{x}_b x_b^{1/2} dx_b = 2\eta'_E \psi \frac{\int_{x_{bm}}^{x_{bt}} \int dx_b x_b^{1/2} dx_b}{\int_{\tau_1}^{\tau_x} dt} = \frac{4\eta'_E \psi (x_{bt}^{5/2} - x_{bm}^{5/2})}{5\tau_2} \quad (4.79)$$

where x_{bt} is the plastic displacement between contacting surfaces in the moment of the time of rebound, when $t = \tau_x$ and $x_c = 0$.

Now, since $\eta'_E = E\tau_E$, the equations for the work of the normal viscoelastic forces in the compression and the restitution can be written as follows:

$$A_{xm} = A_{xcm} + A_{xbm} = \frac{4}{15} \psi E \left(x_{cm}^{5/2} + 3\frac{\tau_E}{\tau_1} x_{bm}^{5/2} \right) \quad (4.80)$$

$$A_{xt} = A_{xct} - A_{xbt} = \frac{4}{15} \psi E \left(x_{cm}^{5/2} - \frac{3\tau_E}{\tau_2} (x_{bt}^{5/2} - x_{bm}^{5/2}) \right) \quad (4.81)$$

Also, taking into account Equations 4.56 and 4.59, we get

$$x_c^{3/2} = 3\tau_E \dot{x}_b x_b^{1/2} \quad (4.82)$$

Further, it is obvious that according to the initial conditions $x_c = 0$, $x_b = 0$, $\dot{x}_c = V_{cx} = V_x$, $\dot{x}_b = 0$, and according to the boundary conditions in the moment of the time $t = \tau_1$ of the maximal elastic deformation, when $x_c = x_{cm}$, $x_b = x_{bm}$, and $\dot{x}_c = 0$, $\dot{x}_b = \dot{x}_{bm} = \dot{x}_m$ [where $\dot{x}_m$ is relative velocity of displacements (deformations) between contacting surfaces in the initial point A in the moment of the time $t = \tau_1$] follows that $x_{cm} = \frac{V_x}{2}\tau_1$ and $x_{bm} = \frac{\dot{x}_{bm}}{2}\tau_1$, and then taking into account Equation 4.79 we get

$$x_{bm} = x_{cm}\left(\frac{\tau_1}{6\tau_E}\right)^{2/3} \tag{4.83}$$

and

$$\dot{x}_{bm} = V_x\left(\frac{\tau_1}{6\tau_E}\right)^{2/3} \tag{4.84}$$

But on the other hand, it is obvious that $x_{bt} = x_{btm} + x_{bm}$, and also we can take that $x_{bm} = \frac{\dot{x}_{bm}}{2}\tau_1$ and $x_{btm} = \frac{\dot{x}_{bm}}{2}\tau_2$ (where x_{btm} is the plastic displacement for the time of the elastic restitution τ_2), and since $\tau_1 + \tau_2 = \tau_x$ we get for the maximum of plastic deformation $x_{bt} = x_{bm}\frac{\tau_x}{\tau_1}$, and then according to Equation 4.83, it follows

$$x_{bt} = x_{cm}\frac{\tau_x}{\tau_1}\left(\frac{\tau_1}{6\tau_E}\right)^{2/3} \tag{4.85}$$

Now, if we take into account Equations 4.83 and 4.85, we can rewrite Equations 4.80 and 4.81 as follows

$$A_{xm} = A_{xcm} + A_{xbm} = \frac{4}{15}\psi E x_{cm}^{5/2}\left(1 + \frac{3}{6^{5/3}}\left(\frac{\tau_1}{\tau_E}\right)^{2/3}\right) \tag{4.86}$$

and

$$A_{xt} = A_{xct} - A_{xbt} = \frac{4}{15}\psi E' x_{cm}^{5/2}\left(1 - \frac{3\tau_E}{\tau_2}\left(\frac{\tau_1}{6\tau_E}\right)^{5/3}\left(\left(\frac{\tau_x}{\tau_1}\right)^{5/2} - 1\right)\right) \tag{4.87}$$

It is obvious that the total dissipative energy of plastic deformation equals the full dissipative work, which can be found by integration of the viscous force in moment of the time $t = \tau_x$ and $x_b = x_{bt}$ as

$$A_{xb} = \int_0^{x_{bt}} F_{bn}\,dx_b = \int_0^{x_{bt}} 2\eta_E'\psi\dot{x}_b x_b^{1/2}\,dx_b = 2\eta_E'\psi\frac{\int_0^{x_{bt}}\int dx_b x_b^{1/2}\,dx_b}{\int_0^{\tau_x} dt} = \frac{4\eta_E'\psi x_{bt}^{5/2}}{5\tau_x} \tag{4.88}$$

Taking into account Equation 4.82, we get the following

$$A_{xb} = \frac{4\eta_E'\psi}{5}x_{cm}^{5/2}\frac{\tau_x^{3/2}\tau_1^{2/3}}{(6\tau_E)^{5/3}} \tag{4.89}$$

4.3.2 WORK OF THE TANGENTIAL FORCES

Taking into account Equations 4.60 and 4.48–4.51, the work at a shear of the tangential viscous and elastic forces $F_{b\tau y}$, $F_{c\tau y}$ and $F_{b\tau z}$, $F_{c\tau z}$ can be found by integration as follows:

$$A_{ycm} = \int_0^{y_{cm}} F_{c\tau y} dy_c = G \int_0^{P_{bxm}} dP_{bx} \int_0^{y_{cm}} y_c dy_c = \frac{G}{2} P_{bxm} y_{cm}^2 \quad (4.90)$$

$$A_{ybm} = \int_0^{y_{bm}} F_{b\tau y} dy_b = \int_0^{P_{bxm}} dP_{bx} \int_0^{y_{bm}} \dot{y}_b dy_b = \eta_G' P_{bxm} \frac{\int_0^{y_{bm}} \int dy_b dy_b}{\int_0^{\tau_1} dt} = \eta_G' P_{bxm} \frac{y_{bm}^2}{2\tau_1} \quad (4.91)$$

$$A_{zcm} = \int_0^{z_{cm}} F_{c\tau z} dz_c = G \int_0^{P_{axm}} dP_{ax} \int_0^{z_{cm}} z_c dz_c = \frac{G}{2} P_{axm} z_{cm}^2 \quad (4.92)$$

$$A_{zbm} = \int_0^{z_{bm}} F_{b\tau y} dz_b = \eta_G' \int_0^{P_{axm}} dP_{ax} \int_0^{z_{bm}} \dot{z}_b dz_b = \eta_G' P_{axm} \frac{\int_0^{z_{bm}} \int dz_b dz_b}{\int_0^{\tau_1} dt} = \eta_G' P_{axm} \frac{z_{bm}^2}{2\tau_1} \quad (4.93)$$

where y_{cm} and z_{cm} are the elastic deformations at the shear between the contacting surfaces along axes Y and Z at the instance of time $t = \tau_1$; y_{bm}, z_{bm} are viscous deformations at the shear between contacting surfaces at the instant of the time $t = \tau_1$, and where $P_{bxm} = k_{ha} x_m + 2k_{pb} R_b^{1/2} x_m^{1/2}$, $P_{axm} = k_{hb} x_m + 2k_{pa} R_a^{1/2} x_m^{1/2}$.

Since at the moment of rebound $t = \tau_x$, the work of all tangential visco-elastic forces $F_{b\tau y}$, $F_{c\tau y}$ and $F_{b\tau z}$, $F_{c\tau z}$ in the phase of the restitution can be found by integration as follows:

$$A_{yct} = -\int_{y_{cm}}^{y_{ct}} F_{c\tau y} dy_c = -G \int_{P_{bxm}}^{0} dP_{bx} \int_{y_{cm}}^{y_{ct}} y_c dy_c = \frac{G}{2} P_{bxm} (y_{ct}^2 - y_{cm}^2) \quad (4.94)$$

$$A_{ybt} = -\int_{y_{bm}}^{y_{bt}} F_{b\tau y} dy_b = -\eta_G' \int_{P_{bxm}}^{0} dP_{bx} \int_{y_{bm}}^{y_{bt}} \dot{y}_b dy_b = \eta_G' P_{bxm} \frac{\int_{y_{bm}}^{y_{bt}} \int dy_b dy_b}{\int_{\tau_1}^{\tau_x} dt} = \eta_G' P_{bxm} \frac{y_{bt}^2 - y_{bm}^2}{2\tau_2} \quad (4.95)$$

$$A_{zct} = -\int_{z_{cm}}^{z_{ct}} F_{c\tau z} dz_c = -G \int_{Q_{axm}}^{0} dP_{ax} \int_{z_{cm}}^{z_{ct}} z_c dz_c = \frac{G}{2} P_{axm} (z_{ct}^2 - z_{cm}^2) \quad (4.96)$$

$$A_{zbt} = -\int_{z_{bm}}^{z_{bt}} F_{b\tau z} dz_b = -\eta_G' \int_{P_{axm}}^{0} dP_{ax} \int_{z_{bm}}^{z_{bt}} \dot{z}_b dz_b = \eta_G' P_{axm} \frac{\int_{z_{bm}}^{z_{bt}} \int dz_b dz_b}{\int_{\tau_1}^{\tau_x} dt} = \eta_G' P_{axm} \frac{z_{bt}^2 - z_{bm}^2}{2\tau_2} \quad (4.97)$$

where y_{ct} and z_{ct} are the elastic deformations at the shear between the contacting surfaces along axes Y and Z at the instance of time $t = \tau_x$; y_{bt}, z_{bt} are viscous deformations at the shear between contacting surfaces at the instance of time $t = \tau_x$. The direction of the displacement in the restitution phase by axis Y is not changing; therefore, the integrals for the works were taken with the plus sign.

The equations for work of the tangential viscoelastic forces can be found as

$$A_{yc} = A_{ycm} + A_{yct} = \frac{G}{2} P_{bxm} y_{ct}^2 \tag{4.98}$$

$$A_{yb} = A_{ybm} + A_{ybt} = \frac{\eta'_G}{2} P_{bxm} \left(\frac{y_{bm}^2 \tau_2 + \tau_1 (y_{bt}^2 - y_{bm}^2)}{\tau_1 \tau_2} \right) \tag{4.99}$$

$$A_{zc} = A_{zcm} + A_{zct} = \frac{G}{2} P_{axm} z_{ct}^2 \tag{4.100}$$

$$A_{zb} = A_{zbm} + A_{zbt} = \frac{\eta'_G}{2} P_{axm} \left(\frac{z_{bm}^2 \tau_2 + \tau_1 (z_{bt}^2 - z_{bm}^2)}{\tau_1 \tau_2} \right) \tag{4.101}$$

The dissipative energy W_{by} at the shear equals to the dissipative work A_b and equals to the sum of the work of the viscous tangential forces, and can be found as follows

$$W_b = A_b = A_{yb} + A_{zb} = \frac{\eta'_G}{2\tau_1 \tau_2} \left(P_{bxm} [y_{bm}^2 \tau_2 + \tau_1 (y_{bt}^2 - y_{bm}^2)] + P_{axm} [z_{bm}^2 \tau_2 + \tau_1 (z_{bt}^2 - z_{bm}^2)] \right) \tag{4.102}$$

4.4 DYNAMICS OF AN ELASTOPLASTIC CONTACT BETWEEN SMOOTH SURFACES OF TWO SPHERICAL BODIES AT IMPACT

4.4.1 NORMAL DISPLACEMENT

4.4.1.1 NORMAL ELASTIC AND VISCOUS FORCES

The case of an elastoplastic contact between two spheres can be considered in a similar way as it was done in Chapter 3 for the viscoelastic impact. Therefore, we can again take that $R_a = R_b = R$, $r_a = r_b = r$, $a = b = 2r$, see Figure 3.1, and that $k_p = k_{pa} = k_{pb}$ and $\psi = 2_{kp} R^{1/2}$, then according to Equations 4.36 and 4.39 we get the next equations for the general elastic forces:

$$F_{cn} = \frac{4}{3} k_p E R^{1/2} x_c^{3/2} \tag{4.103}$$

$$F_{bn} = 4k_p\eta'_E R^{1/2}\dot{x}_b x_b^{1/2} \tag{4.104}$$

On the other hand, according to Newton's Second Law and in the case of elastic-plastic contact for Maxwell model when the elastic and viscous forces are equal, $F_n = F_{cn} = F_{bn} = -m\ddot{x}$ we can write that

$$F_{cn} = -m\ddot{x} = \frac{4}{3}k_p E R^{1/2} x_c^{3/2} \tag{4.105}$$

and

$$F_{bn} = -m\ddot{x} = 4\eta'_E k_p R^{1/2}\dot{x}_b x_b^{1/2} \tag{4.106}$$

4.4.1.2 WORK, ENERGY AND SIZE OF INDENTATION

Since the limits of integration in this case are x_c and x_b, the works A_{xcm} and A_{xbm} at the compression can be found as

$$A_{xcm} = \int_0^{x_{cm}} F_{cn}dx_c = \frac{4}{3}k_p E R^{1/2}\int_0^{x_{cm}} x_c^{3/2}dx_c = \frac{8}{15}k_p E R^{1/2} x_{cm}^{5/2} \tag{4.107}$$

where x_{cm} is the maximum of the elastic displacement between the centres of mass of the contacting bodies

$$A_{xbm} = \int_0^{x_{bm}} F_{bn}dx_b = \int_0^{x_{bm}} 4k_p\eta'_G R^{1/2}\dot{x}_b x_b^{1/2}dx_b = \eta'_E k_p R^{1/2}\frac{\int_0^{x_{bm}}\int dx_b x_b^{1/2}dx_b}{\int_0^{\tau_1} dt} = \frac{8k_p\eta'_E R^{1/2} x_{bm}^{5/2}}{5\tau_1} \tag{4.108}$$

where x_{bm} is the plastic displacement between the centres of mass of the contacting bodies in the moment of time $t = \tau_1$ and when $x_c = x_{cm}$.

The total work in the moment of time $t = \tau_1$ and when $x_c = x_{cm}$ can be expressed as

$$A_{xm} = A_{xcm} + A_{xbm} = W_{xm} = \frac{mV_x^2}{2} - \frac{m(\dot{x}_{bm})^2}{2} \tag{4.109}$$

Using Equations 4.107–4.109, we can write that

$$A_{xm} = \frac{8}{15}k_p R^{1/2}\left(E x_{cm}^{5/2} + \frac{3\eta'_E}{\tau_1}x_{bm}^{5/2}\right) = \frac{m}{2}(V_x^2 - (\dot{x}_{bm})^2) \tag{4.110}$$

where $\dot{x}_{bm}$ is the relative velocity of the plastic displacement between the centres of mass of the contacting bodies in the moment of time $t = \tau_1$ and when $x_c = x_{cm}$.

The full dissipative work between the centres of mass of the contacting bodies in the moment of time $t = \tau_1$ and when $x_b = x_{bt}$ can be found as follows

$$A_{xb} = \int_0^{x_{bt}} F_{bn} dx = \int_0^{x_{bt}} 4k_p \eta'_E \dot{x}_b x_b^{1/2} dx_b = 4k_p \eta'_E \frac{\int_0^{x_{bt}} \int dx_b x_b^{1/2} dx_b}{\int_0^{\tau_x} dt} = \frac{8k_p \eta'_E R^{1/2} x_{bt}^{5/2}}{5\tau_x} \quad (4.111)$$

where x_{bt} is the plastic displacement between the centres mass of the contacting bodies in the moment of the time of rebound, when $t = \tau_x$ and $x_c = 0$.

It is obvious that the full dissipative work is equal to the change of the kinetic energy at the time of impact, and; therefore, we can write that

$$A_{xb} = \frac{mV_x^2}{2} - \frac{mV_{tx}^2}{2} = \frac{mV_x^2}{2}(1 - k_x^2) \quad (4.112)$$

whereas we know k_x is the coefficient of restitution.

Thus, according to Equations 4.111 and 4.112, and taking into account that $\chi = (1 - k_x^2)$ is the coefficient of mechanical losses at impact, it follows

$$A_{xb} = \frac{8k_p \eta'_E R^{1/2} x_{bt}^{5/2}}{5\tau_x} = \frac{mV_x^2}{2}\chi \quad (4.113)$$

Also, taking into account $F_n = F_{cn} = F_{bn} = -m\ddot{x}$ and using Equations 4.105, 4.106, 4.59, we get the same result as in Equation 4.82, $x_c^{3/2} = 3\tau_E \dot{x}_b x_b^{1/2}$. Therefore, we can use Equations 4.83–4.85 in this case, too. It is obvious that these equations can be used for any kind of an elliptical, circular or oval contact.

Thus, using Equation 4.59, we can rewrite Equation 4.110 as

$$A_{xm} = \frac{8}{15} k_p R^{1/2} x_{cm}^{5/2} E\left(1 + \frac{3\tau_E}{\tau_1} x_{bm}^{5/2}\right) = \frac{m}{2}(V_x^2 - (\dot{x}_{bm})^2) \quad (4.114)$$

Then, using Equations 4.83–4.85 and 4.114 we get the formula for calculation of x_{cm} as given below

$$x_{cm} = \left\{\frac{15mV_x^2}{16k_p E R^{1/2}} f_{\tau E}\right\}^{2/5} \quad (4.115)$$

where $f_{\tau E}$ is the dissipative function, which can be written as

$$f_{\tau E} = \left(\frac{1 - \left(\frac{\tau_1}{\tau_E}\right)^{4/1}}{1 + \frac{3}{(6)^{5/3}}\left(\frac{\tau_1}{6\tau_E}\right)^{2/3}}\right) \quad (4.116)$$

Also, according to Equations 4.85 and 4.115, the expression for the maximum of indentation at impact between spherical bodies can be written as

$$x_t = x_{bt} = \left\{ \frac{15mV_x^2}{16k_p ER^{1/2}} f_{\tau E} \right\}^{2/5} \times \frac{\tau_x}{\tau_1}\left(\frac{\tau_1}{6\tau_E}\right)^{2/3} \tag{4.117}$$

Equations 4.115 and 4.117 are valid only in the case of elastic-plastic deformations, when $tg\beta \geq 1/3$ and respectively when $k_x \leq \frac{\pi - 1}{\pi + 1}$.

Now, according to Equations 4.59, 4.85 and 4.113 we get

$$\chi = \frac{16k_p E\tau_E R^{1/2}}{5m\tau_x V_x^2} x_{cm}^{5/2} \left(\frac{\tau_x}{\tau_1}\right)^{5/2} \left(\frac{\tau_1}{6\tau_E}\right)^{5/3} \tag{4.118}$$

Using Equations 4.115 and 4.118, the expression for the coefficient of mechanical losses can be written as

$$\chi = f_{\tau E} \frac{3\tau_x^{3/2}}{(6)^{5/3} \tau_E^{2/3} \tau_1^{5/6}} \tag{4.119}$$

4.1.1.3 APPROXIMATE SOLUTION TO THE DIFFERENTIAL EQUATIONS OF THE NORMAL DISPLACEMENT BY USING THE METHOD OF THE EQUIVALENT WORKS

Using expressions for work A_{xcm} and A_{xbm} from Equations 4.107 and 4.108, the equivalent constant parameters B_x, C_x can be found by the already known way as given below

$$A_{xcm} = C_x \int_0^{x_{cm}} x_c dx_c = \frac{1}{2} C_x x_{cm}^2 = \frac{8}{15} k_p ER^{1/2} x_{cm}^{5/2} \tag{4.120}$$

and

$$A_{xbm} = B_x \int_0^{x_{bm}} \dot{x}_b dx_b = B_x \frac{\int_0^{x_{bm}} x_b dx_b}{\int_0^{\tau_1} dt} = B_x \frac{x_{bm}^2}{2\tau_1} = \frac{8k_p \eta_E' R^{1/2} x_{bm}^{5/2}}{5\tau_1} \tag{4.121}$$

Hence, we can write the expressions for the equivalent constant viscoelasticity parameters as

$$C_x = \frac{16}{15} k_p ER^{1/2} x_{cm}^{1/2} \tag{4.122}$$

$$B_x = \frac{16\eta_E' k_p R^{1/2}}{5} x_{bm}^{1/2} \tag{4.123}$$

Thus, Equations 4.105 and 4.106 with variable parameters can be rewritten as the equations with constant parameters

$$F_{cn} = -m\ddot{x} = C_x x_c \tag{4.124}$$

and

$$F_{bn} = -m\ddot{x} = B_x \dot{x}_b \tag{4.125}$$

Hence, we can write that

$$\dot{x}_b = \frac{C_x}{B_x} x_c \tag{4.126}$$

Or after differentiation, it follows

$$\ddot{x}_b = \frac{C_x}{B_x} \dot{x}_c \tag{4.127}$$

Taking into account that $\ddot{x} = \ddot{x}_c + \ddot{x}_b$, Equation 4.124 can be written now as

$$m\left(\ddot{x}_c + \frac{C_x}{B_x} \dot{x}_C\right) = -C_x x_c \tag{4.128}$$

Thus, the next differentiation equation related to x_c can be expressed as

$$m\ddot{x}_c + B_{Cx}\dot{x}_c + C_x x_c = 0 \tag{4.129}$$

where $B_{Cx} = m\frac{C_x}{B_x}$. Equation 4.129 is the equation of damped oscillations and the solution to this equation is known as

$$x_c = C_1 e^{-\delta_x t} \sin(\omega_x t) + C_2 e^{-\delta_x t} \cos(\omega_x t) \tag{4.130}$$

where $\omega_x = \sqrt{\omega_{0x}^2 - \delta_x^2}$, $\delta_x = \frac{B_{Cx}}{2m} = \frac{C_x}{2B_x}$ is the normal damping factor, $\omega_{0x} = \sqrt{\frac{C_x}{m}}$ is the angular frequency of free harmonic oscillations by axis X. The constants of integration C_1 and C_2, which can be found from the initial conditions, if $t = 0$ follows $C_2 = 0$ and

$$x_c = C_1 e^{-\delta_x t} \sin(\omega_x t) \tag{4.131}$$

Then, following integration we get

$$\dot{x}_c = C_1 e^{-\delta_x t}[\omega_x \cos(\omega_x t) - \delta_x \sin(\omega_x t)] \tag{4.132}$$

According to the initial condition $\dot{x} = V_x$ follows that $C_1 = \frac{V_x}{\omega_x}$, and after substitution follows

$$x_c = \frac{V_x}{\omega_x} e^{-\delta_x t} \sin(\omega_x t) \tag{4.133}$$

and as well we get the equation for the normal velocity between centres of mass of contacting bodies as follows

$$\dot{x}_c = \frac{V_x}{\omega_x} e^{-\delta_x t} [\omega_x \cos(\omega_x t) - \delta_x \sin(\omega_x t)] \tag{4.134}$$

Now, since $V_{tx} = \dot{x}_c\,(t = \tau_x)$ and $\delta_x = \frac{B_{Cx}}{2m} = \frac{C_x}{2B_x}$, the energetic coefficient of restitution can be found as

$$e_x = k_x^2 = \frac{V_{tx}^2}{V_x^2} = \exp\left(-\frac{C_x}{B_x}\tau_x\right) \tag{4.135}$$

4.4.2 *TANGENTIAL DISPLACEMENT*

4.4.2.1 TANGENTIAL ELASTIC AND VISCOUS FORCES AT ROLLING ELASTOPLASTIC CONTACT

As we know, in the case of rolling contact $\dot{y}_r = \frac{x}{l}\dot{y}$, see Equation 3.65, and $y_r = \frac{x}{l} y$, see Equation 3.66, it is obvious that in case of elastoplastic contact the viscous deformation at the rolling shear can be taken analogically as

$$\dot{y}_{br} = \frac{x}{l}\dot{y}_b \tag{4.136}$$

and obviously, we can take the following for the elastic deformation at the rolling shear rolling shear

$$y_{cr} = \frac{x}{l} y_c \tag{4.137}$$

Further, in the case of rolling contact between two spheres, see Figure 3.1, we can take that $F_{c\tau z} = 0$, $F_{b\tau z} = 0$, $P_{ax} = 0$, $P_{bx} = P_x$, $k_{ha} = k_{hb} = k_h$. Then, taking Equations 4.48 and 4.49, that $F_{b\tau} = F_{b\tau y}$, $F_{c\tau} = F_{c\tau y}$, we get next equations for the tangential elastic and viscous forces in the case of rolling contact

$$F_{c\tau r} = G P_x y_{cr} \tag{4.138}$$

$$F_{b\tau r} = \eta'_G P_x \dot{y}_{br} \tag{4.139}$$

where $P_x = \frac{x^{1/2}}{l}(k_h x^{1/2} + 2k_p R^{1/2})$. Taking into account Equations 4.136–4.139, we get the following

$$F_{c\tau r} = G' P_s y_c \tag{4.140}$$

$$F_{b\tau r} = \eta'_G P_s \dot{y}_b \tag{4.141}$$

similar as in Equation 3.69 $P_s = \frac{x}{l} P_x = \frac{x^{3/2}}{l}(k_h x^{1/2} + 2k_p R^{1/2})$.

On the other hand, according to Newton's Second Law, we can write that

$$F_{c\tau r} = -m\ddot{y} = G P_s y_c \tag{4.142}$$

and

$$F_{b\tau r} = -m\ddot{y} = \eta'_G P_s \dot{y}_b \tag{4.143}$$

According to Equation 4.60 $\tau_G = \frac{\eta'_G}{G}$, and as in the case of elastic-plastic contact for Maxwell model when the elastic and the viscous forces are equal, $F_{\tau r} = F_{c\tau r} = F_{b\tau r} = -m\ddot{y}$ using Equations 4.142 and 4.143, we get

$$y_c = \tau_G \dot{y}_b \tag{4.144}$$

4.4.2.2 WORK AND APPROXIMATE SOLUTION TO THE DIFFERENTIAL EQUATIONS OF THE TANGENTIAL DISPLACEMENT

The works A_{ycm} and A_{ybm} of the rolling shear in the moment of time $t = \tau_1$ and $x = x_{cm}$ can be found as given below

$$A_{ycm} = \int_0^{y_{cm}} F_{c\tau y} dy_c = G \int_0^{P_m} dP_s \int_0^{y_{cm}} y_c dy_c = \frac{G}{2} P_m y_{cm}^2 \tag{4.145}$$

where y_{cm} is the maximum of the tangential elastic deformation between contacting bodies and

$$A_{ybm} = \int_0^{y_{bm}} F_{b\tau y} dy_b = \eta'_G \int_0^{P_m} dP_s \int_0^{y_{bm}} \dot{y}_b dy_b = \eta'_G P_m \frac{\int_0^{y_{bm}} \int dy_b dy_b}{\int_0^{\tau_1} dt} = \eta'_G P_m \frac{y_{bm}^2}{2\tau_1} \tag{4.146}$$

where y_{bm} is the plastic deformation of the shear between the surfaces of the contacting bodies along axis Y at the instant of the time $t = \tau_1$ and $x = x_{cm}$, and where

$$P_m = \frac{x_m^{3/2}}{l}(k_h x_m^{1/2} + 2k_p R^{1/2}) \tag{4.147}$$

We can find the approximate solutions for equations of motion in the same manner as for the equations with the equivalent constant viscoelasticity parameters, if we choose the equivalent constant parameters B_y, C_y so that the works A_{ycm} and A_{ybm} with the variable viscoelasticity parameters c_y, b_y will be equal to the work with the constant viscoelasticity parameters. Thus, according to this statement and boundary conditions $t=\tau_1$, $x=x_{cm}$ and $y=y_{cm}$, and also using the expressions for works A_{ycm} and A_{ybm} from Equations 4.145 and 4.146 in the phase of the rolling shear, we get

$$A_{ycm} = C_{yr}\int_0^{y_{cm}} y_c dy_c = \frac{1}{2}C_{yr}y_{cm}^2 = \frac{1}{2}GP_m y_{cm}^2 \tag{4.148}$$

$$A_{ybm} = B_{yr}\int_0^{y_{bm}} \dot{y}_b dy_b = B_{yr}\frac{\int_0^{y_{bm}}\int dy_b dy_b}{\int_0^{\tau_1} dt} = B_{yr}\frac{y_{bm}^2}{2\tau_1} = \frac{\eta_G'}{2\tau_1}P_m y_{bm}^2 \tag{4.149}$$

Hence, we can write the expressions for the equivalent constant viscoelasticity parameters as given below:

$$C_{yr} = GP_m \tag{4.150}$$

$$B_{yr} = \eta_G' P_m \tag{4.151}$$

Thus, Equations 4.142 and 4.143 with variable parameters can be rewritten as the equations with constant parameters as follows

$$F_{c\tau r} = -m\ddot{y} = C_{yr}y_c \tag{4.152}$$

and

$$F_{b\tau r} = -m\ddot{y} = B_{yr}\dot{y}_b \tag{4.153}$$

Hence, we can write that

$$\dot{y}_b = \frac{C_{yr}}{B_{yr}}y_c \tag{4.154}$$

Or after differentiation, it follows

$$\ddot{y}_b = \frac{C_{yr}}{B_{yr}}\dot{y}_c \tag{4.155}$$

Taking into account that $\ddot{y} = \ddot{y}_c + \ddot{y}_b$, Equation 4.152 can be written now as

$$m\left(\ddot{y}_c + \frac{C_{yr}}{B_{yr}}\dot{y}_C\right) = -C_{yr}y_c \tag{4.156}$$

Thus, the next differentiation equation related to y_c can be expressed as

$$m\ddot{y} + B_{Cyr}\dot{y} + C_{yr}y = 0 \tag{4.157}$$

where $B_{Cyr} = m\frac{C_{yr}}{B_{yr}}$. Equation 4.157 is the equation of the damped oscillations and according to the initial condition $y_c = 0$ and $\dot{y}_c = V_y$, the solution to this equation is known as

$$y_c = \frac{V_y}{\omega_{yr}} e^{-\delta_{yr}t} \sin(\omega_{yr}t) \tag{4.158}$$

where $\omega_{yr} = \sqrt{\omega_{0yr}^2 - \delta_{yr}^2}$, $\delta_{yr} = \frac{B_{Cyr}}{2m} = \frac{C_{yr}}{2B_{yr}}$ is the tangential damping factor, $\omega_{0yr} = \sqrt{\frac{C_{yr}}{m}}$ is the angular frequency of the harmonic oscillations by axis *Y*.

As well the equation for the tangential relative velocity between centres of mass of the contacting bodies can be received by differentiation of Equation 4.158 as

$$\dot{y}_c = \frac{V_y}{\omega_{yr}} e^{-\delta_{yr}t} [\omega_{yr}\cos(\omega_{yr}t) - \delta_{yr}\sin(\omega_{yr}t)] \tag{4.159}$$

Now, since $V_{ty} = \dot{y}_c(t = \tau_x)$ and $\delta_{yr} = \frac{C_{yr}}{2B_{yr}}$, the energetic coefficient of restitution for tangential displacement can find

$$e_y = k_y^2 = \frac{V_{ty}^2}{V_y^2} = \exp\left(-\frac{C_{yr}}{B_{yr}}\tau_x\right) \tag{4.160}$$

The full changing of the energy of the dissipative system at the rolling shear can be found as the difference between A_y and A_{yc} as given below

$$\Delta W_y = A_y - A_{yc} = \frac{mV_y^2}{2} - \frac{mV_{ty}^2}{2} - \frac{J_z\omega_t^2}{2} = A_{ybm} + A_{ybt} = A_{yb} \tag{4.161}$$

Now, let us find the works A_{yc} and A_{yb} for the full-time period of contact:

$$A_{yc} = C_{yr}\int_0^{y_{ct}} y_c dy_c = \frac{1}{2}C_{yr}y_{ct}^2 \tag{4.162}$$

$$A_{yb} = Q_{\omega} = B_{yr} \int_{0}^{y_{bt}} \dot{y}_b dy_b = B_{yr} \frac{\int_0^{y_{bt}} \int dy_b dy_b}{\int_0^{\tau_x} dt} = B_{yr} \frac{y_{bt}^2}{2\tau_x} \tag{4.163}$$

As we already know, according to Equation 3.85 the conclusion can be drawn that the work A_{yc} is transformed into the kinetic energy of the relative rotation between the bodies, but on the other hand, the work A_{yb} is transformed into dissipative energy Q_{ω} in the process of the internal friction. According to Equation 3.85, 4.161 and 4.162, it follows

$$A_{yc} = \frac{J_z \omega_t^2}{2} = \frac{1}{2} G' P_m y_{ct}^2 \tag{4.164}$$

Hence, the equation for the relative angular velocity at the time of rebound can be written as follows

$$\omega_t = \left(\frac{G' P_m}{J_z} \right)^{\frac{1}{2}} y_{ct} \tag{4.165}$$

4.4.3 COEFFICIENT OF FRICTION

Since the viscous and elastic forces in the Maxwell model are equal, the coefficient of rolling friction can be expressed by the four variants such as

$$f_r = \frac{F_{\tau r}}{F_n} = \frac{F_{b\tau r}}{F_{cn}} = \frac{F_{b\tau r}}{F_{bn}} = \frac{F_{c\tau r}}{F_{cn}} = \frac{F_{c\tau r}}{F_{bn}} \tag{4.166}$$

Choosing a third of them, we can write that

$$f_r = \frac{F_{c\tau r}}{F_{cn}} = \frac{3G' P_s}{4k_p E R^{1/2}} \times \frac{y_c}{x_c^{3/2}} \tag{4.167}$$

Also since $P_s = \frac{x}{l} P_x = \frac{x^{3/2}}{l} (k_h x^{1/2} + 2k_p R^{1/2})$ we finally get

$$f_r = \frac{F_{c\tau r}}{F_{cn}} = \frac{3G}{4k_p E R^{1/2}} \times \frac{x^{3/2} (k_h x^{1/2} + 2k_p R^{1/2}) y_c}{l x_c^{3/2}} \tag{4.168}$$

4.5 DYNAMICS OF AN ARBITRARY ELASTOPLASTIC COLLISION BETWEEN A SLED, WHICH HAS AN ARBITRARY CURVILINEAR CYLINDRICAL FORWARD SURFACE AND A SEMI-SPACE

4.5.1 NORMAL DISPLACEMENT

4.5.1.1 NORMAL ELASTIC AND VISCOUS FORCES

As we already know, in the case of elastoplastic contacts for the definition of all elastic and viscous forces, the integration for all viscous and elastic differential forces have to be done separately according to the limits of integrations: x_c, x_b, y_c, y_b. Or better say in other words, in the case of elastoplastic contact between two bodies, we get similar equations for all elastic and viscous forces as in a viscoelastic contact, but separately for elastic and viscous displacements.

For example, in the case of an arbitrary elastoplastic collision between a sled, which has an arbitrary curvilinear cylindrical forward surface, see Figure 2.13, and the flat surface of semi-space; the integration of series of Equations 2.550–2.553 by limits x_c, x_b, $\dot{x}_b$ gives us the normal forces similar as in Equations 2.560 and 2561, respectively as

$$F_{cn} = E(L_s + L)D_1 x_c + E\frac{C_r D_1^{1/n}}{n+1} x_c^{\frac{n+1}{n}} \tag{4.169}$$

$$F_{bn} = \eta_E'(L_s + L)D_1 \dot{x}_b + \eta_E' C_r D_1^{1/n} \dot{x}_b x_b^{1/n} \tag{4.170}$$

As we remember, in the case of the Maxwell model, instead of the storage moduli E', E_1', E_2' and G', E_1', G_2', we use the dynamic Young's elasticity moduli E, E_1, E_2 and G, G_1, G_2.

On the other hand, according to Newton's Second Law and since in the case of elastic-plastic contact for Maxwell model when the elastic and viscous forces are equal, $F_n = F_{cn} = F_{bn} = -m\ddot{x}$, we can write that

$$F_{cn} = -m\ddot{x} = E(L_s + L)D_1 x_c + E\frac{C_r D_1^{1/n}}{n+1} x_c^{\frac{n+1}{n}} \tag{4.171}$$

and

$$F_{bn} = -m\ddot{x} = \eta_E'(L_s + L)D_1 \dot{x}_b + \eta_E' C_r D_1^{1/n} \dot{x}_b x_b^{1/n} \tag{4.172}$$

where x_c, x_b are the normal displacement of the centre mass of a sled, and $\dot{x}_c$, $\dot{x}_b$ are the velocities of the normal displacement of the centre mass of a sled.

4.5.1.2 APPROXIMATE SOLUTION TO THE DIFFERENTIAL EQUATIONS OF THE NORMAL DISPLACEMENT

The works A_{xcm} and A_{xbm} at the maximum elastic compression $x_c = x_{cm}$ and $x_b = x_{bm}$ can be found by integration as given below

$$A_{xcm} = \int_0^{x_{cm}} F_{cn} dx_c = D_1 E(L_s + L) \int_0^{x_{cm}} x_c dx_c + E \frac{C_r D_1^{1/n}}{n+1} \int_0^{x_m} x_c^{\frac{n+1}{n}} dx_c$$

$$= E x_{cm}^2 \left(\frac{D_1}{2}(L_s + L) + \frac{n C_r D_1^{1/n}}{(2n+1)(n+1)} x_{cm}^{1/n} \right) \tag{4.173}$$

and

$$A_{xbm} = \int_0^{x_{bm}} F_{bn} dx_b = \eta_E'(L_s + L) D_1 \int_0^{x_{bm}} \dot{x} dx + \eta_E' C_r D_1^{1/n} \int_0^{x_{bm}} \dot{x} x^{1/n} dx$$

$$= \eta_E' \frac{(L_s + L) D_1 \int_0^{x_{bm}} \int dx dx + C_r D_1^{1/n} \int_0^{x_{bm}} \int dx x^{1/n} dx}{\int_0^{\tau_1} dt} = \eta_E' x_{bm}^2 \tag{4.174}$$

$$\left(\frac{D_1}{2}(L_s + L) + \frac{n C_r D_1^{1/n}}{2n+1} x_{bm}^{1/n} \right)$$

Denoting $\frac{D_1}{2}(L_s + L) = \kappa_1$, $\frac{n C_r D_1^{1/n}}{(2n+1)(n+1)} = \kappa_2$ and $\frac{n C_r D_1^{1/n}}{2n+1} = \kappa_3$, the equations for the works can be written as:

$$A_{xcm} = E x_{cm}^2 (\kappa_1 + \kappa_2 x_{cm}^{1/n}) \tag{4.175}$$

$$A_{xbm} = \frac{\eta_E'}{\tau_1} x_{bm}^2 (\kappa_1 + \kappa_3 x_{bm}^{1/n}) \tag{4.176}$$

According to the boundary conditions $t = \tau_1$, $x_c = x_{cm}$ and $x_b = x_{bm}$, using the expressions for works A_{xcm} and A_{xbm} from Equations 4.173 and 4.174, we can write that

$$A_{xcm} = C_x \int_0^{x_{cm}} x_c dx_c = \frac{1}{2} C_x x_{cm}^2 = E x_{cm}^2 (\kappa_1 + \kappa_2 x_{cm}^{1/n}) \tag{4.177}$$

and

$$A_{xbm} = B_x \int_0^{x_{bm}} \dot{x}_b dx_b = B_x \frac{\int_0^{x_{bm}} x_b dx_b}{\int_0^{\tau_1} dt} = B_x \frac{x_{bm}^2}{2\tau_1} = \frac{\eta_E'}{\tau_1} x_{bm}^2 (\kappa_1 + \kappa_3 x_{bm}^{1/n}) \tag{4.178}$$

Hence, we can write the expressions for the equivalent constant viscoelasticity parameters as follows

$$C_x = 2E(\kappa_1 + \kappa_2 x_{cm}^{1/n}) \quad (4.179)$$

$$B_x = 2\eta_E'(\kappa_1 + \kappa_3 x_{bm}^{1/n}) \quad (4.180)$$

However, since practically $\kappa_1 >> \kappa_2 x_{cm}^{1/n}$ and $\kappa_1 >> \kappa_3 x_{bm}^{1/n}$; therefore, we can approximately take that

$$C_x = 2E\kappa_1 = ED_1(L_s + L) \quad (4.181)$$

and

$$B_x = 2\eta_E'\kappa_1 = \eta_E' D_1(L_s + L) \quad (4.182)$$

As we can see, the constant parameters C_x and B_x do not depend on the displacement x, or better to say that, in reality, they weakly depend on x.

Thus, Equations 4.171 and 4.172 with variable parameters can be rewritten as the equations with constant parameters as follows

$$F_{cn} = -m\ddot{x} = \mathrm{C_x} x_c = E'D_1(L_s + L)x_c \quad (4.183)$$

and respectively

$$F_{bn} = -m\ddot{x} = \mathrm{B_x}\dot{x}_b = \eta_E' D_1(L_s + L)\dot{x}_b \quad (4.184)$$

Hence, we can write that

$$\dot{x}_b = \frac{1}{\tau_E} x_c \quad (4.185)$$

Further, respectively, after differentiation, it follows as

$$\ddot{x}_b = \frac{1}{\tau_E} \dot{x}_c \quad (4.186)$$

Taking into account that $\ddot{x} = \ddot{x}_c + \ddot{x}_b$, Equation 4.183 can be written now as

$$m\left(\ddot{x}_c + \frac{1}{\tau_E}\dot{x}_C\right) = -ED_1(L_s + L)x_c \quad (4.187)$$

Thus, the next differentiation equation related to x_c can be expressed as

$$m\ddot{x}_c + \frac{m}{\tau_E}\dot{x}_c + ED_1(L_s + L)x_c = 0 \quad (4.188)$$

or in the canonical form as

$$m\ddot{x}_c + B_{Cx}\dot{x}_c + C_x x_c = 0 \quad (4.189)$$

where

$$B_{Cx} = \frac{m}{\tau_E} \tag{4.190}$$

Equation 4.189 is the equation of damped oscillations and the solution to this equation is known as

$$x_c = \frac{V_x}{\omega_x} e^{-\delta_x t} \sin(\omega_x t) \tag{4.191}$$

where $\omega_x = \sqrt{\omega_{0x}^2 - \delta_x^2}$, $\delta_x = \frac{B_{Cx}}{2m}$ is the normal damping factor, $\omega_{0x} = \sqrt{\frac{C_x}{m}}$ is the angular frequency of free harmonic oscillations by axis X. Then, after integration, we get for the velocity of a sled

$$\dot{x}_c = \frac{V_x}{\omega_x} e^{-\delta_x t} [\omega_x \cos(\omega_x t) - \delta_x \sin(\omega_x t)] \tag{4.192}$$

Now, using Equation 4.192, the time of indentation $t = \tau_1$ can be found according to the condition $\dot{x}_c = 0$, respectively as

$$\tau_1 = \frac{1}{\omega_x} \text{arctg}\left(\frac{\omega_x}{\delta_x}\right) \tag{4.193}$$

Thus, using Equations 4.191 and 4.193, we get for the maximum of an elastic displacement

$$x_{cm} = \frac{V_x}{\omega_x} \exp\left(-\frac{\delta_x}{\omega_x} \text{arctg}\left(\frac{\omega_x}{\delta_x}\right)\right) \times \sin\left(\text{arctg}\left(\frac{\omega_x}{\delta_x}\right)\right) \tag{4.194}$$

Taking into account that $\delta_x = \frac{B_{Cx}}{2m}$, it follows

$$\delta_x = \frac{1}{2\tau_E} \tag{4.195}$$

But on the other hand, according to Equation 4.65 $tg\beta_E = \frac{1}{\omega_x \tau_E}$, and respectively

$$\delta_x = \frac{\omega_x tg\beta_E}{2} \tag{4.196}$$

Substituting $\delta_x = \frac{\omega_x tg\beta_E}{2}$ into Equation 4.194 gives

$$x_{cm} = \frac{V_x}{\omega_x} \exp\left(-\frac{tg\beta_E}{2} \text{arctg}\left(\frac{2}{tg\beta_E}\right)\right) \times \sin\left(\text{arctg}\left(\frac{2}{tg\beta_E}\right)\right) \tag{4.197}$$

Now, since $V_{tx} = \dot{x}_c(t = \tau_x)$, the energetic coefficient of restitution can be found as

$$e_x = k_x^2 = \frac{V_{tx}^2}{V_x^2} = \exp\left(-2\delta_x \tau_x\right) \tag{4.198}$$

Substituting δ_x from Equations 4.196–4.198 and since $\tau_x = \dfrac{\pi}{\omega_x}$, we get finally that

$$e_x = k_x^2 = \frac{V_{tx}^2}{V_x^2} = \exp\left(-\pi tg\beta_E\right) \tag{4.199}$$

It is obvious that according to the initial conditions $x_c = 0, x_b = 0, \dot{x}_c = V_{cx} = V_x$, $\dot{x}_b = 0$, and boundary conditions in the moment of time $t = \tau_1$ of the maximal elastic deformation, when $x_c = x_{cm}$, $x_b = x_{bm}$ and $\dot{x}_c = 0$, $\dot{x}_b = \dot{x}_{bm} = \dot{x}_m$ [where $\dot{x}_m$ is relative velocity of displacements (deformations) between contacting surfaces in the moment of the time ($t = \tau_1$), it follows that $x_{cm} = \dfrac{V_x}{2}\tau_1$ and $x_{bm} = \dfrac{\dot{x}_{bm}}{2}\tau_1$, and taking into account Equation 4.185, we get the following equation

$$x_{bm} = \frac{\tau_1}{2\tau_E} x_{cm} \tag{4.200}$$

and

$$\dot{x}_{bm} = \frac{V_x \tau_1}{2\tau_E} \tag{4.201}$$

But on the other hand, it is obvious that $x_{bt} = x_{btm} + x_{bm}$, and also we can take that $x_{bm} = \dfrac{\dot{x}_{bm}}{2}\tau_1$ and $x_{btm} = \dfrac{\dot{x}_{bm}}{2}\tau_2$ (where x_{btm} is the plastic displacement for the time of the elastic restitution τ_2), and also since $\tau_1 + \tau_2 = \tau_x$ we get for the maximum of plastic deformation $x_{bt} = x_{bm}\dfrac{\tau_x}{\tau_1}$, and taking into account Equation 4.200, it follows

$$x_{bt} = \frac{\tau_x}{2\tau_E} x_{cm} \tag{4.202}$$

The total indentation $x_m = x_{cm} + x_{bm}$ of a sled into semi-space in the moment of the time $t = \tau_1$ can be expressed as

$$x_m = \frac{x_{cm}}{2}\left(2 + \frac{\tau_1}{\tau_E}\right) \tag{4.203}$$

4.5.2 TANGENTIAL DISPLACEMENT

4.5.2.1 TANGENTIAL ELASTIC AND VISCOUS FORCES AT ELASTOPLASTIC CONTACT

In the case of an arbitrary elastoplastic collision between a sled, which has an arbitrary curvilinear cylindrical forward surface, see Figure 2.13, and the flat surface of semi-space, the integration of series of Equations 2.554–2.557 by limits y_c, y_b, $\dot{y}_b$ gives us for the tangential forces similar as in Equations 2.546, 2.547, 3.394 and 3.395

$$F_{c\tau y} = Gy_c P_{Lx} \tag{4.204}$$

$$F_{b\tau y} = \eta'_G \dot{y}_b P_{Lx} \tag{4.205}$$

where $P_{Lx} = L + D_{1x}$, and where, in case of Maxwell model, instead of the effective storage modulus G' we use the effective dynamic Young's elasticity modulus G.

On the other hand, according to Newton's Second Law and since in the case of elastic-plastic contact for Maxwell model when the elastic and viscous forces are equal, $F_\tau = F_{c\tau y} = F_{b\tau y} = -m\ddot{y}$, we can write that

$$F_{c\tau y} = -m\ddot{y} = G' P_{Lx} y_c \tag{4.206}$$

and

$$F_{b\tau y} = -m\ddot{y} = \eta'_G P_{Lx} \dot{y}_b \tag{4.207}$$

Taking into account Equation 4.60 and using Equations 4.206 and 4.207, we get

$$\dot{y}_b = \frac{1}{\tau_G} y_c \tag{4.208}$$

4.5.2.2 WORK AND APPROXIMATE SOLUTION TO THE DIFFERENTIAL EQUATIONS OF THE TANGENTIAL DISPLACEMENT

The works A_{ycm} and A_{ybm} of the rolling shear in the moment of the time $t = \tau_1$ and $x = x_{cm}$, can be found as given below

$$A_{ycm} = \int_0^{y_{cm}} F_{c\tau y} dy_c = G \int_0^{P_{Lxm}} dP_{Lx} \int_0^{y_{cm}} y_c dy_c = \frac{G}{2} P_{Lxm} y_{cm}^2 \tag{4.209}$$

where y_{cm} is the maximum of the tangential elastic deformation between contacting surfaces and respectively

$$A_{ybm}=\int\limits_{0}^{y_{bm}}F_{b\tau y}dy_b=\eta'_G\int\limits_{0}^{P_{Lxm}}dP_{Lx}\int\limits_{0}^{y_{bm}}\dot{y}_b dy_b=\eta'_G P_{Lxm}\frac{\int_0^{y_{bm}}\int dy_b dy_b}{\int_0^{\tau_1}dt}=\eta'_G P_{Lxm}\frac{y_{bm}^2}{2\tau_1}\quad(4.210)$$

where

$$P_{Lxm}=x_m^{1/2}(k_h x_m^{1/2}+2k_p R^{1/2})\quad(4.211)$$

and where y_{bm} is the plastic deformation of the shear between the surfaces of the contacting surfaces along axis Y at the instant of the time $t=\tau_1$ and $x=x_{cm}$.

We can find the approximate solutions for equations of motion in the same manner as for the equations with the equivalent constant viscoelasticity parameters, if we choose the equivalent constant parameters B_y, C_y so that works A_{ycm} and A_{ybm} with the variable viscoelasticity parameters c_y, b_y will be equal to the work with the constant viscoelasticity parameters. Thus, according to this statement and the boundary conditions $t=\tau_1$, $x=x_{cm}$ and $y=y_{cm}$, and also using the expressions for works A_{ycm} and A_{ybm} from Equations 4.209 and 4.210 in the phase of sliding shear, we get

$$A_{ycm}=C_y\int\limits_{0}^{y_{cm}}y_c dy_c=\frac{1}{2}C_y y_{cm}^2=\frac{1}{2}GP_{Lxm}y_{cm}^2\quad(4.212)$$

$$A_{ybm}=B_y\int\limits_{0}^{y_{bm}}\dot{y}_b dy_b=B_y\frac{\int_0^{y_{bm}}\int dy_b dy_b}{\int_0^{\tau_1}dt}=B_y\frac{y_{bm}^2}{2\tau_1}=\frac{\eta'_G}{2\tau_1}P_{Lxm}y_{bm}^2\quad(4.213)$$

Hence, we can write the expressions for the equivalent constant viscoelasticity parameters as follows:

$$C_y=GP_{Lxm}\quad(4.214)$$

$$B_y=\eta'_G P_{Lxm}\quad(4.215)$$

Thus, Equations 4.206 and 4.207 with variable parameters can be rewritten as the equations with constant parameters as follows

$$F_{c\tau y}=-m\dddot{y}=C_y y_c\quad(4.216)$$

and

$$F_{b\tau y}=-m\ddot{y}=B_y\dot{y}_b\quad(4.217)$$

Hence, we can write that

$$\dot{y}_b=\frac{C_y}{B_y}y_c\quad(4.218)$$

Or respectively, after differentiation, it follows

$$\ddot{y}_b = \frac{C_y}{B_y}\dot{y}_c \tag{4.219}$$

Taking into account that $\ddot{y} = \ddot{y}_c + \ddot{y}_b$, Equation 4.216 can be written now as

$$m\left(\ddot{y}_c + \frac{C_y}{B_y}\dot{y}_C\right) = -C_y y_c \tag{4.220}$$

Thus, the next differentiation equation related to y_c can be expressed as

$$m\ddot{y} + B_{Cy}\dot{y} + C_y y = 0 \tag{4.221}$$

where $B_{Cy} = m\frac{C_y}{B_y}$. Equation 4.221 is the equation of the damped oscillations and according to the initial condition $y_c = 0$ and $\dot{y}_c = V_y$, the solutions to this equation is known as

$$y_c = \frac{V_y}{\omega_y}e^{-\delta_y t}\sin(\omega_y t) \tag{4.222}$$

where $\omega_y = \sqrt{\omega_{0y}^2 - \delta_y^2}$, $\delta_y = \frac{B_{Cy}}{2m} = \frac{C_y}{2B_y}$ is the tangential damping factor, $\omega_{0y} = \sqrt{\frac{C_y}{m}}$ is the angular frequency of the harmonic oscillations by axis Y.

The equation for the tangential velocity of the centres of mass of a sled can be received by differentiation of Equation 4.222 as

$$\dot{y}_c = \frac{V_y}{\omega_y}e^{-\delta_y t}[\omega_y\cos(\omega_y t) - \delta_y\sin(\omega_y t)] \tag{4.223}$$

Now, since $V_{ty} = \dot{y}_c(t = \tau_x)$ the energetic coefficient of restitution for tangential displacement can be found respectively as

$$e_y = k_y^2 = \frac{V_{ty}^2}{V_y^2} = \exp\left(-2\delta_y\tau_x\right) \tag{4.224}$$

Taking into account Equation 4.60 and that $\delta_y = \frac{B_{Cy}}{2m}$, it follows

$$\delta_y = \frac{1}{2\tau_G} \tag{4.225}$$

But on the other hand, according to Equation 4.65 $tg\beta_G = \frac{1}{\omega_y \tau_G}$, and respectively

$$\delta_y = \frac{\omega_y tg\beta_G}{2} \tag{4.226}$$

Substituting $\tau_x = \frac{\pi}{\omega_x}$ and $\delta_y = \frac{\omega_y tg\beta_G}{2}$ into Equation 4.224 gives

$$e_y = k_y^2 = \frac{V_{ty}^2}{V_y} = \exp\left(-\frac{\pi\omega_y tg\beta_G}{\omega_x}\right) \tag{4.227}$$

4.5.3 COEFFICIENT OF FRICTION

Since the viscous and elastic forces in the Maxwell model are equal, the coefficient of sliding friction can be expressed by the four variants, such as

$$f_r = \frac{F_\tau}{F_n} = \frac{F_{b\tau y}}{F_{cn}} = \frac{F_{b\tau y}}{F_{bn}} = \frac{F_{c\tau y}}{F_{cn}} = \frac{F_{c\tau y}}{F_{bn}} \tag{4.228}$$

Choosing the third of them, we can write according to Equations 4.206 and 4.183 that in the case of collision between a sled and a semi-space approximately as

$$f_s = \frac{F_{c\tau y}}{F_{cn}} = \frac{GP_{Lx}}{ED_1(L_s + L)} \times \frac{y_c}{x_c} \tag{4.229}$$

or since $P_{Lx} = L + D_1 x$, it follows

$$f_s = \frac{F_{c\tau y}}{F_{cn}} = \frac{G(L + D_1 x)}{ED\ (L_s + L)} \times \frac{y_c}{x_c} \tag{4.230}$$

4.6 DYNAMICS OF AN ARBITRARY ELASTOPLASTIC COLLISION BETWEEN A VESSEL AND SEMI-SPACE

4.6.1 NORMAL DISPLACEMENT

4.6.1.1 NORMAL ELASTIC AND VISCOUS FORCES

In the case of an arbitrary elastoplastic collision between a vessel and the flat surface of semi-space, see Figure 2.14, taking into account that limits of integration are $x_c, x_b, \dot{x}_b$, we get the normal forces similar as in Equations 2.568 and 2.569 as given below

$$F_{cn} = EL_s D_1 x_c + 4E\frac{C_r D_1^{1/n}}{n+1} x_c^{\frac{n+1}{n}} \tag{4.231}$$

$$F_{bn} = \eta_E' L_s D_1 \dot{x}_b + 4\eta_E' C_r D_1^{1/n} \dot{x}_b x_b^{1/n} \tag{4.232}$$

Where, in the case of Maxwell model, instead of the storage modulus E' we use the dynamic Young's elasticity modulus E.

On the other hand, according to the Newton's Second Law and since in the case of elastic-plastic contact for Maxwell model when the elastic and the viscous forces are equal, $F_n = F_{cn} = F_{bn} = -m\ddot{x}$, we can write that

$$F_{cn} = -m\ddot{x} = E' L_s D_1 x_c + 4E' \frac{C_r D_1^{1/n}}{n+1} x_c^{\frac{n+1}{n}} \tag{4.233}$$

and

$$F_{bn} = -m\ddot{x} = \eta_E' L_s D_1 \dot{x}_b + 4\eta_E' C_r D_1^{1/n} \dot{x}_b x_b^{1/n} \tag{4.234}$$

where x_c,x_b are the normal displacement of the centre of mass of a vessel, and $\dot{x}_c$, $\dot{x}_b$ are the velocities of the normal displacement of the centre of mass of a vessel.

4.6.1.2 APPROXIMATE SOLUTION TO THE DIFFERENTIAL EQUATIONS OF THE NORMAL DISPLACEMENT

The works A_{xcm} and A_{xbm} at the maximum elastic compression $x_c = x_{cm}$ and $x_b = x_{bm}$, can be found by integration in a similar way like it was done in Equations 4.110 and 4.111, respectively

$$\begin{aligned} A_{xcm} &= \int_0^{x_{cm}} F_{cn} dx_c = D_1 E L_s \int_0^{x_{cm}} x_c dx_c + 4E\frac{C_r D_1^{1/n}}{n+1} \int_0^{x_m} x_c^{\frac{n+1}{n}} dx_c \\ &= E x_{cm}^2 \left(\frac{D_1}{2} L_s + \frac{4nC_r D_1^{1/n}}{(2n+1)(n+1)} x_{cm}^{1/n} \right) \end{aligned} \tag{4.235}$$

and

$$\begin{aligned} A_{xbm} &= \int_0^{x_{bm}} F_{bn} dx_b = \eta_E' L_s D_1 \int_0^{x_{bm}} \dot{x} dx + 4\eta_E' C_r D_1^{1/n} \int_0^{x_{bm}} \dot{x} x^{1/n} dx \\ &= \eta_E' \frac{L_s D_1 \int_0^{x_{bm}} \int dx dx + 4C_r D_1^{1/n} \int_0^{x_{bm}} \int dx x^{1/n} dx}{\int_0^{\tau_1} dt} = \frac{\eta_E'}{\tau_1} x_{bm}^2 \left(\frac{D_1}{2} L_s + \frac{4nC_r D_1^{1/n}}{2n+1} x_{bm}^{1/n} \right) \end{aligned} \tag{4.236}$$

Denoting $\frac{D_1}{2}L_s = q_1$, $\frac{4nC_rD_1^{1/n}}{(2n+1)(n+1)} = q_2$ and $\frac{4nC_rD_1^{1/n}}{2n+1} = q_3$, the equations for the works can be written respectively as:

$$A_{xcm} = Ex_{cm}^2(q_1 + q_2x_{cm}^{1/n}) \tag{4.237}$$

$$A_{xbm} = \frac{\eta'_E}{\tau_1}x_{bm}^2(q_1 + q_3x_{bm}^{1/n}) \tag{4.238}$$

According to the boundary conditions $t = \tau_1$, $x_c = x_{cm}$ and $x_b = x_{bm}$, and using the expressions for works A_{xcm} and A_{xbm} from Equations 4.237 and 4.238, we can write that

$$A_{xcm} = C_x\int_0^{x_{cm}} x_c dx_c = \frac{1}{2}C_x x_m^2 = Ex_{cm}^2(q_1 + q_2x_{cm}^{1/n}) \tag{4.239}$$

and

$$A_{xbm} = B_x\int_0^{x_{bm}} \dot{x}_b dx_b = B_x\frac{\int_0^{x_{bm}} x_b dx_b}{\int_0^{\tau_1} dt} = B_x\frac{x_{bm}^2}{2\tau_1} = \frac{\eta'_E}{\tau_1}x_{bm}^2(q_1 + q_3x_{bm}^{1/n}) \tag{4.240}$$

Hence, we can write the expressions for the equivalent constant viscoelasticity parameters as follows

$$C_x = 2E(q_1 + q_2x_{cm}^{1/n}) \tag{4.241}$$

$$B_x = 2\eta'_E(q_1 + q_3x_{bm}^{1/n}) \tag{4.242}$$

But, since practically $q_1 >> q_2x_m^{1/n}$ and $q_1 >> q_3x_m^{1/n}$, we can approximately take that

$$C_x = 2Eq_1 = ED_1L_s \tag{4.243}$$

and

$$B_x = 2\eta'_E q_1 = \eta'_E D_1L_s \tag{4.244}$$

Thus, Equations 4.233 and 4.234 with variable parameters can be rewritten as the equations with constant parameters as follows

$$F_{cn} = -m\ddot{x} = C_x x_c = ED_1L_s x_c \tag{4.245}$$

and respectively

$$F_{bn} = -m\ddot{x} = B_x\dot{x}_b = \eta'_E D_1L_s\dot{x}_b \tag{4.246}$$

Since $F_{cn} = F_{bn}$, we can write that

$$\dot{x}_b = \frac{1}{\tau_E}x_c \tag{4.247}$$

Further, after differentiation, it follows

$$\ddot{x}_b = \frac{1}{\tau_E}\dot{x}_c \tag{4.248}$$

Taking into account that $\ddot{x} = \ddot{x}_c + \ddot{x}_b$, Equation 4.245 can be written now as

$$m\left(\ddot{x}_c + \frac{1}{\tau_E}\dot{x}_C\right) = -ED_1L_sx_c \tag{4.249}$$

Thus, the next differentiation equation related to x_c can be expressed as

$$m\ddot{x}_c + \frac{m}{\tau_E}\dot{x}_c + ED_1L_sx_c = 0 \tag{4.250}$$

or in the canonical form

$$m\ddot{x}_c + B_{Cx}\dot{x}_c + C_xx_c = 0 \tag{4.251}$$

where respectively

$$B_{Cx} = \frac{m}{\tau_E} \tag{4.252}$$

Equation 4.251 is the equation of damped oscillations and the solution to this equation is already known as

$$x_c = \frac{V_x}{\omega_x}e^{-\delta_x t}\sin(\omega_x t) \tag{4.253}$$

where $\omega_x = \sqrt{\omega_{0x}^2 - \delta_x^2}$, $\delta_x = \dfrac{B_{Cx}}{2m}$ is the normal damping factor, $\omega_{0x} = \sqrt{\dfrac{C_x}{m}}$ is the angular frequency of free harmonic oscillations by axis X. Then, after integration of Equation 4.253, for the velocity of vessel, we get the following

$$\dot{x}_c = \frac{V_x}{\omega_x}e^{-\delta_x t}[\omega_x\cos(\omega_x t) - \delta_x\sin(\omega_x t)] \tag{4.254}$$

Now, using Equation 4.254, the time of indentation $t = \tau_1$ can be found according to the condition $\dot{x}_c = 0$ as

$$\tau_1 = \frac{1}{\omega_x}\operatorname{arctg}\left(\frac{\omega_x}{\delta_x}\right) \tag{4.255}$$

Thus, using Equations 4.253 and 4.255, we get for the maximum of an elastic displacement

$$x_{cm} = \frac{V_x}{\omega_x} \exp\left(-\frac{\delta_x}{\omega_x} \operatorname{arctg}\left(\frac{\omega_x}{\delta_x}\right)\right) \times \sin\left(\operatorname{arctg}\left(\frac{\omega_x}{\delta_x}\right)\right) \tag{4.256}$$

Taking into account that $\delta_x = \frac{B_{Cx}}{2m}$, it follows

$$\delta_x = \frac{1}{2\tau_E} \tag{4.257}$$

But on the other hand, since $tg\beta_E = \frac{1}{\omega_x \tau_E}$, see Equation 4.65, it follows respectively

$$\delta_x = \frac{\omega_x tg\beta_E}{2} \tag{4.258}$$

Substituting $\delta_x = \frac{\omega_x tg\beta_E}{2}$ into Equation 4.256 gives the following equation

$$x_{cm} = \frac{V_x}{\omega_x} \exp\left(-\frac{tg\beta_E}{2} \operatorname{arctg}\left(\frac{2}{tg\beta_E}\right)\right) \times \sin\left(\operatorname{arctg}\left(\frac{2}{tg\beta_E}\right)\right) \tag{4.259}$$

Now, since $V_{tx} = \dot{x}_c\,(t = \tau_x)$, the energetic coefficient of restitution can be found as

$$e_x = k_x^2 = \frac{V_{tx}^2}{V_x^2} = \exp(-2\delta_x \tau_x) \tag{4.260}$$

Substituting δ_x from Equation 4.258 into 4.260 and since $\tau_x = \frac{\pi}{\omega_x}$, we finally get that

$$e_x = k_x^2 = \frac{V_{tx}^2}{V_x} = \exp(-\pi tg\beta_E) \tag{4.261}$$

It is obvious that, in this case, we can use Equation 4.202 for the maximum of plastic deformation x_{bt} and Equation 4.203 for the elastic indentation x_m.

4.6.2 TANGENTIAL DISPLACEMENT

4.6.2.1 TANGENTIAL ELASTIC AND VISCOUS FORCES

In the case of an arbitrary elastoplastic collision between a vessel and the flat semi-space, see Figure 2.14, since limits of integration are $y_c, y_b, \dot{y}_b$, the tangential forces should be taken similarly as it has been presented in Equations 2.404 and 2.405, respectively as

$$F_{c\tau y} = F_{b\tau} = GP_{xn}y_c \tag{4.262}$$

$$F_{b\tau y} = F_{b\tau} = \eta_G' P_{xn}\dot{y}_b \tag{4.263}$$

where $P_{xn} = 2C_r D_1^{1/n} x^{1/n} + D_1 x$, see Equation 2.406, and where, in case of Maxwell model, instead of the effective storage modulus G' we use the effective dynamic Young's elasticity modulus G.

On the other hand, according to Newton's Second Law and since in the case of elastic-plastic contact for Maxwell model when the elastic and viscous forces are equal, $F_\tau = F_{c\tau} = F_{b\tau} = -m\ddot{y}$, we can write that

$$F_{c\tau y} = -m\ddot{y} = GP_{xn}y_c \tag{4.264}$$

and

$$F_{b\tau y} = -m\ddot{y} = \eta_G' P_{xn}\dot{y}_b \tag{4.265}$$

Taking into account Equation 4.60 and using Equations 4.264 and 4.265, we get

$$\dot{y}_b = \frac{1}{\tau_G} y_c \tag{4.266}$$

4.6.2.2 WORK AND APPROXIMATE SOLUTION TO THE DIFFERENTIAL EQUATIONS OF THE TANGENTIAL DISPLACEMENT

The works of the rolling shear in the moment of time $t = \tau_1$ and $x = x_{cm}$, we can find as follows

$$A_{ycm} = \int_0^{y_{cm}} F_{c\tau y} dy_c = G' \int_0^{P_{xnm}} dP_{xn} \int_0^{y_{cm}} y_c dy_c = \frac{G}{2} P_{xnm} y_{cm}^2 \tag{4.267}$$

where y_{cm} is the maximum of the tangential elastic deformation between contacting surfaces and

$$A_{ybm} = \int_0^{y_{bm}} F_{b\tau y} dy_b = \eta_G' \int_0^{P_{xnm}} dP_{xn} \int_0^{y_{bm}} \dot{y}_b dy_b = \eta_G' P_{xnm} \frac{\int_0^{y_{bm}} \int dy_b dy_b}{\int_0^{\tau_1} dt} = \eta_G' P_{xnm} \frac{y_{bm}^2}{2\tau_1} \tag{4.268}$$

where

$$P_{xnm} = 2C_r D_1^{1/n} x_m^{1/n} + D_1 x_m \tag{4.269}$$

and where y_{bm} is the plastic deformation of shear between the surfaces of the contacting surfaces along axis Y at the instance of time $t = \tau_1$ and $x = x_{cm}$.

Now, according to the boundary conditions $t = \tau_1$ and $x = x_{cm}$ and $y = y_{cm}$, and also using the expressions for works A_{ycm} and A_{ybm} from Equations 4.267 and 4.268 in the phase of sliding shear, thus, we can write

$$A_{ycm} = C_y \int_0^{y_{cm}} y_c dy_c = \frac{1}{2} C_y y_{cm}^2 = \frac{1}{2} G P_{xnm} y_{cm}^2 \tag{4.270}$$

$$A_{ybm} = B_y \int_0^{y_{bm}} \dot{y}_b dy_b = B_y \frac{\int_0^{y_{bm}} \int dy_b dy_b}{\int_0^{\tau_1} dt} = B_y \frac{y_{bm}^2}{2\tau_1} = \frac{\eta_G'}{2\tau_1} P_{xnm} y_{bm}^2 \tag{4.271}$$

Hence, we can write the expressions for the equivalent constant viscoelasticity parameters as:

$$C_y = G P_{xnm} \tag{4.272}$$

$$B_y = \eta_G' P_{xnm} \tag{4.273}$$

Thus, Equations 4.264 and 4.265 with variable parameters can be rewritten as the equations with constant parameters as follows

$$F_{c\tau y} = -m\ddot{y} = C_y y_c \tag{4.274}$$

and

$$F_{b\tau y} = -m\ddot{y} = B_y \dot{y}_b \tag{4.275}$$

Hence, we can write that

$$\dot{y}_b = \frac{C_y}{B_y} y_c \tag{4.276}$$

Or respectively, after differentiation, it follows

$$\ddot{y}_b = \frac{C_y}{B_y} \dot{y}_c \tag{4.277}$$

Taking into account that $\ddot{y} = \ddot{y}_c + \ddot{y}_b$, Equation 4.274 can be written now as

$$m\left(\ddot{y}_c + \frac{C_y}{B_y} \dot{y}_C\right) = -C_y y_c \tag{4.278}$$

Thus, the next differentiation equation related to y_c can be expressed as

$$m\ddot{y} + B_{Cy}\dot{y} + C_y y = 0 \tag{4.279}$$

where

$$B_{Cy} = m\frac{C_y}{B_y} = m\frac{G}{\eta'_G} \tag{4.280}$$

Equation 4.279 is the equation of the damped oscillations and according to the initial condition $y_c = 0$ and $\dot{y}_c = V_y$, the solutions to this equation is known as

$$y_c = \frac{V_y}{\omega_y} e^{-\delta_y t} \sin(\omega_y t) \tag{4.281}$$

where $\omega_y = \sqrt{\omega_{0y}^2 - \delta_y^2}$, $\delta_y = \frac{B_{Cy}}{2m} = \frac{C_y}{2B_y}$ is the tangential damping factor, $\omega_{0y} = \sqrt{\frac{C_y}{m}}$ is the angular frequency of the harmonic oscillations by axis Y.

As well the equation for the tangential velocity of the centres of mass of a vessel can be received by differentiation of Equation 4.281 as

$$\dot{y}_c = \frac{V_y}{\omega_y} e^{-\delta_y t} [\omega_y \cos(\omega_y t) - \delta_y \sin(\omega_y t)] \tag{4.282}$$

Now, since $V_{ty} = \dot{y}_c (t = \tau_x)$, the energetic coefficient of restitution for tangential displacement can be found as

$$e_y = k_y^2 = \frac{V_{ty}^2}{V_y^2} = \exp\left(-2\delta_y \tau_x\right) \tag{4.283}$$

Taking into account Equation 4.60 and that $\delta_y = \frac{B_{Cy}}{2m}$, it follows

$$\delta_y = \frac{1}{2\tau_G} \tag{4.284}$$

But on the other hand, according to Equation 4.65 $tg\beta_G = \frac{1}{\omega_y \tau_G}$, and

$$\delta_y = \frac{\omega_y tg\beta_G}{2} \tag{4.285}$$

Substituting $\tau_x = \frac{\pi}{\omega_x}$ and $\delta_y = \frac{\omega_y tg\beta_G}{2}$ into Equation 4.283 gives

$$e_y = k_y^2 = \frac{V_{ty}^2}{V_y^2} = \exp\left(-\frac{\pi\omega_y tg\beta_G}{\omega_x}\right) \tag{4.286}$$

4.6.3 COEFFICIENT OF FRICTION

Since the viscous and elastic forces in the Maxwell model are equal, the coefficient of sliding friction can be expressed by the four variants such as

$$f_r = \frac{F_\tau}{F_n} = \frac{F_{b\tau y}}{F_{cn}} = \frac{F_{b\tau y}}{F_{bn}} = \frac{F_{c\tau y}}{F_{cn}} = \frac{F_{c\tau y}}{F_{bn}} \tag{4.287}$$

If we choose the third of them, we can write according to Equations 4.264 and 4.247 that approximately in the case of collision between a vessel and a semi-space

$$f_s = \frac{F_{c\tau y}}{F_{cn}} = \frac{GP_{xn}}{ED_1L_s} \times \frac{y_c}{x_c} \tag{4.288}$$

or since $P_{xn} = 2C_r D_1^{1/n} x^{1/n} + D_1 x$, see Equation 2.406, it follows

$$f_s = \frac{F_{c\tau y}}{F_{cn}} = \frac{G(2C_r D_1^{1/n} x^{1/n} + D_1 x)}{ED_1L_s} \times \frac{y_c}{x_c} \tag{4.289}$$

4.7 DYNAMICS OF AN ARBITRARY ELASTOPLASTIC COLLISION BETWEEN CYLINDRICAL WHEEL AND SEMI-SPACE

4.7.1 NORMAL DISPLACEMENT

4.7.1.1 NORMAL ELASTIC AND VISCOUS FORCES

In the case of an arbitrary elastoplastic collision between a cylindrical wheel and the flat surface of semi-space, see Figure 2.12, taking into account that limits of integration are $x_c, x_b, \dot{x}_b$, we get similar results as in Equations 2.506 and 2.507 for the normal forces as given below

$$F_{cn} = \frac{2}{3} k_p E R_w^{1/2} x_c^{3/2} + E' L_w x_c \tag{4.290}$$

$$F_{bn} = 2k_p \eta'_E R_w^{1/2} \dot{x}_b x_b^{1/2} + 2\eta'_E L_w \dot{x}_b \tag{4.291}$$

Where, in the case of the Maxwell model, we use the effective dynamic Young's elasticity modulus E. On the other hand, according to Newton's Second Law and since in the case of elastic-plastic contact for the Maxwell model when the elastic and viscous forces are equal, $F_n = F_{cn} = F_{bn} = -m\ddot{x}$, we can write that

$$F_{cn} = -m\ddot{x} = \frac{2}{3}k_p E R_w^{1/2} x_c^{3/2} + E L_w x_c \tag{4.292}$$

and

$$F_{bn} = -m\ddot{x} = 2\eta_E' k_p R_w^{1/2} \dot{x}_b x_b^{1/2} + 2\eta_E' L_w \dot{x}_b \tag{4.293}$$

where respectively, x_c, x_b are the normal displacement of the centre mass of a wheel, and $\dot{x}_c, \dot{x}_b$ are the velocities of the normal displacement of the centre mass of a wheel.

4.7.1.2 APPROXIMATE SOLUTION TO THE DIFFERENTIAL EQUATIONS OF THE NORMAL DISPLACEMENT

The works A_{xcm} and A_{xbm} at the maximum elastic compression, $x_c = x_{cm}$ and $x_b = x_{bm}$, can be found by integration in a similar way like it was done in Equations 3.589 and 3.590

$$A_{xcm} = \int_0^{x_{cm}} F_{cn} dx_c = E \int_0^{x_{cm}} \left(\frac{2}{3} k_p R_w^{1/2} x_c^{3/2} + L_w x_c \right) dx_c = E x_{cm}^2 \frac{1}{2} \left(\frac{8}{15} k_p R_w^{1/2} x_{cm}^{1/2} + L_w \right) \tag{4.294}$$

and

$$\begin{aligned} A_{xbm} &= \int_0^{x_{bm}} F_{bn} dx_b = 2\eta_E' \int_0^{x_{bm}} \left(k_p R_w^{1/2} x_b^{1/2} + L_w \right) \dot{x}_b dx_b \\ &= 2\eta_E' \frac{\int_0^{x_{bm}} \int dx_b \left(k_p R_w^{1/2} x_b^{1/2} + L_w \right) dx_b}{\int_0^{\tau_1} dt} = \frac{\eta_E'}{\tau_1} x_{bm}^2 \left(\frac{4}{5} k_p R_w^{1/2} x_{bm}^{1/2} + L_w \right) \end{aligned} \tag{4.295}$$

But, practically since $L_w >> \frac{8}{15} k_p R_w^{1/2} x_{cm}^{1/2}$ and $L_w >> \frac{4}{5} k_p R_w^{1/2} x_{bm}^{1/2}$; therefore, we can approximately take that:

$$A_{xcm} = \frac{1}{2} E x_{cm}^2 L_w \tag{4.296}$$

and

$$A_{xbm} = \frac{\eta_E'}{\tau_1} x_{bm}^2 L_w \tag{4.297}$$

According to the boundary conditions $t = \tau_x$, $x_c = x_{cm}$ and $x_b = x_{bm}$, and using the expressions for works A_{xcm} and A_{xbm} from Equations 4.296 and 4.297, we can write that

$$A_{xcm} = C_x \int_0^{x_{cm}} x_c dx_c = \frac{1}{2} C_x x_{cm}^2 = \frac{1}{2} E x_{cm}^2 L_w \tag{4.298}$$

and

$$A_{xbm} = B_x \int_0^{x_{bm}} \dot{x}_b dx_b = B_x \frac{\int_0^{x_{bm}} x_b dx_b}{\int_0^{\tau_1} dt} = B_x \frac{x_{bm}^2}{2\tau_1} = \frac{\eta_E'}{\tau_1} x_{bm}^2 L_w \tag{4.299}$$

Hence, we can write the expressions for the equivalent constant viscoelasticity parameters as

$$C_x = EL_w \tag{4.300}$$

and

$$B_x = 2\eta_E' L_w \tag{4.301}$$

Thus, Equations 4.292 and 4.293 with variable parameters can be rewritten as the equations with constant parameters as follows

$$F_{cn} = -m\ddot{x} = C_x x_c = EL_w x_c \tag{4.302}$$

and respectively

$$F_{bn} = -m\ddot{x} = B_x \dot{x}_b = 2\eta_E' L_w \dot{x}_b \tag{4.303}$$

Hence, since $F_{cn} = F_{bn}$ and $\tau_E = \frac{\eta_E'}{E}$, see Equation 4.59, we can write that

$$\dot{x}_b = \frac{1}{2\tau_E} x_c \tag{4.304}$$

Further, after differentiation, it follows

$$\ddot{x}_b = \frac{1}{2\tau_E} \dot{x}_c \tag{4.305}$$

Taking into account that $\ddot{x} = \ddot{x}_c + \ddot{x}_b$, Equation 4.302 can be written now as

$$m\left(\ddot{x}_c + \frac{1}{2\tau_E} \dot{x}_C\right) = -EL_w x_c \tag{4.306}$$

Thus, the next differentiation equation related to x_c can be expressed as

$$m\ddot{x}_c + \frac{m}{2\tau_E} \dot{x}_c + ED_1 L_s x_c = 0 \tag{4.307}$$

or in the canonical form respectively

$$m\ddot{x}_c + B_{Cx} \dot{x}_c + C_x x_c = 0 \tag{4.308}$$

whereas in this case, respectively

$$B_{Cx} = \frac{m}{2\tau_E} \tag{4.309}$$

The Equation 4.308 the solution to this equation can be expressed respectively as

$$x_c = \frac{V_x}{\omega_x} e^{-\delta_x t} \sin(\omega_x t) \tag{4.310}$$

where $\omega_x = \sqrt{\omega_{0x}^2 - \delta_x^2}$, $\delta_x = \frac{B_{Cx}}{2m}$ is the normal damping factor, $\omega_{0x} = \sqrt{\frac{C_x}{m}}$ is the angular frequency of free harmonic oscillations by axis X. Then after integration, we get for the velocity of wheel

$$\dot{x}_c = \frac{V_x}{\omega_x} e^{-\delta_x t} [\omega_x \cos(\omega_x t) - \delta_x \sin(\omega_x t)] \tag{4.311}$$

Using Equation 4.311, the time of indentation $t = \tau_1$ can be found according to the condition $\dot{x}_c = 0$, respectively as

$$\tau_1 = \frac{1}{\omega_x} \operatorname{arctg}\left(\frac{\omega_x}{\delta_x}\right) \tag{4.312}$$

Thus, we get for the maximum of an elastic displacement

$$x_{cm} = \frac{V_x}{\omega_x} \exp\left(-\frac{\delta_x}{\omega_x} \operatorname{arctg}\left(\frac{\omega_x}{\delta_x}\right)\right) \times \sin\left(\operatorname{arctg}\left(\frac{\omega_x}{\delta_x}\right)\right) \tag{4.313}$$

Taking into account Equation 4.309 and that $\delta_x = \frac{B_{Cx}}{2m}$, it follow

$$\delta_x = \frac{1}{4\tau_E} \tag{4.314}$$

But on the other hand, since according to $tg\beta_E = \frac{1}{\omega_x \tau_E}$, see Equation 4.65, it follows in this case respectively that

$$\delta_x = \frac{\omega_x tg\beta_E}{4} \tag{4.315}$$

Substituting $\delta_x = \frac{\omega_x tg\beta_E}{4}$ into Equation 4.313 gives

$$x_{cm} = \frac{V_x}{\omega_x} \exp\left(-\frac{tg\beta_E}{4} \operatorname{arctg}\left(\frac{4}{tg\beta_E}\right)\right) \times \sin\left(\operatorname{arctg}\left(\frac{4}{tg\beta_E}\right)\right) \tag{4.316}$$

Now, since $V_{tx} = \dot{x}_c(t = \tau_x)$, the energetic coefficient of restitution can be found respectively as

$$e_x = k_x^2 = \frac{V_{tx}^2}{V_x^2} = \exp(-2\delta_x\tau_x) \tag{4.317}$$

Substituting δ_x from Equation 4.315 in Equation 4.317 and since $\tau_x = \dfrac{\pi}{\omega_x}$, we get finally that

$$e_x = k_x^2 = \frac{V_{tx}^2}{V_x^2} = \exp\left(-\frac{\pi tg\beta_E}{2}\right) \tag{4.318}$$

Further, according to the initial conditions $x_c = 0, x_b = 0, \dot{x}_c = V_{cx}, \dot{x}_b = 0$, and the boundary conditions in the moment of the time $t = \tau_1$ of the maximal elastic deformation, when $x_c = x_{cm}, x_b = x_{bm}$ and $\dot{x}_c = 0, \dot{x}_b = \dot{x}_{bm} = \dot{x}_m$, it follows that $x_{cm} = \dfrac{V_x}{2}\tau_1$ and $x_{bm} = \dfrac{\dot{x}_{bm}}{2}\tau_1$, and taking in account Equation 4.304, we get

$$x_{bm} = \frac{\tau_1}{4\tau_E}x_{cm} \tag{4.319}$$

and

$$\dot{x}_{bm} = \frac{V_x}{4\tau_E}\tau_1 \tag{4.320}$$

But on the other hand, since $x_{bt} = x_{btm} + x_{bm}$, and since we can take that $x_{bm} = \dfrac{\dot{x}_{bm}}{2}\tau_1$ and $x_{btm} = \dfrac{\dot{x}_{bm}}{2}\tau_2$ (where x_{btm} is the plastic displacement for the time of the elastic restitution τ_2), and also since $\tau_1 + \tau_2 = \tau_x$ respectively we get for the maximum of plastic deformation $x_{bt} = x_{bm}\dfrac{\tau_x}{\tau_1}$, and then according to the Equation 4.319, it follows in this case that

$$x_{bt} = \frac{\tau_x}{4\tau_E}x_{cm} \tag{4.321}$$

The total indentation $x_m = x_{cm} + x_{bm}$ of a sled into semi-space in the moment of the time $t = \tau_1$, respectively can be expressed as

$$x_m = \frac{x_{cm}}{4}\left(4 + \frac{\tau_1}{\tau_E}\right) \tag{4.322}$$

4.7.2 TANGENTIAL DISPLACEMENT

4.7.2.1 TANGENTIAL ELASTIC AND VISCOUS FORCES

It is obvious that, in the case of an elastoplastic collision between cylindrical wheel and the flat semi-space, the contact of rolling shear has place. Also, since limits of integration are y_c, y_b, $\dot{y}_b$, the tangential forces should be taken similar like they have presented in Equations 3.292 and 3.293, respectively as

$$F_{c\tau y} = G' P_{xw} y_{cr} \tag{4.323}$$

$$F_{b\tau y} = \eta'_G P_{xw} \dot{y}_{br} \tag{4.324}$$

where $P_{xw} = D_1 x + L_w$, and where, in the case of the Maxwell model, instead the effective storage modulus G' we use the effective dynamic Young's elasticity modulus G.

Further, since in the case of a rolling contact, according to Equations 3.65 and 3.66 $\dot{y}_r = \frac{x}{l}\dot{y}$, $y_r = \frac{x}{l} y$ and since in this case $l = R_w$, see Figure 2.12, the equations for the tangential viscoelastic forces for the rolling contact, can be rewritten as

$$F_{c\tau y} = G P_{rw} y_c \tag{4.325}$$

$$F_{b\tau y} = P_{rw} \dot{y}_b \tag{4.326}$$

where (you can see as well Eq. 3.302), respectively for cylindrical wheel

$$P_{rw} = \frac{x}{R_w}(D_1 x + L_w) \tag{4.327}$$

On the other hand, according to the Newton's Second Law and since in the case of elastoplastic contact for the Maxwell model when the elastic and the viscous forces are equal, $F_\tau = F_{c\tau} = F_{b\tau} = -m\ddot{y}$, we can write that

$$F_{c\tau y} = -m\ddot{y} = G P_{rw} y_c \tag{4.328}$$

and

$$F_{b\tau y} = -m\ddot{y} = \eta'_G P_{rw} \dot{y}_b \tag{4.329}$$

Taking into account Equations 4.328, 4.329 and 4.60, we get

$$\dot{y}_b = \frac{1}{\tau_G} y_c \tag{4.330}$$

4.7.2.2 WORK AND APPROXIMATE SOLUTION TO THE DIFFERENTIAL EQUATIONS OF THE TANGENTIAL DISPLACEMENT

The works of the rolling shear in the moment of the time $t = \tau_1$ and $x = x_{cm}$, we can find respectively as

$$A_{ycm} = \int_0^{y_{cm}} F_{c\tau y} dy_c = G \int_0^{P_{rwm}} dP_{rw} \int_0^{y_{cm}} y_c dy_c = \frac{G}{2} P_{rwm} y_{cm}^2 \tag{4.331}$$

where y_{cm} is the maximum of the tangential elastic deformation between contacting surfaces, and respectively

$$A_{ybm} = \int_0^{y_{bm}} F_{b\tau y} dy_b = \eta'_G \int_0^{P_{rwm}} dP_{rw} \int_0^{y_{bm}} \dot{y}_b dy_b = \eta'_G P_{rwm} \frac{\int_0^{y_{bm}} \int dy_b dy_b}{\int_0^{\tau_1} dt} = \frac{\eta'_G}{2} P_{rwm} \frac{y_{bm}^2}{\tau_1} \tag{4.332}$$

where

$$P_{rwm} = \frac{x_m}{R_w}(D_1 x_m + L_w) \tag{4.333}$$

and where y_{bm} is the plastic deformation of the shear between the surfaces of the contacting surfaces along axis Y at the instant of the time $t = \tau_1$ and $x = x_{cm}$.

Now, according to the boundary conditions $t = \tau_1$, $x = x_{cm}$ and $y = y_{cm}$, and also using the expressions for work A_{ycm} and A_{ybm} from Equations 4.331 and 4.332 in the phase of rolling shear, respectively we can write

$$A_{ycm} = C_{yr} \int_0^{y_{cm}} y_c dy_c = \frac{1}{2} C_{yr} y_{cm}^2 = \frac{1}{2} G P_{rwm} y_{cm}^2 \tag{4.334}$$

and

$$A_{ybm} = B_{yr} \int_0^{y_{bm}} \dot{y}_b dy_b = B_{yr} \frac{\int_0^{y_{bm}} \int dy_b dy_b}{\int_0^{\tau_1} dt} = B_{yr} \frac{y_{bm}^2}{2\tau_1} = \frac{\eta'_G}{2\tau_1} P_{rwm} y_{bm}^2 \tag{4.335}$$

Hence, we can write the expressions for the equivalent constant viscoelasticity parameters, respectively as:

$$C_{yr} = G P_{rwm} \tag{4.336}$$

$$B_{yr} = \eta'_G P_{rwm} \tag{4.337}$$

Thus, Equations 4.328 and 4.329 with variable parameters can be rewritten as the equations with constant parameters as follows

$$F_{c\tau y} = -m\ddot{y} = C_{yr} y_c \tag{4.338}$$

and

$$F_{b\tau y} = -m\ddot{y} = B_{yr} \dot{y}_b \tag{4.339}$$

Hence, we can write that

$$\dot{y}_b = \frac{C_{yr}}{B_{yr}} y_c \tag{4.340}$$

Or respectively, after differentiation, it follows

$$\ddot{y}_b = \frac{C_{yr}}{B_{yr}} \dot{y}_c \tag{4.341}$$

Taking into account that $\ddot{y} = \ddot{y}_c + \ddot{y}_b$, the Equation 4.338 can be rewritten as

$$m\left(\ddot{y}_c + \frac{C_{yr}}{B_{yr}} \dot{y}_C \right) = -C_{yr} y_c \tag{4.342}$$

Thus, the next differentiation equation related to y_c can be expressed as

$$m\ddot{y} + B_{Cyr}\dot{y} + C_{yr} y = 0 \tag{4.343}$$

where $B_{Cyr} = m\frac{C_{yr}}{B_{yr}}$. The Equation 4.343 is the equation of the damped oscillations and according to the initial condition $y_c = 0$ and $\dot{y}_c = V_y$, the solutions to this equation is known as

$$y_c = \frac{V_y}{\omega_{yr}} e^{-\delta_{yr} t} \sin(\omega_y t) \tag{4.344}$$

where $\omega_{yr} = \sqrt{\omega_{0yr}^2 - \delta_{yr}^2}$, $\delta_{yr} = \frac{B_{Cyr}}{2m} = \frac{C_{yr}}{2B_{yr}}$ is the tangential damping factor, $\omega_{0yr} = \sqrt{\frac{C_{yr}}{m}}$ is the angular frequency of the harmonic oscillations by axis Y.

The equation for the tangential velocity of the centres of mass of a cylindrical wheel can be received by differentiation of Equation 4.344 respectively as

$$\dot{y}_c = \frac{V_y}{\omega_{yr}} e^{-\delta_{yr} t} [\omega_{yr} \cos(\omega_{yr} t) - \delta_{yr} \sin(\omega_{yr} t)] \tag{4.345}$$

Now, since $V_{ty} = \dot{y}_c(t = \tau_x)$, the energetic coefficient of restitution for tangential displacement can be found respectively as

$$e_y = k_y^2 = \frac{V_{ty}^2}{V_y^2} = \exp\left(-2\delta_y \tau_x\right) \tag{4.346}$$

Taking into account Equation 4.60 and that $\delta_{yr} = \dfrac{C_{yr}}{2B_{yr}}$, it follows

$$\delta_{yr} = \frac{1}{2\tau_G} \tag{4.347}$$

But on the other hand, according to Equation 4.65, we can write that in this case $tg\beta_G = \dfrac{1}{\omega_{yr}\tau_G}$, and respectively, it follows that

$$\delta_{yr} = \frac{\omega_{yr} tg\beta_G}{2} \tag{4.348}$$

Substituting $\tau_x = \dfrac{\pi}{\omega_x}$ and $\delta_{yr} = \dfrac{\omega_{yr} tg\beta_G}{2}$ into Equation 4.346 gives

$$e_y = k_y^2 = \frac{V_{ty}^2}{V_y^2} = \exp\left(-\frac{\pi\omega_{yr} tg\beta_G}{\omega_x}\right) \tag{4.349}$$

The full changing of the energy of the dissipative system at the rolling shear can be found as the difference between A_y and A_{yc} as follows

$$\Delta W_y = A_y - A_{yc} = \frac{mV_y^2}{2} - \frac{mV_{ty}^2}{2} - \frac{J_z\omega_t^2}{2} = A_{ybm} + A_{ybt} = A_{yb} \tag{4.350}$$

Now, let us find the work A_{yc} and A_{yb} for the full period time of contact:

$$A_{yc} = C_{yr}\int_0^{y_{ct}} y_c dy_c = \frac{1}{2}C_{yr} y_{ct}^2 \tag{4.351}$$

$$A_{yb} = Q_\omega = B_{yr}\int_0^{y_{bt}} \dot{y}_b dy_b = B_{yr}\frac{\int_0^{y_{bt}}\int dy_b dy_b}{\int_0^{\tau_x} dt} = B_{yr}\frac{y_{bt}^2}{2\tau_x} \tag{4.352}$$

As we already know, according to the Equation 3.85 the conclusion can be drawn that the work A_{yc} is transformed into the kinetic energy of rotation of a wheel, but on the other hand the work A_{yb} is transformed into dissipative energy Q_ω in the process of the internal friction. Accordingly to Equation 3.85 and Equations 4.350 and 4.351, it follows

$$A_{yc} = \frac{J_z \omega_t^2}{2} = \frac{1}{2} G P_{rwm} y_{ct}^2 \tag{4.353}$$

Hence, the equation for the relative angular velocity at the instant time of rebound can be written as follows

$$\omega_t = \left(\frac{G P_{rwm}}{J_z} \right)^{\frac{1}{2}} y_{ct} \tag{4.354}$$

It is obvious that, in a moment of the time $t = \tau_x$, $y_c = y_{ct}$ and respectively we can write that

$$y_{ct} = \frac{V_y}{\omega_{yr}} e^{-\delta_{yr}\tau_x} \sin(\omega_{yr}\tau_x) \tag{4.355}$$

Taking into account that $\tau_x = \frac{\pi}{\omega_x}$ and $\delta_{yr} = \frac{\omega_{yr} tg \beta_G}{2}$, we get

$$y_{ct} = \frac{V_{yr}}{\omega_{yr}} \exp\left(-\frac{\pi \omega_{yr} tg \beta_G}{2\omega_x} \right) \times \sin\left(\frac{\pi \omega_{yr}}{\omega} \right) \tag{4.356}$$

4.7.3 COEFFICIENT OF FRICTION

We can write according to Equations 4.292 and 4.323 that, in the case of collision between a cylindrical wheel and a semi-space approximately

$$f_s = \frac{F_{c\tau y}}{F_{cn}} = \frac{G P_{rw}}{E L_w} \times \frac{y_c}{x_c} \tag{4.357}$$

or since $P_{rw} = \frac{x}{R_w}(D_1 x + L_w)$, it follows

$$f_s = \frac{F_{c\tau y}}{F_{cn}} = \frac{G x (D_1 x + L_w)}{R_w E L_w} \times \frac{y_c}{x_c} \tag{4.358}$$

As we can see that, in case if $L_w >> x$, it follows

$$f_s = \frac{F_{c\tau y}}{F_{cn}} = \frac{G'(x_c + x_b)}{R_w E'} \times \frac{y_c}{x_c} \tag{4.359}$$

4.8 DYNAMICS OF AN ARBITRARY ELASTOPLASTIC COLLISION BETWEEN TOROIDAL WHEEL AND SEMI-SPACE

4.8.1 NORMAL DISPLACEMENT

4.8.1.1 NORMAL ELASTIC AND VISCOUS FORCES

As we already know, in the case of an arbitrary elastoplastic collision between a toroidal surface of a wheel and the flat surface of semi-space, the contact area has an elliptical form, see Figure 2.12. Also, in this case, since we can take that $R_a = R_w$, $r_a = r_y$, $R_b = R_p$ and $r_b = r_z$, and take into account that limits of integration are x_c,x_b,$\dot{x}_b$, we get for the normal forces similar as in Equations 2.517 and 2.518 respectively as

$$F_{cn} = \frac{2}{3} E \psi_w x_c^{3/2} \tag{4.360}$$

and

$$F_{bn} = 2\eta'_E \psi_w \dot{x}_b x_b^{1/2} \tag{4.361}$$

whereas in this case

$$\psi_w = (k_{pa} R_w^{1/2} + k_{pb} R_p^{1/2}) \tag{4.362}$$

On the other hand, according to Newton's Second Law and since in the case of elastic-plastic contact for the Maxwell model when the elastic and the viscous forces are equal, $F_n = F_{cn} = F_{bn} = -m\ddot{x}$, we can write that

$$F_{cn} = -m\ddot{x} = \frac{2}{3} \psi_w E x_c^{3/2} \tag{4.363}$$

and

$$F_{bn} = -m\ddot{x} = 2\eta'_E \psi_w \dot{x}_b x_b^{1/2} \tag{4.364}$$

where respectively, x_c, x_b are the normal displacement of the centre mass of a wheel, and $\dot{x}_c$, $\dot{x}_b$ are the velocities of the normal displacement of the centre mass of a wheel.

Taking into account Equations 4.59 and 4.364, 4.365, we get the same result as in Equation 4.82, respectively

$$x_c^{3/2} = 3\tau_E \dot{x}_b x_b^{1/2} \tag{4.365}$$

Thus, in the dynamic elastoplastic contact between a toroidal wheel and the flat semi-space we can use Equations 4.80–4.82 as well. It is obvious that these equations can be used for any kind of elliptical, circular or oval contacts.

4.8.1.2 WORK OF THE NORMAL FORCES

Since the limits of integration in this case are x_c and x_b, the work A_{xcm} of elastic force F_{cn}, see Equation 4.363, in the compression phase can be found respectively as

$$A_{xcm} = \int_0^{x_{cm}} F_{cn} dx_c = \frac{2}{3} E\psi_w \int_0^{x_{cm}} x_c^{3/2} dx_c = \frac{4}{15} E\psi_w x_{cm}^{5/2} \tag{4.366}$$

The work A_{xbm} of viscous force F_{bn} at the compression can be found respectively as

$$A_{xbm} = \int_0^{x_{bm}} F_{bn} dx_b = \int_0^{x_{bm}} 2\eta_E' \psi_w \dot{x}_b x_b^{1/2} dx_b = 2\eta_E' \psi_w \frac{\int_0^{x_{bm}} \int dx_b x_b^{1/2} dx_b}{\int_0^{\tau_1} dt} = \frac{4\eta_E' \psi_w x_{bm}^{5/2}}{5\tau_1} \tag{4.367}$$

The total work in the moment of the time $t = \tau_1$ and when $x_c = x_{cm}$ can be expressed as

$$A_{xm} = A_{xcm} + A_{xbm} = W_{xm} = \frac{mV_x^2}{2} - \frac{m(\dot{x}_{bm})^2}{2} \tag{4.368}$$

Using Equations 4.366–4.368 and 4.83–4.85, and since $\eta_E' = E\tau_E$ we can write respectively that

$$A_{xm} = \frac{4}{15} \psi_w E x_{cm}^{5/2} \left(1 + \frac{3}{6^{5/3}} \left(\frac{\tau_1}{\tau_E} \right)^{2/3} \right) = \frac{mV_x^2}{2} \left(1 - \left(\frac{\tau_1}{6\tau_E} \right)^{4/3} \right) \tag{4. 369}$$

Thus, we get the formula for calculation x_{cm}, respectively

$$x_{cm} = \left\{ \frac{15mV_x^2}{8\psi_w E} f_{\tau E} \right\}^{2/5} \tag{4.370}$$

where $f_{\tau E}$ is the dissipative function, which is the same as in Equation 4.116

Also, according to Equations 4.85 and 4.118, the expression for the maximum of indentation at impact between spherical bodies can be respectively written as

$$x_t = x_{bt} = \left\{ \frac{15mV_x^2}{8\psi_w E} f_{\tau E} \right\}^{2/5} \times \frac{\tau_x}{\tau_1} \left(\frac{\tau_1}{6\tau_E} \right)^{2/3} \tag{4.371}$$

The Equations 4.370 and 4.371 are valid only in the case of elastic-plastic deformations, when $tg\beta \geq 1/3$ and respectively when $k_x \leq \frac{\pi - 1}{\pi + 1}$.

On the other hand, the full dissipative work of the contacting bodies in the moment of the time $t = \tau_x$ and when $x_b = x_{bt}$, can be found as follows

$$A_{xb} = \int_0^{x_{bt}} F_{bn} dx_b = \int_0^{x_{bt}} 2\eta'_E \psi \dot{x}_b x_b^{1/2} dx_b = 2\eta'_E \psi \frac{\int_0^{x_{bt}} \int dx_b x_b^{1/2} dx_b}{\int_0^{\tau_x} dt} = \frac{4\eta'_E \psi x_{bt}^{5/2}}{5\tau_x} \quad (4.372)$$

where x_{bt} is the plastic displacement between the centres mass of the contacting bodies in the moment of the time of rebound, when $t = \tau_x$ and $x_c = 0$. Taking into account Equation 4.82, we get that

$$A_{xb} = \frac{4\eta'_E \psi_w}{5} x_{cm}^{5/2} \frac{\tau_x^{3/2} \tau_1^{2/3}}{(6\tau_E)^{5/3}} \quad (4.373)$$

It is obvious that the full dissipative work is equal to the change of the kinetic energy in the time of impact, and we can write, therefore, that

$$A_{xb} = \frac{mV_x^2}{2} - \frac{mV_{tx}^2}{2} = \frac{mV_x^2}{2}(1 - k_x^2) \quad (4.374)$$

whereas we know k_x is the coefficient of restitution.

Thus according to Equations 4.374 and 4.373, and taking into account that $\chi = (1 - k_x^2)$ is the coefficient of mechanical losses at impact, it follows

$$A_{xb} = \frac{4\eta'_E \psi_w}{5} x_{cm}^{5/2} \frac{\tau_x^{3/2} \tau_1^{2/3}}{(6\tau_E)^{5/3}} = \chi \frac{mV_x^2}{2} \quad (4.375)$$

Now, according to Equations 4.59, 4.85 and 4.113, we get respectively

$$\chi = \frac{8\eta'_E \psi_w}{5mV_x^2} x_{cm}^{5/2} \frac{\tau_x^{3/2} \tau_1^{2/3}}{(6\tau_E)^{5/3}} \quad (4.376)$$

Using Equations 4.376 and 4.370, and since $\eta'_E = E\tau_E$, the expression for the coefficient of mechanical losses can be written respectively as

$$\chi = \frac{3\tau_x^{3/2}}{6^{5/3}} \left(\frac{\tau_1}{\tau_E} \right)^{2/3} \times f_{\tau E} \quad (4.377)$$

4.8.1.3 APPROXIMATE SOLUTION TO THE DIFFERENTIAL EQUATIONS OF THE NORMAL DISPLACEMENT BY USING THE METHOD OF EQUIVALENT WORKS

Using expressions for work A_{xcm} and A_{xbm} from Equations 4.366 and 4.367, the equivalent constant parameters B_x, C_x can be found by the already known way, respectively as

$$A_{xcm} = C_x \int_0^{x_{cm}} x_c dx_c = \frac{1}{2} C_x x_{cm}^2 = \frac{4}{15} E\psi_w x_{cm}^{5/2} \tag{4.378}$$

and

$$A_{xbm} = B_x \int_0^{x_{bm}} \dot{x}_b dx_b = B_x \frac{\int_0^{x_{bm}} x_b dx_b}{\int_0^{\tau_1} dt} = B_x \frac{x_{bm}^2}{2\tau_1} = \frac{4\eta_E' \psi_w x_{bm}^{5/2}}{5\tau_1} \tag{4.379}$$

Hence, we can write the expressions for the equivalent constant viscoelasticity parameters, respectively as

$$C_x = \frac{8}{15} E\psi_w x_{cm}^{1/2} \tag{4.380}$$

$$B_x = \frac{8\eta_E' \psi_w}{5} x_{bm}^{1/2} \tag{4.381}$$

Thus, Equations 4.363 and 4.364 with variable parameters can be rewritten as the equations with constant parameters as follows

$$F_{cn} = -m\ddot{x} = C_x x_c \tag{4.382}$$

and respectively

$$F_{bn} = -m\ddot{x} = B_x \dot{x}_b \tag{4.383}$$

Hence, since $F_{cn} = F_{bn}$, we can write that

$$\dot{x}_b = \frac{C_x}{B_x} x_c \tag{4.384}$$

Or respectively, after differentiation, it follows

$$\ddot{x}_b = \frac{C_x}{B_x} \dot{x}_c \tag{4.385}$$

Taking into account that $\ddot{x} = \ddot{x}_c + \ddot{x}_b$, the Equation 4.382 can be written now as

$$m\left(\ddot{x}_c + \frac{C_x}{B_x}\dot{x}_C\right) = -C_x x_c \tag{4.386}$$

Thus, the next differentiation equation related to x_c can be expressed as

$$m\ddot{x}_c + B_{Cx}\dot{x}_c + C_x x_c = 0 \tag{4.387}$$

where respectively

$$B_{Cx} = m\frac{C_x}{B_x} = \frac{mE}{3\eta_E'}\frac{x_{cm}^{1/2}}{x_{bm}^{1/2}} \tag{4.388}$$

Taking into account Equations 4.59 and 4.83, it follows

$$B_{Cx} = \frac{m}{3\tau_E^{2/3}}\left(\frac{6}{\tau_1}\right)^{1/3} \tag{4.389}$$

The Equation 4.387 is the equation of damped oscillations and the solution to this equation is known as

$$x_c = \frac{V_x}{\omega_x}e^{-\delta_x t}\sin(\omega_x t) \tag{4.390}$$

where $\omega_x = \sqrt{\omega_{0x}^2 - \delta_x^2}$, $\delta_x = \frac{B_{Cx}}{2m} = \frac{C_x}{2B_x}$ is the normal damping factor, $\omega_{0x} = \sqrt{\frac{C_x}{m}}$ is the angular frequency of free harmonic oscillations by axis X. After integration Equation 4.390, we get the equation for the normal velocity of centres of mass of a wheel

$$\dot{x}_c = \frac{V_x}{\omega_x}e^{-\delta_x t}[\omega_x\cos(\omega_x t) - \delta_x\sin(\omega_x t)] \tag{4.391}$$

Now, since $V_{tx} = \dot{x}_c\ (t = \tau_x)$ the energetic coefficient of restitution can find respectively as

$$e_x = k_x^2 = \frac{V_{tx}^2}{V_x^2} = \exp(-2\delta_x\tau_x) \tag{4.392}$$

Also, since $\delta_x = \frac{B_{Cx}}{2m}$, taking into account Equation 4.389, we can write Equation 4.392 finally as

$$e_x = k_x^2 = \frac{V_{tx}^2}{V_x^2} = \exp\left(-\frac{2\tau_x}{3\tau_E^{2/3}}\left(\frac{6}{\tau_1}\right)^{1/3}\right) \quad (4.393)$$

4.8.2 TANGENTIAL DISPLACEMENT

4.8.2.1 TANGENTIAL ELASTIC AND VISCOUS FORCES

Taking into account that limits of integration are $y_c, y_b, \dot{y}_b$, we get for the normal forces similar as in Equations 2.523 and 2.524, respectively as

$$F_{c\tau y} = GP_{bx}y_c \quad (4.394)$$

$$F_{b\tau y} = \eta'_G P_{bx}\dot{y}_b \quad (4.395)$$

where

$$P_{bx} = D_1 x + 2k_{pb}R_p^{1/2}x^{1/2} \quad (4.396)$$

As we know, in the case of rolling contact $\dot{y}_r = \frac{x}{l}\dot{y}$, see Equation 3.65, and $y_r = \frac{x}{l}y$, see Equation 3.66, and since in this case $l = R_w$, see Figure 2.12, the equations for the tangential viscoelastic forces for the rolling contact, can be rewritten as

$$F_{c\tau y} = G'P_{br}y_c \quad (4.397)$$

$$F_{b\tau y} = \eta'_G P_{br}\dot{y}_b \quad (4.398)$$

where $P_{br} = \frac{x^{3/2}}{R_w}(D_1 x^{1/2} + 2k_p R^{1/2})$, and where, in case of the Maxwell model, instead of the effective storage modulus G' we use the effective dynamic Young's elasticity modulus G.

On the other hand, according to the Newton's Second Law and since in the case of elastic-plastic contact for Maxwell model when the elastic and the viscous forces are equal, $F_{\tau r} = F_{c\tau r} = F_{b\tau r} = -m\ddot{y}$, we can write that

$$F_{c\tau r} = -m\ddot{y} = GP_{br}y_c \quad (4.399)$$

and

$$F_{b\tau r} = -m\ddot{y} = \eta'_G P_{br}\dot{y}_b \quad (4.400)$$

4.8.2.2 WORK AND APPROXIMATE SOLUTION TO THE DIFFERENTIAL EQUATIONS OF THE TANGENTIAL DISPLACEMENT

The works of the rolling shear in the moment of the time $t = \tau_1$ and $x = x_{cm}$, we can find respectively as

$$A_{ycm} = \int_0^{y_{cm}} F_{c\tau y} dy_c = G \int_0^{P_{brm}} dP_{br} \int_0^{y_{cm}} y_c dy_c = \frac{G}{2} P_{brm} y_{cm}^2 \tag{4.401}$$

where y_{cm} is the maximum of the tangential elastic deformation between contacting bodies, and respectively

$$A_{ybm} = \int_0^{y_{bm}} F_{b\tau y} dy_b = \eta_G' \int_0^{P_{brm}} dP_{br} \int_0^{y_{bm}} \dot{y}_b dy_b = \eta_G' P_{brm} \frac{\int_0^{y_{bm}} \int dy_b dy_b}{\int_0^{\tau_1} dt} = \eta_G' P_{brm} \frac{y_{bm}^2}{\tau_1} \tag{4.402}$$

where

$$P_{brm} = \frac{x_m^{3/2}}{R_w} (D_1 x_m^{1/2} + 2k_p R^{1/2}) \tag{4.403}$$

and where y_{bm} is the plastic deformation of the shear between the surfaces of the contacting bodies along axis Y at the instant of the time $t = \tau_1$ and $x = x_{cm}$.

Further, according to the boundary conditions $t = \tau_1$, $x = x_{cm}$ and $y = y_{cm}$, and also using the expressions for work A_{ycm} and A_{ybm} from Equations 4.401 and 4.402 in the phase of the rolling shear, respectively we get

$$A_{ycm} = C_{yr} \int_0^{y_{cm}} y_c dy_c = \frac{1}{2} C_{yr} y_{cm}^2 = \frac{1}{2} G P_{brm} y_{cm}^2 \tag{4.404}$$

$$A_{ybm} = B_{yr} \int_0^{y_{bm}} \dot{y}_b dy_b = B_{yr} \frac{\int_0^{y_{bm}} \int dy_b dy_b}{\int_0^{\tau_1} dt} = B_{yr} \frac{y_{bm}^2}{2\tau_1} = \frac{\eta_G'}{2\tau_1} P_{brm} y_{bm}^2 \tag{4.405}$$

Hence, we can write the expressions for the equivalent constant viscoelasticity parameters, respectively as:

$$C_{yr} = G P_{brm} \tag{4.406}$$

$$B_{yr} = \eta_G' P_{brm} \tag{4.407}$$

Thus, Equations 4.399 and 4.400 with variable parameters can be rewritten as the equations with constant parameters as follows

$$F_{c\tau r} = -m\ddot{y} = C_{yr} y_c \tag{4.408}$$

and

$$F_{b\tau r} = -m\ddot{y} = B_{yr} \dot{y}_b \tag{4.409}$$

Hence, we can write that

$$\dot{y}_b = \frac{C_{yr}}{B_{yr}} y_c \tag{4.410}$$

Or respectively, after differentiation, it follows

$$\ddot{y}_b = \frac{C_{yr}}{B_{yr}} \dot{y}_c \tag{4.411}$$

Taking into account that $\ddot{y} = \ddot{y}_c + \ddot{y}_b$, the Equation 4.408 can be written now as

$$m\left(\ddot{y}_c + \frac{C_{yr}}{B_{yr}} \dot{y}_C\right) = -C_{yr} y_c \tag{4.412}$$

Thus, the next differentiation equation related to y_c can be expressed as

$$m\ddot{y} + B_{Cyr}\dot{y} + C_{yr} y = 0 \tag{4.413}$$

where $B_{Cyr} = m\frac{C_{yr}}{B_{yr}}$. The Equation 4.413 is the equation of the damped oscillations and according to the initial condition $y_c = 0$ and $\dot{y}_c = V_y$, the solutions to this equation is known as

$$y_c = \frac{V_y}{\omega_{yr}} e^{-\delta_{yr} t} \sin(\omega_{yr} t) \tag{4.414}$$

where $\omega_{yr} = \sqrt{\omega_{0yr}^2 - \delta_{yr}^2}$, $\delta_{yr} = \frac{B_{Cyr}}{2m} = \frac{C_{yr}}{2B_{yr}}$ is the tangential damping factor, $\omega_{0yr} = \sqrt{\frac{C_{yr}}{m}}$ is the angular frequency of the harmonic oscillations by axis *Y*.

As well the equation for the tangential relative velocity of centres of mass of a toroidal well can be received by differentiation of Equation 4.414 respectively as

$$\dot{y}_c = \frac{V_y}{\omega_{yr}} e^{-\delta_{yr}t}[\omega_{yr}\cos(\omega_{yr}t) - \delta_{yr}\sin(\omega_{yr}t)] \tag{4.415}$$

Now, since $V_{ty} = \dot{y}_c(t = \tau_x)$, the energetic coefficient of restitution for tangential displacement can find respectively as

$$e_y = k_y^2 = \frac{V_{ty}^2}{V_y^2} = \exp\left(-2\delta_y \tau_x\right) \tag{4.416}$$

Taking into account Equation 4.60 and that $\delta_{yr} = \frac{C_{yr}}{2B_{yr}}$, it follows

$$\delta_{yr} = \frac{1}{2\tau_G} \tag{4.417}$$

But on the other hand, according to Equation 4.65, we can write that in this case $tg\beta_G = \frac{1}{\omega_{yr}\tau_G}$, and respectively, it follows that

$$\delta_{yr} = \frac{\omega_{yr} tg\beta_G}{2} \tag{4.418}$$

Substituting $\tau_x = \frac{\pi}{\omega_x}$ and $\delta_{yr} = \frac{\omega_{yr} tg\beta_G}{2}$ into Equation 4.416 gives

$$e_y = k_y^2 = \frac{V_{ty}^2}{V_y^2} = \exp\left(-\frac{\pi\omega_{yr} tg\beta_G}{\omega_x}\right) \tag{4.419}$$

The full changing of the energy of the dissipative system at the rolling shear can be found as the difference between A_y and A_{yc} as follows

$$\Delta W_y = A_y - A_{yc} = \frac{mV_y^2}{2} - \frac{mV_{ty}^2}{2} - \frac{J_z\omega_t^2}{2} = A_{ybm} + A_{ybt} = A_{yb} \tag{4.420}$$

Now, let us find the work A_{yc} and A_{yb} for the full period time of contact:

$$A_{yc} = C_{yr}\int_0^{y_{ct}} y_c dy_c = \frac{1}{2} C_{yr} y_{ct}^2 \tag{4.421}$$

$$A_{yb} = Q_\omega = B_{yr}\int_0^{y_{bt}} \dot{y}_b dy_b = B_{yr}\frac{\int_0^{y_{bt}}\int dy_b dy_b}{\int_0^{\tau_x} dt} = B_{yr}\frac{y_{bt}^2}{2\tau_x} \tag{4.422}$$

As we already know, according to the Equation 3.85 the conclusion can be drawn that the work A_{yc} is transformed into the kinetic energy of rotation of a wheel, but on the other hand, the work A_{yb} is transformed into dissipative energy Q_{ω} in the process of the internal friction. Accordingly to Equation 3.85 and Equations 4.420 and 4.421, it follows

$$A_{yc} = \frac{J_z \omega_t^2}{2} = \frac{1}{2} G P_{rwm} y_{ct}^2 \tag{4.423}$$

Hence, the equation for the relative angular velocity at the instant time of rebound can be written as follows

$$\omega_t = \left(\frac{G P_{rwm}}{J_z} \right)^{\frac{1}{2}} y_{ct} \tag{4.424}$$

It is obvious that in the moment of time $t = \tau_x$, $y_c = y_{ct}$, we can write that

$$\dot{y}_{ct} = \frac{V_y}{\omega_{yr}} e^{-\delta_{yr}\tau_x} [\omega_{yr} \cos(\omega_{yr}\tau_x) - \delta_{yr} \sin(\omega_{yr}\tau_x)] \tag{4.425}$$

Taking into account that $\tau_x = \frac{ð}{\omega_x}$ and $\delta_{yr} = \frac{\omega_{yr} tg \beta_G}{2}$, we get

$$y_{ct} = \frac{V_{yr}}{\omega_{yr}} \exp\left(-\frac{\pi \omega_{yr} tg \beta_G}{2\omega_x} \right) \times \sin\left(\frac{\pi \omega_{yr}}{\omega} \right) \tag{4.426}$$

4.8.3 COEFFICIENT OF FRICTION

Since the viscous and elastic forces in the Maxwell model are equal, the coefficient of rolling friction can be expressed by the four variants, see Equation 4.168. But for example, if to choose the third of them, where $F_{c\tau r} = -m\ddot{y} = G P_{br} y_c$, see Equation 4.399, and $F_{cn} = -m\ddot{x} = \frac{2}{3} \psi_w E x_c^{3/2}$, see Equation 4.360, we get the following equation

$$f_r = \frac{F_{c\tau r}}{F_{cn}} = \frac{3 G P_{br}}{2 \psi_w E} \times \frac{y_c}{x_c^{3/2}} \tag{4.427}$$

Also since $P_{br} = \frac{x^{3/2}}{R_w} (D_1 x^{1/2} + 2k_p R^{1/2})$, finally we get

$$f_r = \frac{F_{c\tau r}}{F_{cn}} = \frac{3G}{2\psi_w E} \times \frac{x^{3/2} (D_1 x^{1/2} + 2k_p R^{1/2}) y_c}{x_c^{3/2} l} \tag{4.428}$$

KEYWORDS

- **elastoplastic contacts**
- **specific forces**
- **Young's elasticity modulus**
- **Maxwell model**
- **tangential forces**

CHAPTER 5

DEFINITION OF THE TIME OF FATIGUE LIFE AND DYNAMIC MECHANICAL PROPERTIES OF MATERIALS OF THE CONTACTING SURFACES

ABSTRACT

The basic principle of the definition of the time of fatigue life prior to the beginning of destruction of a material of the contacting surface, using the high acceleration fatigue life test and the stress-number of cycles (S-N) Voeller diagrams are described in this chapter. The equations for the definition of the time of fatigue life of cylindrical gears and the surface of a semi-space in the flow of rigid bodies have been derived. The experimental techniques for the determination of dynamics elasticity and viscosity modulus by using the method of the 'temperature–time superposition' have been considered.

5.1 TIME OF FATIGUE LIFE

As we already know, also it has been studied, for example by Johnson (1985); Alfredsson (2000); Golenko (2010) and many others, that in the process of cyclic contact between contacting bodies, particularly at impact, their surfaces are subjected to quickly change cyclical deformations and stresses. Therefore, it is obvious that in some moment of the time, which is called the time of fatigue life, the destruction of the contact surface will happen. Usually, in this moment of time, the fatigue crack will arise, which will develop very quickly. Therefore, it is very important to define the time of fatigue life prior to the beginning of the destruction of a material of the contacting surface. For example, for the definition of the time of fatigue life, we can use the high acceleration fatigue life test, using the Stress-Number of cycles (S-N) Voeller diagram (see Golenko, 2010), the illustration is given in Figure 5.1.

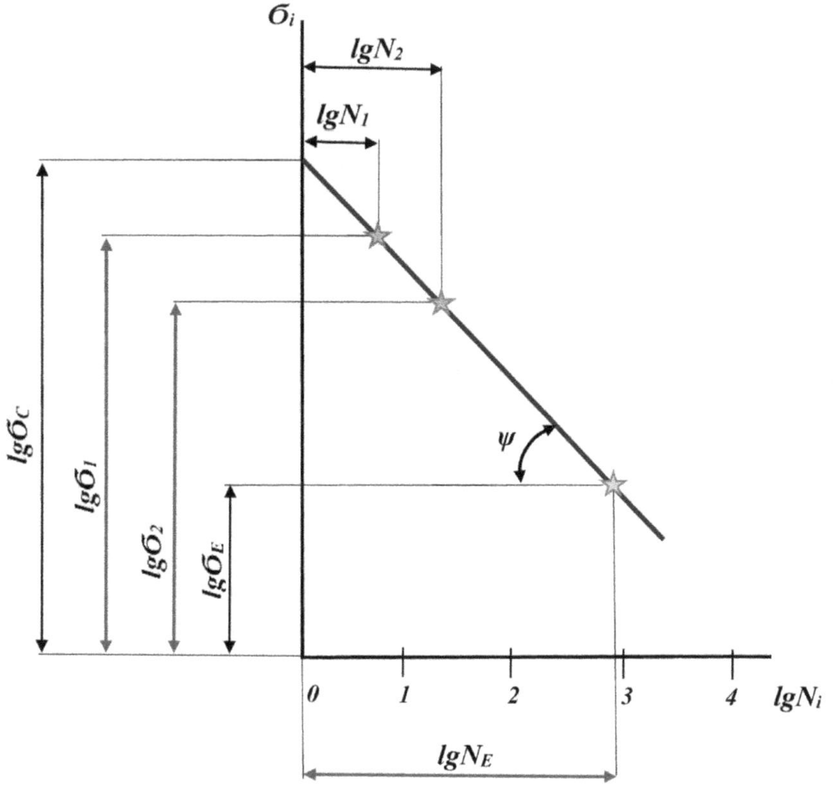

FIGURE 5.1 (See color insert.) Illustration of the stress-number of cycles (S-N) Voeller diagram of the high acceleration fatigue life test for the definition of the exponent λ of the curve of a fatigue life.

In a general way, the time of fatigue life τ_f can be defined as follows

$$\tau_f = \frac{N_E}{v_i} \tag{5.1}$$

where v_i denotes the frequency of contact and N_E is the effective number of cycles of the contact loading prior to the beginning of the destruction of materials of the contacting surfaces. N_E can be calculated by using the following formula.

$$N_E = \left(\frac{\sigma_c}{\sigma_E} \right)^{\lambda} \tag{5.2}$$

where σ_c is the limiting strength at $N_i=1$, see Figure 5.1

The equation of an exponent of curve of fatigue in the logarithmic coordinates for any $\sigma_i < \sigma_c$ and for $1 < N_i$ (Figure 5.1) looks as

$$\lg \sigma_i = \lg \sigma_c - \mathrm{tg}\psi \lg N_i \tag{5.3}$$

were here respectively $tg\psi = 1/\lambda$, and hence

$$\lambda = \frac{\lg N_i}{\lg \sigma_c - \lg \sigma_i} \tag{5.4}$$

where, see Figure 5.1, N_i is the limit of number of cycles of deformation of testing at the moment of the tear of sample; σ_i is the stress of deformation of testing at the moment of the tear of sample; λ is an exponent of curve of fatigue life. For example, we can use the result of test when $\sigma_i = \sigma_1$ and $N_i = N_1$. Practically, for definition of the time of fatigue life of a new material, we have to use the experimental number N_1 and N_2 cycles of the tear or appearing of cracks of destruction on the contacting surface, and then by using statistical analysis we can calculate λ.

Also, it is obvious that the main problem is the calculation of the effective contact *stresses* σ_E.

5.2 DEFINITION OF THE TIME OF FATIGUE LIFE OF THE CYLINDRICAL GEARING

First of all, let us consider an example of the kinematics and dynamics of the cylindrical involute gearing, which are well known (Budynas and Nisbett, 2011; Golenko (2010), Singh (2012). The schematic illustration of the contact at impact between involute surfaces of two teeth of two gears is shown in Figure 5.2.

As we can see here, the gears depicted in the process of their driving with angular velocities ω_1 and ω_1. The teeth of gears enter into contact at impact at the initial point B and then the contact zone moves to the point *A* of the maximum approach between contacting surfaces. At the point C contacting surfaces get out of contact. The line *BC* represents the direction of action of the normal forces, is called the line of action or the pressure line. Point A is called as the pitch point.

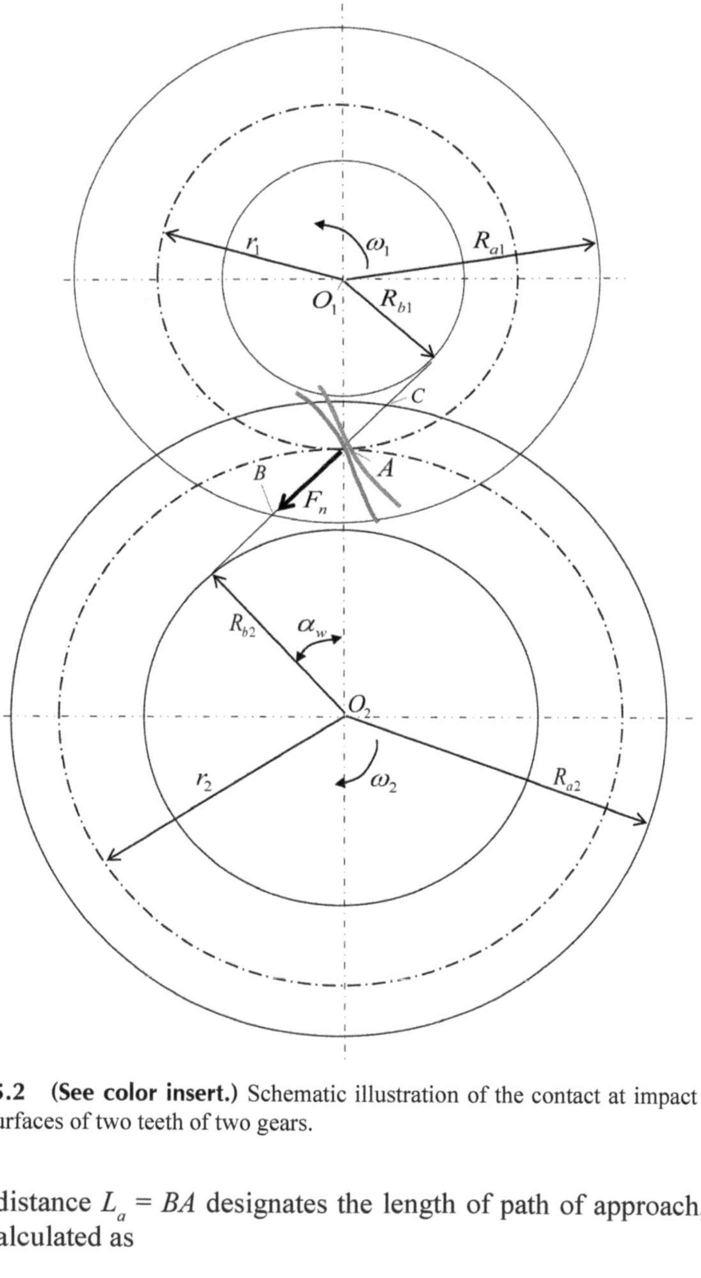

FIGURE 5.2 (See color insert.) Schematic illustration of the contact at impact between involute surfaces of two teeth of two gears.

The distance $L_a = BA$ designates the length of path of approach, and it can be calculated as

$$L_a = (R_{a1}^2 - R_{b1}^2)^{1/2} - r_1 \sin\alpha_w \tag{5.5}$$

But on the other hand, $L_r = AC$ is the length of path of recess and it can be calculated as

$$L_r = (R_{a2}^2 - R_{b2}^2)^{1/2} - r_2 \sin\alpha_w \tag{5.6}$$

where R_{a1}, R_{a2} are the radii of addendum circles; R_{b1}, R_{b2} are the radii of base circles; r_1,r_2 are the radii of pitch circles; α_w is the pressure angle.

The initial maximal velocity of 'sliding', which is indeed the relative velocity of the rolling motion between two cylindrical involute surfaces at the point B (see for example, Budynas and Nisbett (2011), Golenko (2000), Singh (2012)) can be expressed as

$$V_{yB} = (\omega_2 + \omega_1)L_a \tag{5.7}$$

It is known that the velocity of sliding at the point A is equal zero.

It is obvious that the time of contact τ_c of the one pair of teeth can be found as

$$\tau_c = \frac{T_{\omega1}}{z_1} = \frac{T_{\omega2}}{z_2} \tag{5.8}$$

where $T_{\omega1} = \dfrac{2\pi}{\omega_1}$ is the period of rotation of the first gear, $T_{\omega2} = \dfrac{2\pi}{\omega_2}$ is the period of rotation of the second gear, z_1 is the number of the teeth of the first gear, z_2 is the number of the teeth of the second gear. Also, we can take that time of compression $\tau_s = \tau_c / 2$.

As we know, since usually the length of contact area (or line) between two cylinders in many times bigger than the indentation, that is, $L >> k_h x$ and as well it is obvious that $L >> \frac{2}{3} k_p R^{1/2} x^{1/2}$; therefore, for finding viscoelastic forces we can use Equations 2.384 and 2.385. Also, since it is the rolling contact, taking into account Equations 3.65 and 3.66, it follows that

$$F_n = L(2\eta_E' \dot{x} + E'x) \tag{5.9}$$

$$F_\tau = L\frac{x}{l}(\eta_G' \dot{y} + G'y) \tag{5.10}$$

where L is the length of contact line, which is equal to the length of tooth of the second gear and since $\bar{\omega} = \bar{\omega}_1 + \bar{\omega}_2$ we can write, see Figure 2.8, that according to Equation 3.162 $l \approx \frac{\bar{V}_y R_1 R_2}{R_1 \bar{V}_{2y} + R_2 \bar{V}_{1y}}$. In this case, $\bar{V}_{2y} = \bar{V}_{1y} = \omega_1 r_1 = \omega_2 r_2$, $\bar{V}_y = \bar{V}_{1y} + \bar{V}_{2y}$, it follows that $l = 2R = \frac{2R_1 R_2}{R_1 + R_2}$, where $R_1 = r_1$ sin α_w and $R_2 = r_2$ sin α_w are the radii of involutes of the contacting surfaces.

As we can see, and this is obvious, that the contact between two teeth is similar as the dynamic collision at sliding impact between two cylindrical bodies with parallel axes. Also we can see that we get the dangerous point during impact between teeth at the moment of the maximum of compression $x = x_m$ and the tangential motion $y = y_m$ at the point A, when $t = \tau_s = \tau_c / 2$, $\dot{y} = 0$, and $\dot{x} = 0$. Thus, since $\tau_s = \tau_c / 2 = \frac{\pi}{\omega_1 z_1}$ and $V_{yB} = (\omega_2 + \omega_1)L_a$, and since we can take that $y_m = \frac{V_{yB}}{2}\tau_s$, the maximum of relative tangential displacement at the point A can be found as

$$y_m = \frac{\pi}{2\omega_1 z_1}(\omega_1 + \omega_2)L_a \tag{5.11}$$

Thus, taking into account Equations 5.9–5.11 and that $l = 2R$ the expressions for the viscoelastic forces at the moment of maximal compression $x = x_m$ and $y = y_m$ at the point A can be written as

$$F_n = LE'x_m \tag{5.12}$$

$$F_\tau = \frac{\pi x_m}{4R\omega_1 z_1} LL_a(\omega_1 + \omega_2)G' \tag{5.13}$$

The friction coefficient at the moment of the maximum compression and the shear at the point A can be expressed respectively

$$f_{rA} = \frac{F_\tau}{F_n} = \frac{\pi}{4} \frac{L_a G'(\omega_1 + \omega_2)}{RE'\omega_1 z_1} \tag{5.14}$$

Now, for the definition of the time of fatigue life of the cylindrical gears, we have to find the effective contact stresses σ_E at the moment of maximum compression and shear. Since the area of contact between two cylindrical surfaces at the maximum of compression is equal $S_{xm} = a_m L$, where $a_m = 2r_m = 2k_p R^{1/2} x_m^{1/2}$, see Equation 2.154, it follows $S_{xm} = 2k_p L R^{1/2} x_m^{1/2}$, and we can find the maximum of the normal stresses, see Equation 2.326, as given below

$$\sigma_n = \frac{3F_n}{2S_x} = \frac{3E'x_m^{1/2}}{4k_p R^{1/2}} \tag{5.15}$$

On the other hand, we can find the maximal tangential stress as $\sigma_{\tau c} = \frac{3}{2}\sigma_\tau$, see Equation 2.352 and since

$$\sigma_\tau = \frac{\pi x_m^{1/2}}{8k_p R^{3/2}\omega_1 z_1} L_a(\omega_1 + \omega_2)G' \tag{5.16}$$

we get

$$\sigma_{\tau c} = \frac{3\pi x_m^{1/2}}{16k_p R^{3/2}\omega_1 z_1} L_a(\omega_1 + \omega_2)G' \tag{5.17}$$

Also, since the maximal normal force can be written as, see for example Budynas and Nisbett (2011), Golenko (2010), Singh (2012),

$$F_n = LE'x_m = \frac{M_1}{r_1 \cos\alpha_w} \tag{5.18}$$

where M_1 is the torsional moment on the first gear, we can write that

$$x_m = \frac{M_1}{LE'r_1 \cos\alpha_w} \tag{5.19}$$

and also we can rewrite Equations 5.15 and 5.17 as given below

$$\sigma_n = \frac{3F_n}{2S_x} = \frac{3(E')^{1/2}}{4k_p R^{1/2}}\left(\frac{M_1}{Lr_1 \cos\alpha_w}\right)^{1/2} \tag{5.20}$$

$$\sigma_{\tau c} = \frac{3\pi}{16k_p R^{3/2}\omega_1 z_1}\left(\frac{M_1}{Lr_1 E' \cos\alpha_w}\right)^{1/2} L_a(\omega_1 + \omega_2)G' \tag{5.21}$$

According to the hypothesis of maximum tangential stresses the equation for the general contact stresses of viscoelasticity in the centre of the contact area can be expressed as

$$\sigma_E = \sqrt{\sigma_n^2 + 4\sigma_{\tau c}^2} \tag{5.22}$$

where σ_n is the normal contact stresses of viscoelasticity and $\sigma_{\tau c}$ is the tangential contact stresses of viscoelasticity.

Substitution of σ_n and $\sigma_{\tau c}$ into Equation 5.22 gives us

$$\sigma_E = \sqrt{\frac{M_1}{Lr\cos\alpha_{w1}k_p^2}\left(\frac{3E'}{4R}+\frac{4\pi^2}{E'}\left(\frac{3G'}{16R^{3/2}\omega_1 z_1}L_a(\omega_1+\omega_2)\right)^2\right)} \tag{5.23}$$

Since the frequency of loading of the contacting surface of the first gear $\omega_1 = v_i$ and if the exponent λ of curve of a fatigue life and the limiting strength σ_c were found experimentally using Equations 5.1, 5.2 and 5.23, the time of fatigue life can be calculated finally as

$$\tau_f = \frac{\sigma_c^\lambda}{\omega_1}\left(\frac{M_1}{k_p^2 Lr_1\cos\alpha_w}\left(\frac{3E'}{4R}+\frac{4\pi^2}{E'}\left(\frac{3G'}{16R^{3/2}\omega_1 z_1}L_a(\omega_1+\omega_2)\right)^2\right)\right)^{-\frac{\lambda}{2}} \tag{5.24}$$

5.3 DEFINITION OF THE TIME OF FATIGUE LIFE OF THE SURFACE OF SEMI-SPACE IN THE FLOW OF RIGID BODIES

5.3.1 INTRODUCTION

As we know, in nature all solid bodies have diversified forms, but we can calculate the average statistical radius and mass of a solid body and use them in the design and calculations as it has place for a spherical body.

Let a spherical body having the average statistical mass m_2, the average statistical radius R and the initial velocity V_0 come into viscoelastic contact under an arbitrary angle of attack α to the surface of semi-space at the initial instance of time $t=0$, at the initial point of contact 0 (Figure 5.3).

And let the vectors of velocities be applied to the centre of mass of the body (the point C_0). Also, in Figure 5.3 the following are designated: $V_{0x} = V_0 \sin\alpha$, $V_{0y} = V_0 \cos\alpha$ denote the initial normal and tangential velocities of a body; $M = F_\tau l_x$ is the reactive moment; ω is the angular velocity and ε is the angular acceleration around of the centre of mass of a body; V_d is the volume of deformations, which is forming the course of contact.

The contact configuration between a spherical body and a semi-space is depicted in the Figure 5.4.

It is seen here (Figure 5.4) that at the initial instance of the time, the body with the centre of mass at the point C_0 comes into contact with the surface of semi-space at the initial point of the contact 0 with coordinates $x=0$ and $y=0$, but at the instance of time t, the centre of mass of a body (the point C_t) takes the position with coordinates X and Y. We take here that x and y are the displacements of the centre mass of a body (point C_t) equal to the distances of the mutual approach between a body and a semi-space in the coordinates X and Y. Also, it is shown here that x_1 is the normal deformation of the surface of semi-space and also it is the distance between the initial point of contact 0 and the current point of contact A. The initial point of contact 0 coincides with the current contact point A only at the instant of the time $t=0$. The distance x_2 is the normal deformation of the surface of a body. We can see here that $x = h = x_1 + x_2$ is the distance of the mutual approach (the total deformation) between a body and a semi-space, and as well, in the same instance of time, it is the displacement of the centre of mass of a body relative to the initial point of contact 0. The current point of contact A is displacing in the process of contact and it is always placed in the middle of the contact zone. Also, the geometry of the surface of the contact zone is characterized by the geometrical parameters such as, see Figure 5.4: a is the diameter of the contact area; r is the radius of thc contact arca; h_x is the depth of the contact surface, or in other words it is the depth of indentation of a body into the surface of semi-space.

The electronic-mechanical device (Figure 5.5) works in the following way:

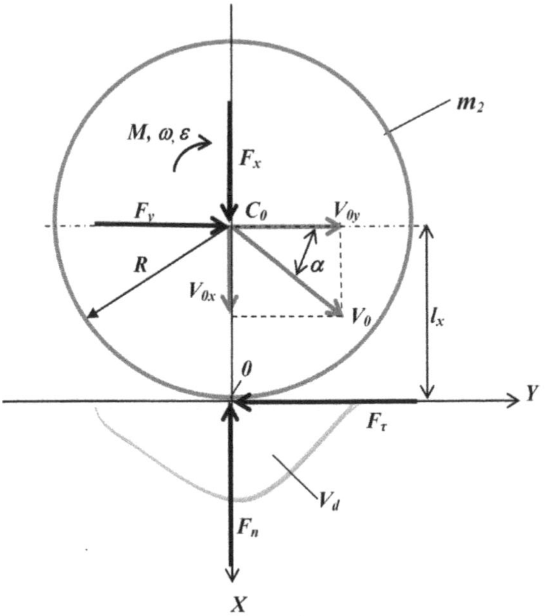

FIGURE 5.3 **(See color insert.)** Schematic illustration of the mechanics of a viscoelastic contact between a spherical solid body and a semi-space at impact.

Source: Reprinted from Goloshchapov, 2015a, with permission from SAGE Publications. http://journals.sagepub.com/doi/abs/10.1177/1056789514560912).

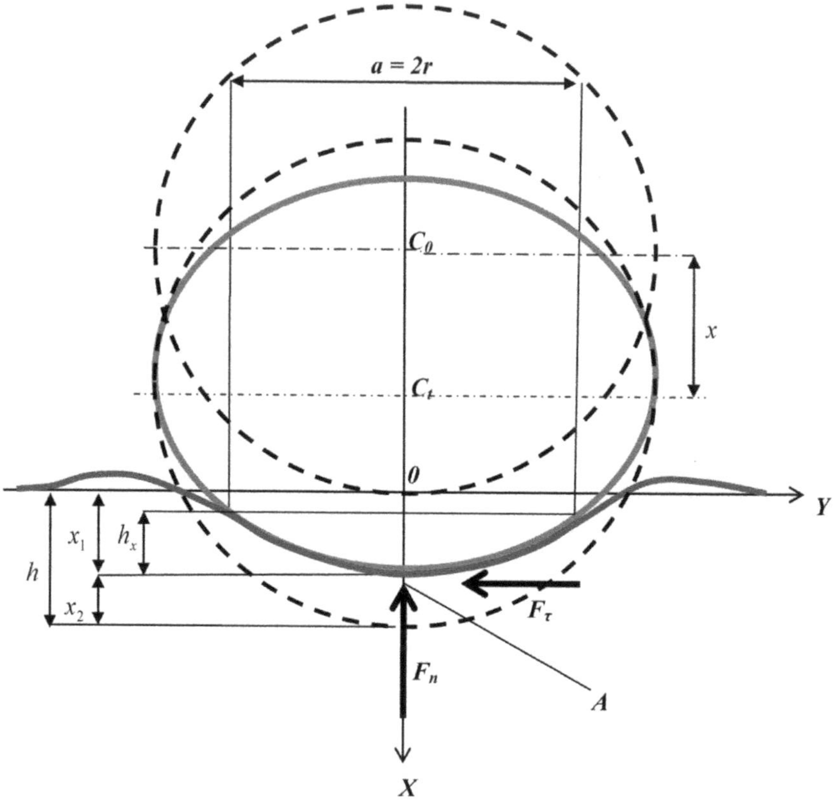

FIGURE 5.4 (See color insert.) The schematic illustration of the indentation of a spherical body into a semi-space.

Source: Reprinted from Goloshchapov, 2015a, with permission from SAGE Publications. http://journals.sagepub.com/doi/abs/10.1177/1056789514560912).

As we already know that in case of impact between two colliding bodies the effective mass m is entered such as for the third body, and also the movement (the displacement) x of the centre of mass of this third body is taken equal to the distance h of the mutual approach (or a compression) of the colliding bodies. In the case of the collision between a body and a semi-space, it is simply necessary to prove that $x = h = x_1 + x_2$, too. At the impact of two bodies, according to the second law of Newton, we can write that $m\frac{dV_x}{dt} = m_1\frac{dV_{1x}}{dt} = m_2\frac{dV_{2x}}{dt}$, where $V_x = V_{1x} + V_{2x}$. These equations are valid only for the movement of the centres of mass of the bodies. All authors who use these equations consider that x is the displacement of centre of mass of this third body and x is the mutual approach (a compression, an overlapping) between the bodies, too. These two expressions follow that $\frac{1}{m}=\frac{1}{m_1}+\frac{1}{m_2}$ and as in the case of the collision of a body and a semi-space $m_1 >> m_2$, and; hence, we can take that $m = m_2$. Hence, it is proved that in the case of collision between a body and a semi-space, the mass of a body m_2 is equal to the effective mass m. Further, in this chapter, the mass of a body is designated by the symbol m. Consequently, the distance h of the mutual approach between a body and a semi-space is equal to the displacement x of the centre of mass of a body.

5.3.2 DEFINITION OF THE MAXIMUM OF FREQUENCY OF CONTACTING

The frequency of contact at any point of surface on a semi-space by a flow of rigid bodies without an influence of the shielding effect (without taking into account the influence of collision between coming nearer bodies and rebounding bodies) can be defined as follows

$$\nu_i = S_{xm} V_0 C \sin\alpha \tag{5.25}$$

where C is the concentration of particles approaching to the surface of contact, V_0 is the average velocity of a body in the flow, α angle of attack between the flow of abrasive particles and a semi-space; $S_{xm} = \pi r_m^2$ is the maximum magnitude of the contact surface and where r_m is the maximum magnitude of the radius of the contact area. Since $r_m^2 = k_p^2 R x_m$, see Equations 2.129 and 2.130, where R is the radius of particle, x_m denotes the maximum indentation of a particle into a semi-space, it follows $S_x = \pi k_p^2 R x_m$, and consequently, we can write for the maximum of frequency of contacting as follows

$$\nu_i = \pi k_p^2 C V_0 R x_m \,\mathrm{Sin}\,\alpha \tag{5.26}$$

5.3.3 VISCOELASTIC STRESSES IN THE CONTACT AREA

It is obvious that the stresses at the point A reach the maximum magnitude at the moment $t=\tau_1$, $x=x_m$, $y_d=y_{dm}$ and when the velocity of compression reaches the minimum $\dot{x} = 0$; therefore, for calculation of the maximum stresses we can use the expression in Equation 3.176 as

$$\sigma_{Am} = \sigma_E = \sqrt{\left(\frac{2E'x_m^{1/2}}{\pi k_p R^{1/2}}\right)^2 + 4\left(\frac{3G'x_m^{1/2}}{2lR^{1/2}}\varsigma_y y_m\right)^2} \tag{5.27}$$

In the case of contact between a spherical body and a flat semi-space, when $R_1 = \infty$ and $\bar{V}_{0y} = \bar{V}_y = \bar{V}_{2y}$, it follows that $l = l_x = R_2 = R$ and $k_h = D_1$, see Section 3.1.6.

Also according to Equation 3.60, since $V_x = V_{0x} = V_0 \sin\alpha$, the maximal indentation of a body into a semi-space can be found as

$$x_m = \left[\frac{15mV_{0x}^2 \sin^2\alpha}{16k_p E'R^{1/2}} \times \frac{(\pi - 3tg\beta_E)}{(\pi + 3tg\beta_E)}\right]^{2/5} \tag{5.28}$$

On the other hand, using Equation 3.94 and since $V_y = V_{0y} = V_0 y \cos\alpha$ the maximum of the rolling shear at the moment of time $t = \tau_1$ can be expressed as

$$y_m = \frac{V_{0y}\cos\alpha}{\omega_{yr}} \exp(-\delta_{yr}\tau_1)\sin(\omega_{yr}\tau_1) \tag{5.29}$$

where $\omega_{yr} = \sqrt{\omega_{0yr}^2 - \delta_{yr}^2}$, $\delta_{yr} = \frac{B_y}{2m}$ is the tangential damping factor, $\omega_{0yr} = \sqrt{\frac{C_{yr}}{m}}$ is the angular frequency of the harmonic oscillations by axis Y. Also, according to Equations 3.88, 3.91, 3.92, and since $k_h = D_1$ and $l = R$, it follows as

$$C_{yr} = G'\frac{x_m^{3/2}}{R}(D_1 x_m^{1/2} + 2k_p R^{1/2}) \tag{5.30}$$

$$B_{yr} = \frac{G''}{\omega_{yr}}\frac{x_m^{3/2}}{R}(D_1 x_m^{1/2} + 2k_p R^{1/2}) \tag{5.31}$$

As we know, see Section 3.1, $\omega_x = \sqrt{\omega_{0x}^2 - \delta_x^2}$, $\delta_{yr} = \frac{B_x}{2m}$ is the tangential damping factor; $\omega_{0x} = \sqrt{\frac{C_x}{m}}$ is the angular frequency of the harmonic oscillations by axis X.

Since at the time of the maximum indentation $t = \tau_1$ it follows that $\dot{x} = 0$; therefore, using Equation 3.56 we get

$$\tau_1 = \frac{1}{\omega_x}\operatorname{arctg}\left(\frac{\omega_x}{\delta_x}\right) \tag{5.32}$$

Thus, by substituting τ_1 from Equation 5.32 into 5.29, the maximum of the rolling shear can be expressed as

$$y_m = \frac{V_{0y}\cos\alpha}{\omega_{yr}} \exp\left(-\frac{\delta_{yr}}{\omega_x}\operatorname{arctg}\left(\frac{\omega_x}{\delta_x}\right)\right)\sin\left(\frac{\omega_{yr}}{\omega_x}\operatorname{arctg}\left(\frac{\omega_x}{\delta_x}\right)\right) \quad (5.33)$$

5.3.4 THE TIME OF FATIGUE LIFE OF THE CONTACT AREA

Finally, using Equations 5.1, 5.2, 5.26, 5.27 and since $l = R$, the time of fatigue life can be expressed as

$$\tau_f = \frac{\sigma_c^{\lambda}}{\pi k_p C V_0 R x_m \sin\alpha}\left(\left(\frac{2E' x_m^{1/2}}{\pi k_p R^{1/2}}\right)^2 + 4\left(\frac{3G' x_m^{1/2}}{2R^{3/2}}\varsigma_y y_m\right)^2\right)^{-\frac{\lambda}{2}} \quad (5.34)$$

If the exponent λ of fatigue life curve and the limiting strength σ_c were found experimentally, using Equations 5.1, 5.2 and 5.23, we can calculate the time of fatigue life simply substituting into Equation 5.34 all magnitudes x_m and y_m, which can be calculated by using Equations 5.33 and 5.28.

5.4 EXPERIMENTAL TECHNIQUES FOR THE DETERMINATION OF THE DYNAMICS MODULI BY THE METHOD OF THE 'TEMPERATURE–TIME SUPERPOSITION'

Using Equation 3.64, the expression for the calculation of the effective dynamic elasticity module can be written as

$$E' = \frac{5m}{8k_p V_x^{1/2} R^{1/2} \tau_x^{5/2}} \times \left(\frac{(-2\ln k_x)(1+k_x)}{k_x^{1/5}(1-k_x)}\right)^{5/4} \quad (5.35)$$

and since $tg\beta_E = \frac{\pi}{3}\times\frac{(1-k_x)}{(1+k_x)} = \frac{E''}{E'}$ is known from Equation 3.58, we get

$$E'' = \frac{15\pi m}{24k_p V_x^{1/2} R^{1/2} \tau_x^{5/2}} \times (-2\ln k_x)^{5/4}\left(\frac{(1+k_x)}{k_x(1-k_x)}\right)^{1/4} \quad (5.36)$$

The dynamic elasticity and viscosity modules for high velocities of the collision can be found, if we follow the principles of the 'temperature–time superposition' according to the equation of the 'WLF—Williams–Landel–Ferry or Arrhenius' (Ferry, 1963; Van Krevelen, 1972; Moore, 1975; Nilsen and Landel, 1994). First of all, we have to define experimentally the effect of temperature for the period of the contact time τ_x, and for the coefficient of restitution k_x at the fixed initial velocity of impact. For example, if we define these parameters for velocity at 2 m/s, then using the principles of the 'time–temperature superposition' we can determine their values for any velocities interesting for us, for example for velocity 100 m/c and for temperature 100°C. After this, when τ_x and k_x will be known, we can find the value of $tg\beta$ and the dynamic modules E'' and E'.

According to the principle temperature–time (velocity or frequency) superposition, the increase of the velocity or the frequency of loading is equivalent to decreasing the temperature of testing. For description of behavior of amorphous polymers and elastomers usually the WLF equation is used

$$\lg a_T = -\frac{C_1(T_{M.i} - T_g)}{C_2 + (T_{M.i} - T_g)} \tag{5.37}$$

where $T_{M.i}$ denotes the temperature of mechanical glazing, T_g is the temperature of structural glazing, which is the constant of the examinee of material, constant $C_1 = 17.44$, constant $C_2 = 51.6$, and $\lg a_T = \lg \frac{V_g}{V_i}$, where V_g is the velocity of structural glazing (see Goloshchapov, 1989), when processes of the mechanical and the structural glazing coincidence, V_i denotes the initial velocity of impact under a test. Index $i = 1,2,3,.$ for any $V_i \geq V_g$.

On the other hand, for description of the behavior of amorphous polymers and elastomers usually the Arrhenius' equation is used

$$\lg a_T = \frac{U_a}{2.3R_g}\left(\frac{1}{T_{M.i}} - \frac{1}{T_g}\right) \tag{5.38}$$

where U_a is the energy of activation of process of mechanical glazing, R_g is the universal gas constant.

For the definition of viscoelastic properties of materials of the contacting surfaces at impact, we can use the electronic-mechanical device, the schema of which is depicted in Figure 5.5. It has the two-channel service oscilloscope, piezo-detectors, thermo-chamber, pendulum elastomer, thermocouples and digital milli-voltmeter have been used. This equipment permitted the determination of the thermokinetic dependences for the moment of impact τ_x and the coefficient k_x with the initial velocities $V_i \leq 2\ [m/s]$.

The given device (see Figure 5.5) allows for the definition of elasticity on a rebound e, the coefficient of restitution k_x and the time of impact τ_x depending on the temperature of the sample.

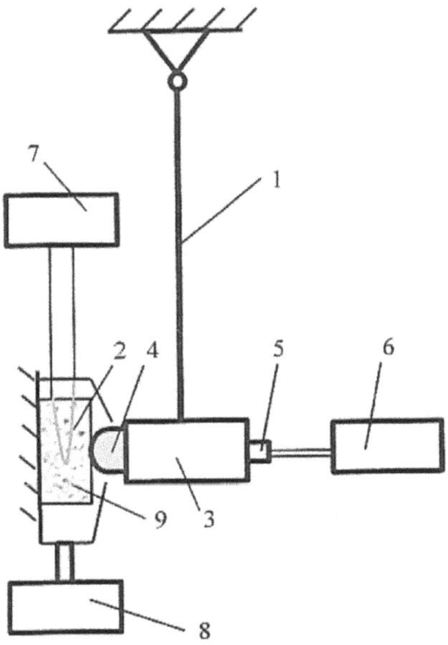

FIGURE 5.5 The schema of the electronic-mechanical device for the definition of the dynamic mechanical properties of materials at impact.

The electronic-mechanical device (see Figure 5.5) works in the following way: The hammer 3 is placed on the end of a pendulum. The sample 9 is placed in a clip of elasticity meter. The volume temperature of the sample is measured by means of the thermocouple 2 and digital milli voltmeter 7 to within 0,05°C with a milli-voltmeter resolution of 10^{-6}V. An electric signal is generated in the piezo-sensor 5 during the impact by the indenter 4 of the hammer 3 at the surface of the sample 9. The form of the signal (see Figure 5.9) is registered by the oscilloscope 6. Elasticity on a rebound is defined mechanically by the size of the rebound of the hammer 3 and is fixed on a scale elasticity meter. A time of impact τ_x is measured at the same time by the oscilloscope and an electronic frequency meter in a mode of measurement of current's impulse time, arising at impact in a piezo-element chain, and also in an electric chain the indenter-sample at contact of the hammer's indenter with the sample. Heating of the sample is carried out by hot air, cooling by streams of nitrogen using a heater and a cryogenic refrigerator 8. The thermocouple 2 is entered in the sample directly ahead of the beginning of tests. The thermocouple practically has no heat exchange with the outside environment. The temperature of the sample was registered at the moment when the indenter 4 started to penetrate into the sample 9.

The example of the theoretically obtained curves for the viscoelastic forces are depicted in Figure 5.6. By comparison, Figure 5.7 shows a photo of the experimental oscilloscopic curve of the piezo-electric signal, which has been obtained as a result of the collision of the spherical steel indenter and the rubber sample.

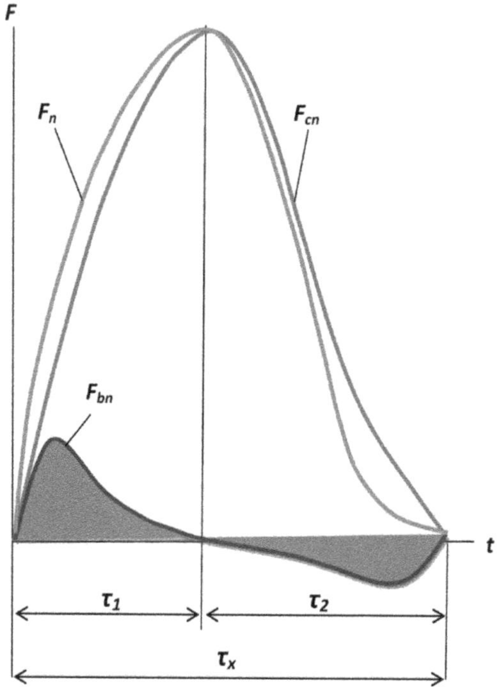

FIGURE 5.6 (See color insert.) Graphical illustration of the results and the forms of curves obtained theoretically for viscoelastic forces.

The power of the signal, which is generated by the piezo-detector placed inside the indenter, is directly proportional to the viscoelastic force. In the result of the graphical comparison of these curves, which are presented in Figure 5.6 and 5.7, we can see that the curves have very similar forms and; hence, this, once more, confirms that the theoretical solutions were made in the correct manner.

In conclusion, first of all, we have to mark here, that the method of specific viscoelastic forces presented in this book can be used for finding the viscoelastic forces, which act in the contact area. This method can be used in designing of wear-resistant elements and coverings for components of machines and equipment, which are working in harsh conditions where they are subjected to the action of flow or jet abrasive particles.

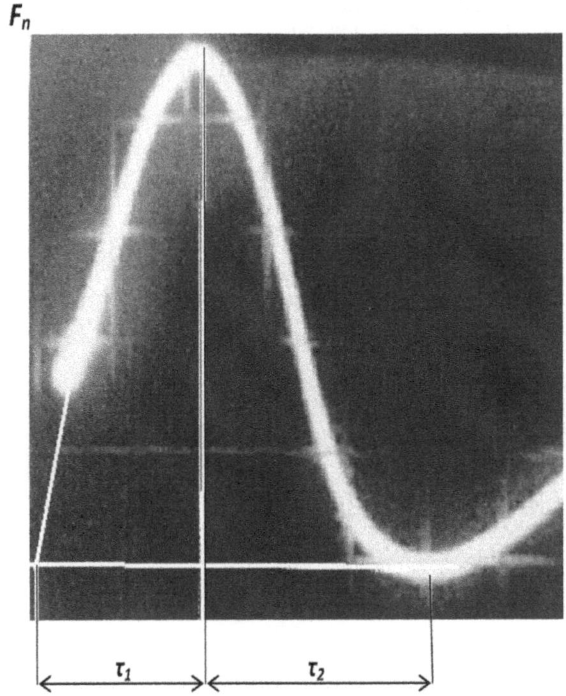

FIGURE 5.7 Photo of the experimental oscilloscopic curve of the piezoelectric signal.

Also, the results of the experimental and theoretical research and the method of the differential specific forces presented in this book can be used for the determination of the viscoelastic forces, contact stresses, durability and fatigue life for a wide spectrum of the tasks relevant to collisions between solid bodies under different loading conditions. Opportunities to use the results exist practically in the design and development of new advanced materials, wear-resistant elastic coatings and elements for pneumatic and hydraulic systems, stop valves, fans, centrifugal pumps, injectors, valves, gate valves and in other installations. This method is already being used as an instrument in research of the processes of erosion and disintegration of polymers and elastomers and other materials being in gas jet and in gas flow of abrasive particles under high and low temperatures.

The use of this theory gives an opportunity for the development of analytical and experimental methods allowing optimizing the basic dynamic and mechanical qualities of viscoelastic materials, for example elastomers (rubbers), for exploitation in gas abrasive substances and for different temperatures and velocities of loading.

As a result of research, based on the theoretical principals given in this book, new materials can be designed and used for designing of advanced wear-resistant elements for grinding mills, for example for pneumatic and hydraulic equipment, and in many other applications.

KEYWORDS

- **time of fatigue life**
- **stress-number of cycles**
- **viscoelastic force**
- **compression**
- **temperature–time superposition**

REFERENCES

Alfredsson, B. A Study on Contact Fatigue Mechanism. Doctoral Thesis, Royal Institute of Technology, Sweden, 2000. http://www.nmpa.ir/files/150.pdf

Antypov, D.; Elliott, J. A.; Bruno, C.; Hancock, B. C. Effect of Particle Size on Energy Dissipation in Viscoelastic Granular Collisions. *Phys. Rev. E* **2011**, *84,* 021303. http:// Phys Rev zE.84.021303.

Archard, J. F. Elastic Deformation and the Laws of Friction. *Proc. R. Soc. London Ser. A* **1957**, *243*(1233), 190–205.

Barber, J. R.; Ciavarella, M. Contact Mechanics. *Int. J. Solids Struct.* **2000**, *37*. http:/ doi:10.1016/S0020–7683(99)00075-X.

Becker, V.; Schwager, T.; Pöschel, T. Coefficient of Tangential Restitution for the Linear Dashpot Model. *Phys. Rev. E* **2008**, *77,* 011304. http:/doi: 10.1103/PhysRevE.77.011304

Biryukov, A. A.; Vorobyova, E. N.; Semchinova, I. N. Mechanical Oscillations. Publishing House Samara University, 2011; p 167. mechanic/meh_kolebaniya.pdf

Bordbar, M. H.; Hyppänen, T. Modeling of Binary Collision between Multisize Viscoelastic Spheres. *J. Numer. Anal. Sci. Eng. (ESCMSE)* **2007**, *2*(3–4), 115–128.

Boussinesq, J. *Application des potentiels à l'étude de l'équilibre et du mouvement des solides élastiques;* Gauthier-Villars: Paris, 1885.

Bowden, F. P.; Tabor, D. The Area of Contact between Stationary and between Moving Surfaces. *Proc. R. Soc. London Ser. A* **1939**, *169*(938), 391–413.

Brilliantov, N. V.; Spahn, F.; Hertzsch, J.-M.; Pöschel, T. Model for Collisions in Granular Gases. *Phys. Rev. E* **1996**, *53,* 5382. from arxiv.org

Budynas, R. G.; Nisbett, J. K. *Shigley's Mechanical Engineering Design,* 9th ed.; (Series in Mechanical Engineering), McGraw-Hill: New Delhi, 2011; p 1109.

Bush, A. W.; Gibson, R. D.; Thomas, T. R. The Elastic Contact of a Rough Surface. *Wear* **1975**, *35*(1), 87–111.

Carbone, G.; Putignano, C. A Novel Methodology to Predict Sliding/Rolling Friction in Viscoelastic Materials: Theory and Experiments. *J. Mech. Phys. Solids* **2013**, http://doi: 10.1016/j.jmps.2013.03.005

Carbone, G.; Lorenz, B.; Persson, B. N. J.; Wohlers, A. Contact Mechanics and Rubber Friction for Randomly Rough Surfaces with Anisotropic Statistical Properties. *Eur. Phys. J. E Soft Matter* **2009**, *29*(3), 275–284. http://dx.doi.org/10.1140/epje/i2009–10484–8

Chenelera, D.; Mehrbanb, N.; Bowenb, J. Spherical Indentation Analysis of Stress Relaxation for Thin Film Viscoelastic Materials. Springer-Verlag, 2013. http://dx.doi.org/doi:10.1007/ s00397–013–0707–5

Cheng, L.; Xia, X.; Scriven, L. E.; Gerberich, W. W. Spherical-Tip Indentation of Viscoelastic Materials. *Mech. Mater.* **2005**, *37,* 213–226. http://www.sciencedirect

Cheng, N.; Subic, A.; Takla, M. Development of a Fast-Solving Numerical Model for the Structural Analysis of Cricket Balls. *Sports Technol.* **2008**, *1*(2–3), 132–144. http:// publication/227591545

Cummins, S. J.; Thornton, C.; Cleary, P. W. Contact Force Models in Inelastic Collisions. *Ninth International Conference on CFD in the Minerals and Process Industries CSIRO,* Melbourne, Australia, 2012. http://www.cfd.com.au/cfd_conf12/PDFs/175CUM.pdf

Cundall, P. A.; Strack, O. D. L. A Discrete Numerical Model for Granular Assemblies. *Geotechnique* **1979**, *29,* 47–65.

Derjaguin, B. V.; Muller, V. M.; Toporov, Y. P. Effect of Contact Deformations on the Adhesion of Particles. *J. Colloid Interface Sci* **1975**, *53*(2), 314–326.

Dintwa, E. Development of Accurate Contact Force Models for use with Discrete Element Method (DEM) Modelling of Bulk Fruit Handling Processes. Dissertationes de Agricultura, Doctoraatsproefschrift nr. 726 aan de faculteit Bio-ingenieurswetenschappen van de K.U. Leuven, Katholieke Universiteit Leuven, Faculteit Bio-ingenieurswetenschappen, Katholieke, 2006.

Farine, M. Instrumented Indentation of Soft Materials and Biological Tissues. Dissertation ETH No. 2156 for Doctor of Sciences. ETH Zurich, Germany, 1985. http://e-collection.library.ethz.ch/eserv/eth:7912/eth-7912-02.pdf

Ferry, J. D. Viscoelastic Properties of Polymer Solutions. *J. Res. Natl. Bur. Stand.* Research Paper RP1903, **1948**, *41,* 53–62.

Ferry, J. D. *Viscoelastic properties of Polymers;* John Wiley & Sons, Inc.: New York, 1963.

Flügge, W. *Viscoelasticity;* Springer-Verlag, Berlin Heidelberg GmbH: Berlin, 1975.

Galin, L. A. *Contact Problems in the Theory of Elasticity;* Dept. of Mathematics, School of Physical Sciences and Applied Mathematics, North Carolina State College, 1961.

Goldsmith, W. *Impact. The Theory and Physical Behaviour of Colliding Solids;* Adward Arnold Publisher Ltd: 1960; p 371.

Golenko, A. *Fundamentals of Machine Design;* Wroclaw University of Technology, 2010; p 175. http://www.dbc.wroc.pl/Content/7154/Golenko_Fundamentals%20of%20Machine%20Design.pdf

Goloshchapov, N. Determination Methods of Cycle-Thermo-Kinetics Viscoelastic and Relaxation Characteristic of Elastomers Under Impact. Cycles, *Materials of the 3-d International Conference,* North-Caucasian State Technical University, Stavropol, Russia, 2001.

Goloshchapov, N. Tribo-Cyclicity. Cycles. Materials of 5th International Conference, North-Caucasian State Technical University, Russia, 2003a.

Goloshchapov, N. Processes of Abrasive Wear Elastomers and their Connection with Viscoelasticity Qualities. *2-d International Scientific Conference "Tribology in Environmental Design", Collected Articles,* Professional Engineering Publishing, London, UK, 2003b.

Goloshchapov, N. Application of the Method of the Specific Forces in the Mechanics of a Viscoelastic Collision between a Solid Body and a Semi-Space. *Int. J. Damage Mech.* August **2015a**, *24*(6), 915–943. http://ijd.sagepub.com/content/24/6/915

Goloshchapov, N. Solutions of the Problems of a Viscoelastic Dynamic Contact between Smooth Curvilinear Surfaces of Two Solid Bodies by the Application of the "Method of the Specific Forces". *Mech. Eng. Res.* **2015b**, *5*(2). http://dx.doi.org/10.5539/mer.v5n2p59

Graham, G. A. C. The Contact Problem in the Linear Theory of Viscoelasticity. *Int. J. Eng. Sci.* **1965**, *3,* 27. http://www.sciencedirect.com/science/article/pii/0020722565900182

Greenwood, J. A.; Williamson, J. B. P. Contact of Nominally Flat Surfaces. *Proc. R. Soc. London Ser. A* **1966**, *295,* 300.

Harrass, M.; Friedrich, K.; Almajid, A. A. Tribological Behavior of Selected Engineering Polymers under Rolling Contact. *Tribol. Int.* **2010**, *43,* 635–646.

Heβ, M. On the Reduction Method of Dimensionality: The Exact Mapping of Axisymmetric Contact Problems with and Without Adhesion. *Phys. Mesomachanics* **2012**, *15*(4), 19–24.

Hertz, H. Über Die Berührung Fester Elastischer Körper (On the Contact of Elastic Solids). *J. Reine Angewandte Mathematik* **1882**, *92*, 156–171.

Hertz, H. Über Die Berührung Fester Elastischer Körper (On the Contact of Rigid Elastic Solids)". In *Miscellaneous Papers;* Jones, D. E., Schott, G. A., Eds.; (J. reine und angewandte Mathematik 92), Macmillan: London, 1896; p 156.

Hosford, W. F. *Mechanical Behaviour of Materials;* Cambridge University Press: New York, 2005. http://assets.cambridge.org/97805218/46707/frontmatter/9780521846707_frontmatter.pdf

Huang, G. Measurements of Viscoelastic Properties by Nanoindentation. Ph.D. Dissertation, Graduate College of the Oklahoma State University, USA, **2007.** http://citeseerx.ist.psu.edu/viewdoc/download?doi=10.1.1.428.156&rep=rep1&type=pdf

Hunter, S. C. The Hertz Problem for a Rigid Spherical Indenter and a Viscoelastic Half Space. *J. Mech. Phys. Solids* **1960**, *8,* 219–234.

Hyun, S.; Robbins, M. O. Elastic Contact between Rough Surfaces. Effect of Roughness at Large and Small Wavelengths. *Trobol. Int.* **2007**, *40,* 1413–1422.

Johnson, K. L. *Contact Mechanics;* Cambridge University Press, 1985.

Johnson, K. L.; Kendall, K.; Roberts, A. D. Surface Energy and the Contact of Elastic Solids. *Proc. R. Soc. London A* **1971**, *324,* 301–313. http://rspa.royalsocietypublishing.org/content/324/1558/30

Kerr, A. D. Elastic and Viscoelastic Foundation Models. *J. Appl. Mech.* **1964**, *31*(3), 491–498. DOI:10.1115/1.3629667.

Lakes, R. S. *Viscoelastic Solids;* II Series: Mechanical Engineering, CRC Press, 1998.

Landau, L. D.; Lifshitz, E. M. Theory of Elastisity. In *Theoretical Physics*; Pergamon Press: Moscow, 1944, vol. 4.

Landau, L. D.; Lifshitz, E. M. *Theory of Elasticity;* Oxford University Press: Oxford, 1965.

Laursen, T. A. *Computational Contact and Impact Mechanics. Fundamentals of Modeling Interfacial Phenomena in Nonlinear Finite Element Analysis;* Springer Verlag: New York, 2002.

Lee, E. H. Viscoelasticity. In *Handbook of Engineering Mechanics;* Flugge, W., Ed.; McGraw-Hill: New York, **1962.**

Lee, E. H.; Radok, J. R. M. The Contact Problem for Viscoelastic Bodies. *J. Appl. Mech.* **1960**, *30,* 438–444.

Li, H. Impact of Cohesion Forces on Practical Mixing and Segregation. Doctoral Dissertation, University of Pittsburgh, 2006. http://d-scholarship.pitt.edu/9699/

Makse, H. A.; Gland, N.; Johnson, D. L.; Schwartz, L. Granular Packings: Nonlinear Elasticity, Sound Propagation and Collective Relaxation Dynamics. *Phys. Rev. E* **2004**, *70,* 061302

McCrum, N. G.; Buckley, C. P.; Bucknall, C. B. *Principles of Polymer Engineering;* Oxford University Press, 1997; p 447.

Menard, K. P. *Dynamic Mechanical Analysis. A Practical Introduction;* CRC Press LLC, 1999.

Menga, N.; Putignano, C.; Carbone, G.; Demelio, G. P. The Sliding Contact of a Rigid Wavy Surface with a Viscoelastic Half-Space. *Proc. R. Soc. A* **2014**, *470.* http://rspa.royalsocietypublishing.org/

Meyers, M. A. *Dynamic Behavior of Materials;* Wiley, 1994; p 668.

Mindlin, R. D. Compliance of Elastic Bodies in Contact. *Trans. ASME J. Appl. Mech.* **1949**, *16,* 259–268.

Moore, D. F. *Principles and Application of Tribology;* Technology & Engineering, Pergamon Press: New York, 1975.

Moore, D. F. *Principles and Application of Tribology;* Mir: Moscow, Russia, 1978.

Nilsen, L. *Mechanical Properties of Polymers and Polymeric Compositions;* Moscow, 1978; p 312.

Nilsen, L. E.; Landel, R. F. *Mechanical Properties of Polymers and Polymeric Compositions,* 2nd ed.; (Rev. Series: Mechanical Engineering). Marcel Dekker, Ink.: New York, 1994, p 531.

Padovan, J.; Paramadilok, O. Transient and Steady State Viscoelastic Rolling Contact. *Comput. Struct.* **1984**, *20,* 545–553.

Peng, G.; Zhang, T.; Feng, Y.; Yong Huan, Y. Determination of Shear Creep Compliance of Linear Viscoelastic-Plastic Solids by Instrumented Indentation. *Polym. Test.* **2012**, *31,* 1038–1044, (Elsevier). http://www.sciencedirect.com/science/article/pii/S0142941812001274

Persson, B. N. J. Rolling Friction for Hard Cylinder and Sphere on Viscoelastic Solid. *Eur. Phys. J. E* **2010**, *33,* 327–333. http://link.springer.com/article/10.1140%2Fepje%2Fi2010–10678-y

Persson, B. N. J.; Bucher, F.; Chiaia, B. Elastic Contact between Randomly Rough Surfaces. Comparison of Theory with Numerical Results. *Phys. Rev. B* **2002**, *65*(18), 184106.

Popov, V. L. *Contact Mechanics and Friction: Physical Principals and Application;* Springer, 2010.

Popov, V. L.; Heβ, M. *Method of Dimensionality Reduction in Contact Mechanics and Friction;* Springer, 2015. http://www.twirpx.com/file/1756685/

Putignano, C.; Carbone, G.; Daniele, D. Theory of Reciprocating Contact for Viscoelastic Solids. *Phys. Rev. E* **2016**, *93*(4), 043003. DOI: 10.1103/PhysRevE.93.043003 www.researchgate.net/publica

Radok, J. R. M. Viscoelastic Stress Analysis. *Q. App. Math.* **1957**, *15,* 198–202.

Rodríguez, P. N. V. Contact and Friction in Systems with Fibre Reinforced Elastomers. Ph.D. Thesis, University of Twente, Enschede, The Netherlands, 2012. DOI 10.3990./1.9789036534543.

Ramírez, R.; Poeschel, T.; Brilliantov, N. V.; Schwager, T. Coefficient of Restitution of Colliding Viscoelastic Spheres. *Phys. Rev. E* **1999**, *60.* http://www.ncbi.nlm.nih.gov/pubmed/11970301

Roylance, D. *Engineering Viscoelasticity;* Department of Materials Science and Engineering, Massachusetts Institute of Technology: Cambridge, MA, 2001; 02139. http://ocw.mit.edu/courses/materials-science-and-engineering/3–11-mechanics-of-materials-fall-1999/modules/visco.pdf

Schafer, J.; Dippel, S.; Wolf, D. E. Force Schemes in Simulations of Granular Materials. *J. Phys I* **1996**, *6,* 5–20. https://hal.archives-ouvertes.fr/jpa-00247176/document

Schwager, T.; Poschel, T. Coefficient of Restitution and Linear-Dashpot Model Revisited. *Granular Matter* **2007**, *9,* 465–469. http://link.springer.com/article/10.1007%2Fs10035–007–0065-z

Schwager, T.; Poschel, T. Coefficient of Restitution for Viscoelastic Spheres: The Effect of Delayed Recovery. *Phys. Rev. E* **2008**, *78,* 051304. http://journals.aps.org/pre/abstract/10.1103/PhysRevE.78.051304

Simon, R. Development of a Mathematical Tool for Evaluating Golf Club Performance. *ASME Design Engineering Congress*, New York, 1967.

Singh, S. *Theory of Machines,* 3rd ed.; Pearson, **2012;** p 1369.

Sneddon, I. N. The Relation between Load and Penetration in the Axisymmetric Boussinesq Problem for a Punch of Arbitrary Profile. *Int. J. Eng. Sci.* **1965**, *3,* 47–57. FWM/Papers/Sneddon1965.pdf

Stronge, W. J. *Impact Mechanics;* Cambridge University Press: Cambridge, 2000; p 280.

Tabor, D. Surface Forces and Surface Interactions. *J Colloid Interface Sci* **1977a**, *58*(1), 2–13.

Tabor, D. The Hardness of Solids. *J. Colloid Interface Sci.* **1977b**, *58,* 145–179.

Thornton, C. Special Issue on the 4th International Conference on Discrete Element Methods. *Powder Technol.* **2009**, *193,* 216–336.

Thornton, C.; Cummins, S. J.; Cleary, P. W. An Investigation of the Comparative Behaviour of Alternative Contact Force Models During Inelastic Collisions. *Powder Technol.* **2012**, *233,* 30–46.

Timoshenko, S.; Goodier, J. N. *Theory of Elasticity;* Mcgraw Hill: N.Y., 1951.

Ting, T. The Contact Stresses between a Rigid Indenter and a Viscoelastic Half Space. *J. Appl. Mech.* **1966**, *33,* 845.

Van Krevelen, D. W. *Properties of Polymers Correlations with Chemical Structure;* Esliier Publishing: Amsterdam-London-New York, 1972.

Van Zeebroeck, M. The Discrete Element Method (DEM) to Simulate Fruit Impact Damage during Transport and Handling. Dissertationes de Agricultura. Proefschrift voorgedragen tot het behalen van de graad van Doctor in de Toegepaste Biologische Wetenschappen door. 2005. https://lirias.kuleuven.be/bitstream/1979/46/2/doctoraatmichaelvanzeebroeck.pdf

Votsios, V. Contact Mechanics and Impact Dynamics of Non-Conforming Elastic and Viscoelastic—Semi-infinite or Thin Bonded Layered Solids. Ph.D. Thesis, Loughborough University, UK, 2003. http://creativecommons.org/licenses/by-nc-nd/2.5/

Webster, M. N.; Sayles, R. S. A Numerical Model for the Elastic Frictionless Contact of Real Rough Surfaces. *ASME* **1986**, *108,* 314–320.

Wriggers, P. *Computational Contact Mechanics;* Springer, 2006; p 514.

Zhou, S. New Solutions of Half-Space Contact Problems Using Potential Theory, Surface Elasticity and Strain Gradient Elasticity. Ph.D. Dissertation, Texas A&M University, 2011.

Zukas, J. A. *Impact Dynamics;* Technology & Engineering, Wiley, 1982; p 452.

INDEX

E

W

Y